**GEO**

07/05

# WEAPONS GRADE

## HOW MODERN WARFARE
## GAVE BIRTH TO
## OUR HIGH-TECH WORLD

# WEAPONS GRADE

## HOW MODERN WARFARE
## GAVE BIRTH TO
## OUR HIGH-TECH WORLD

## David Hambling

CARROLL & GRAF PUBLISHERS
New York

Carroll & Graf Publishers
An imprint of Avalon Publishing Group, Inc.
245 W. 17th Street
New York
NY 10011–5300
www.carrollandgraf.com

AVALON
publishing group incorporated

First published in the UK by Constable,
an imprint of Constable & Robinson Ltd, 2005

First Carroll & Graf edition, 2005

ISBN 0-7867-1561-8

Printed and bound in the EU

This book is dedicated to my parents,
who taught me a love of both science and writing.

# Contents

# Illustrations

# Acknowledgements

Thanks are due to everyone who helped with this book in greater or lesser ways.

Dan Hind at Constable & Robinson
Tim Radford at the *Guardian*
Graham Warwick, Andrew Doyle and Guy Norris at *Flight International*
Jurgen Altmann of the Bochum Verification Project
Justo Miranda at Reichdreams Research Services
Brian 'Megazone' Bikowicz
David Sutton, Paul Sieveking and Bob Rickard at *Fortean Times*
Robin Ramsay of *Lobster* magazine
Not to mention Peter Greenway at ABS, Richard Alexander, the great and astonishingly well-informed community that is Slashdot.org, the usual suspects on the Compuserve UFO forum, Colin Barker of *Computing* magazine, Ron Ardell at ERA technology, Ian Wright of Omega, and the numerous and various folk at Boeing, Lockheed, DARPA, Qinetiq and the Ministry of Defence who fielded queries of all shapes and sizes.

Special thanks to my wife Caroline for editing and support beyond the call of duty.

And to La Princesa Gatuna, whose grasp of computing, laser technology and stealth is a continuing source of wonder.

# List of Acronyms

| | |
|---|---|
| AEC | Atomic Energy Commission |
| AFRL | Air Force Research Laboratory |
| AFSP | Air Force Special Platform |
| AI | Artificial Intelligence |
| AK | Avtomat Kalashnikova = Kalashnikov automatic (rifle) |
| AM | Amplitude Modulation |
| ANP | Aircraft Nuclear Propulsion |
| APTI | Advanced Power Technologies Inc |
| ARPA | Advanced Research Projects Agency |
| ATL | Advanced Tactical Laser |
| ATM | Automated Teller Machine |
| BLU | Bomb, Live Unit |
| BMDO | Ballistic Missile Defence Organization |
| CAV | Common Aero Vehicle |
| CCD | Charge Coupled Device |
| CD | Compact Disc |
| CEH | Centre Electronique Horloger |
| COBOL | COmmon Business Oriented Language |
| CRT | Cathode Ray Tube |
| DA | DiphenylchloroArsine |
| DARPA | Defense Advanced Research Projects Agency |
| DERA | Defence Evaluation & Research Agency |
| DSR | Directed Stick Radiator |
| DVD | Digital Versatile Disc |

| | |
|---|---|
| EELD | Evidence Extraction and Link Discovery |
| EELV | Evolved Expendable Launch Vehicle |
| EMP | ElectroMagnetic Pulse |
| ENIAC | Electronic Numerical Integrator And Computor |
| FCS | Future Combat System |
| FM | Frequency Modulation |
| FORTRAN | FORmula TRANslation |
| GPR | Ground Penetrating Radar |
| GPS | Global Positioning System |
| HAARP | High Frequency Active Auroral Research Program |
| HALSOL | High Altitude SOLar |
| HAPP | High Altitude Powered Platform |
| HCV | Hypersonic Cruise Vehicle |
| HEAT | High Explosive Anti-Tank |
| HUD | Head-Up Display |
| IAO | Information Awareness Office |
| IBM | International Business Machines |
| IC | Integrated Circuit |
| ICT | Institute of Creative Technology |
| IDCSP | Initial Defense Communication Satellite Program |
| IMP | Interface Message Processor |
| INMARSAT | International Maritime Satellite Organization |
| INS | Inertial Navigation System |
| INTELSAT | INternational TELcommunications SATellite Organization |
| ISN | Institute of Soldier Nanotechnology |
| JDAM | Joint Direct Attack Munition |
| Laser | Light Amplification by Stimulated Emission of Radiation |
| LASIK | Laser-Assisted In Situ Keratomileusis |
| LCD | Liquid Crystal Display |
| LISP | LISt Processing |
| LOCAAS | Low Cost Autonomous Attack System |
| MANPAD | MAN-Portable Air Defence System |
| MARAUDER | Magnetically Accelerated Ring to Achieve Ultra-high Directed Energy and Radiation |
| MASER | Microwave Amplification by the Stimulated Emission of Radiation |

| | |
|---|---|
| MAV | Micro-unmanned Air Vehicle |
| MEMS | Micro-Electro-Mechanical Systems |
| MLRS | Multiple Launch Rocket System |
| MMW | Millimetre Waves |
| MOAB | Massive Ordnance Air Blast/Mother Of All Bombs |
| MoD | Ministry of Defence |
| MOSAIC | Multifunctional On-the-move Secure Adaptive Integrated Communications |
| NACA | National Advisory Committee for Aeronautics |
| NASA | National Aeronautics and Space Administration |
| NEPA | Nuclear Energy for Propulsion of Aircraft |
| NORC | Naval Ordnance Research Calculator |
| OECD | Organization for Economic Cooperation and Development |
| OSI | Office of Strategic Influence |
| PC | Personal Computer |
| PDE | Pulse Detonation Engine |
| PDEW | Penetrating Direct Energy Weapon |
| PEP | Pulsed Energy Projectile |
| PIKL | Pulsed Impulsive Kill Laser |
| PPS | Pulsed Periodic Stimuli |
| Radar | RAdio Detection And Ranging |
| RF | Radio Frequency |
| RLG | Ring Laser Gyro |
| RPG | Rocket-Propelled Grenade |
| rTMS | repeated Transcranial Magnetic Stimulation |
| SA | Selective Availability |
| SABRE | Semi-Automated Business Research Environment |
| SACLOS | Semi-Automated Command to Line Of Sight |
| SAM | Surface-to-Air Missile |
| SCORE | Signal Communication by Orbiting Relay Equipment |
| SDI | Strategic Defense Initiative |
| SPRscan | Surface Penetrating Radar scanner |
| TACOM ASCO | Tactical Command Advanced Systems Concept Office |
| Teflon | Polytetrafluoroethylene |
| TEMPEST | Telecommunications Electronics Material Protected From Emanating Spurious Transmissions |

| | |
|---|---|
| TIA | Total Information Awareness |
| TIROS | Television InfraRed Observation Satellite |
| | |
| UAV | Unmanned Airborne Vehicle |
| UCAV | Unmanned Combat Air Vehicle |
| UCLA | University of California at Los Angeles |
| USAF | United States Air Force |
| | |
| VfR | Verein fur Raumschiffahrt (Society for Space travel) |
| VMADS | Vehicle Mounted Active Denial System |
| VRD | Virtual Retinal Display |
| VTOL | Vertical Take-off and Landing |

# Introduction

*And he shall judge among the nations, and shall rebuke many people: and they shall beat their swords into ploughshares, and their spears into pruning hooks: nation shall not lift up sword against nation, neither shall they learn war any more.*

Isaiah 2:4

Every day, without realizing it, we use military technology. When you turn on a computer, get on an aircraft, play a CD or watch satellite television, you are using something originally intended for war. Many people know that the Internet had its origins in a military project from the Cold War, but few realize that everything from microwave ovens to mobile phones to laser eye surgery, come from similar sources.

Sometimes the route taken is a roundabout one; the Sony Walkman, for example, owes its existence to the Manhattan Project which built the first atomic bomb. Sometimes the connection is more obvious, like the evolution of the jet airliner from military jets, or the walkie-talkie to the mobile phone, but even then few people appreciate the closeness of the connection.

The spread and depth of the technology involved is considerable, encompassing some of the most significant developments of the modern world. Often the technology ends up having more effect in the civilian sphere than in the military one it was built for, as the Internet shows. And if you want to know where the technology of the future will come from and what the twenty-first century will look like, then the military labs of today offer some valuable clues.

This stealthy infiltration has happened because little attention is paid to where technology comes from. When you buy a PC or step on to an airliner, what matters is that it works. So long as computers and aircraft do not crash, we are happy to take advantage of them whatever their ultimate origin. There is no great conspiracy to mask the military origins of so much everyday technology; it is simply that there is a pervasive ignorance about the source of all technology.

If we do happen know about of the origin of a particular invention it is often a matter of national pride rather than strict historical accuracy. Ask anyone in Britain who invented the television and they will say it was John Logie Baird; but if you are an American, the answer is more likely to be Philo Farnsworth or Vladimir Zworykin (a naturalized American). Russians might mention Zworykin but may give precedence to Boris Rosing, who transmitted images through the air in 1922.

In America, the Wright Brothers are considered the leading pioneers of manned flight, but as far as the French are concerned it was the Montgolfier brothers' first balloon ascent in 1783 which was the real landmark.

Otherwise it's a matter of generally inaccurate folk history, which often invokes the magical name of the space agency NASA. The Teflon non-stick saucepan and Velcro fasteners are often cited as examples of spin-offs from NASA's space program. Actually, neither of them is. The chemical company DuPont developed Teflon in 1945; Velcro fasteners were the invention of a Swiss engineer in 1948 who had become fascinated by how sticky burrs could attach themselves to clothing.

Stories about Teflon and Velcro may have been started to boost NASA's image, but other stories work against them. One such is the urban myth that they spent $12 million working on a pen that would work in zero-gravity, only to find that the Russians dealt with the same problem by using pencils. In fact the 'space pen' was developed by the Fisher Pen Company at a rather more modest cost, and adopted by NASA. Whereas the NASA connection was a useful marketing ploy for the space pen, the Russians made do with normal ballpoints.

At least with Teflon it is easy to identify a specific inventor at a definite time and place – we can pin the discovery down to Dr Roy J. Plunkett on 6 April 1938. Other inventions like the computer went through many different stages, gradually evolving from abacus to Apple, with no single inventor being able to claim all the credit. Asked who invented the computer, a class of American schoolchildren gave answers ranging from 'Bill Gates' to 'Radio Shack'. Perhaps the key invention that made the modern computer possible is the silicon chip or integrated circuit, another innovation popularly attributed to NASA. In fact this too came from a military research program, as did many others of the elements that make up modern computing.

From the period of the Second World War onwards, military research has been a major driving force of technology. As we shall see, it has been responsible for the digital computers and the Internet as well as satellites and all that they bring, from communication to weather forecasting to navigation. There are also jet aircraft, lasers and nuclear power. The story behind these technologies will be examined in detail in the first part of this book.

The second part of the book looks at the current direction of military research, and what it means for the future of civilian technology. This includes not only areas like aircraft and computers but also more exotic areas like vortex rings and nanotechnology.

## The Second World War: Technology at War

The Second World War was the first in which technology played a key role from the start. Work on radar and rockets was being carried out before the war, but did not mature until its onset. Other developments started in earnest as the war came in sight: the jet engine, digital computers and nuclear power all existed as ideas on paper in 1939, but by 1945 they had all become reality.

The drive for technological superweapons was particularly strong in Nazi Germany. In the opening years of the war, the Wehrmacht swept everything before it, brushing aside the British and French armies and conquering Europe. The numerically superior Red Army was all but destroyed in the initial German assault during Operation Barbarossa; but although they gave ground, the Russians did not surrender. Stalin succeeded in reconstituting Soviet industry, and it was soon building tanks faster than the Germans could destroy them. The entry of the US into the war on the Allied side meant that the Germans were facing enemies in both East and West with vastly greater industrial resources.

Hitler realized that unless Germany could gain an advantage, the war would inevitably turn against her as the greater numbers of the Allies started to count and attrition gradually ate away the German war machine. In 1941, development had been slowed down because the end of the war was in sight, but in 1943 the German Armaments Minister, Albert Speer, persuaded Hitler that advanced technology weapons were the only way to win the war.

'Wonder weapons' were to be Germany's salvation. Hitler poured resources into the V-1 flying bomb and V-2 rocket, and development of the first jet aircraft was accelerated. In addition to these well-known projects, there were hundreds of less successful attempts to produce new weapons: everything from death rays to whirlwind guns, jet packs and rocket planes. Few of these saw action in spite of the time and effort spent on developing them.

Hitler was convinced that these high-tech weapons would turn the tide. In 1944 he told Mussolini of a new 'technical weapon' which would 'transform London into a heap of ruins': the V-1 flying bomb. The V-2 rocket promised to be even more effective – at one point Hitler believed it was the weapon which would win the war – and it would be followed by the V-3, a long-range gun which would rain down shells on London at a tremendous rate. Meanwhile new jet fighters would stem the tide of Allied attacks and reclaim the skies for the Luftwaffe.

None of this worked out. After encouraging initial reports, the V-1 was nicknamed Versager-1 (Disappointment-1) by the German public. The V-2 had even less effect, and the V-3 gun proved impractical and never fired a shot. The small numbers of jet fighters that made it into service caused little impact on the massed fleets of Allied bombers.

The strategy of pushing for high technology was ultimately a failure. The more advanced the plans, the less chance they had of being carried out during the war, and even the most successful products like the Me-262 jet fighter were immature and unreliable by the end of 1945. Whether Germany would have been any more successful by concentrating on less ambitious schemes is a matter for debate.

The Nazi work on secret weapons has now reached legendary status. They are credited, frequently inaccurately, with everything from stealth bombers to air-to-air missiles. The capabilities of German weapons are frequently overstated by enthusiasts who tend to overlook the more important aspects and admire the raw numbers. It is true that the rocket-powered Komet interceptor could fly at 600 miles an hour; but we will see why this did not make it such a technical marvel as some claim. We will also look at the whole mythology of Nazi 'flying saucer' development which tries to credit them with science far in advance of anything that exists fifty years later. In the following chapters we will repeatedly come across ideas which were first given a hearing in Germany in the Second World War and which have still not yet proven practicable in spite of sixty years of technical advances.

In spite of this tendency to mythologize the evil Nazi genius, it was the Allies who had the most notable successes. A tremendous scientific and engineering effort in the US and Britain created triumphs including the development of radar and computers. While the V-2, the very apex of rocket science, had little effect on the course of the war, the same cannot be said of the atomic bomb. It was the Allies, aided greatly by Jewish exiles from Hitler's Germany, who produced the ultimate secret weapon, a fact generally overlooked by fans of the Third Reich.

In the post-war world, wartime technology soon found new uses. Germany itself was wrecked, but much German technology was salvaged – or plundered – and in some cases physically shipped back to the US and Russia, along with the scientists who had developed it. Military systems like the digital computer, the ballistic missile and the jet fighter quickly went from prototype to mature technology, and in the process there were many benefits for the civilian sector.

The era also saw the rise of what President Eisenhower called 'the military industrial complex' driven by the new technology. As he pointed out in a speech in 1960, the backyard inventor had been superseded by the huge research program and the huge investment needed meant that the federal budget had become an essential part of the process of technological advancement.

'This conjunction of an immense military establishment and a large arms industry is new in the American experience,' said Eisenhower. 'We recognize the imperative need for this development. Yet we must not fail to comprehend its grave implications.'

Commentators have been debating the military-industrial complex ever since. But whether it is seen as an essential part of national security or a threat to a free society, there can be no doubt about how important it has been in shaping technology.

If we can see the seeds of the modern world in Second World War technology, it follows that the first shoots of the future may be visible in the technology being developed for the military today.

Whereas this book can only take in the highlights of many thousands of projects, it will take in enough of the significant vistas to get some idea of the whole. Obviously there will be gaps, and if the reader misses a discussion of the Swiss Army Knife (invented by Carl Elsener in 1897 as a multi-purpose tool for the Swiss Army) or flat-panel speakers (from DERA, the British Defence Evaluation and Research Agency) it shows the immense breadth of this field. In other areas, such as the evolution of explosives, the influence of the military is readily apparent and hardly needs further explanation. Further chapters could be written on everything from medicine to catering to pest control (DDT, for example, was developed for the US military), but in these cases the technology is generally doing exactly what it was designed for. What is more important are the ways in which military technology appears unexpectedly in everyday life, how jet fighters turn into Boeing airliners and code-cracking machinery becomes a PC.

From the 1950s to the 1980s, the drive in the West was to build weapons to win the Cold War, or the hot war that might follow it. The anticipated scenario was a massed assault by Soviet armour though the Fulda Gap in Germany, backed by immense numbers of troops and aircraft. The idea then was to follow a similar strategy to the Nazis, defeating a more numerous enemy by the use of high technology and superior quality. Aerospace and electronics were the key.

Air power was always seen as the future, and much emphasis was placed on improving aircraft and missiles to outmatch their Soviet counterparts. This led to 'smart bombs' and advanced missiles which would make up for low numbers with lethal effect. Rather than using a dozen aircraft to drop a hundred dumb bombs on a target, says this argument, a single sophisticated aircraft could do the same job with just one smart bomb. As we shall see, the reality may be more complex.

The current technological drive is directed to the 'war against terror'. It is a different kind of war, and one that calls for different types of weapons. Instead of the earlier emphasis on weapons to take on battalions of tanks, the new opposition is 'asymmetrical': guerrillas and terrorists who will use stealth and surprise to launch their attacks.

The requirement now is on intelligence systems and sensors capable of finding hidden enemies, and weapons to attack them in deep bunkers and caves. Communications and networking become ever more vital because of the need to reduce the 'sensor to shooter' time, the delay between an opponent being sighted and being fired on. A battalion of tanks will not get away easily, but a few guerrillas may easily have vanished before an airstrike can reach them.

New priorities also mean that Allied casualties have become much less acceptable, so there is a drive to keep soldiers as far out of harm's way as possible, filling the front line with robots and remote control systems instead. Body armour and improved vehicle protection are also getting attention.

As in the past, the big spenders in military research and development are in the aerospace sector: high-tech aircraft and missiles are still the most important programs in financial terms. This shows up clearly when you look at the top five defence contractors and their share of the procurement budget:

Lockheed Martin, $15 bn: aircraft including the F-22 Raptor and missiles

Boeing $13 bn: aircraft (e.g. B-52 bomber) and missiles

Northrop Grumman $11 bn: aircraft (e.g. B-2 stealth bomber) and missiles

Raytheon $5 bn: missiles, including the Patriot

General Dynamics $5 bn: aircraft (e.g. F-16) and missiles

The US Air Force deploys aircraft on a massive scale and is the world's biggest buyer in financial terms. But the US Navy is also a major purchaser, having relied on aircraft rather than battleships for most of their firepower since the Second World War. Their fleet of 1,600 carrier-based combat aircraft is the fourth largest air force in the world (the RAF has less than 300). The US Army does not operate fixed-wing combat aircraft, but has over 5,000 helicopters – the largest such force in the world. Even the US Marine Corps has an air force of its own with over 200 Harrier jets (compared to the RAF's sixty).

Whether this emphasis on air power is justified has long been a matter of debate. Its advocates say that airpower gives the US the power to deliver massive force anywhere within hours without risking US soldiers in combat. Critics argue that some jobs can only be carried out by troops on the ground and that air power is a very blunt instrument. But for the meantime, air power rules, and this shows up in the spending figures. In 2003, Operation Iraq Freedom started with a bombardment of over 700 cruise missiles at a cost of up to $1.3 million each. The Air-Launched Cruise Missile is made by Boeing, the Sea-Launched version by Hughes.

This results in a skewing towards aviation projects, which is reflected in this book and the technology spin-offs that appear. Certainly we will see more developments in aerospace, but other areas are involved even in air power. Aircraft need communications, sensors and computers at the very least.

The end products of this new technology drive will filter through to the civilian sector. When they do, they may bring changes at least as dramatic as those brought about by the computer, the Internet, and the jet airliner.

## The Limits of Military Technology

Much of the literature of military technologies is a species of techno porn, with full-colour spreads and exploded diagrams showing the inner workings of military hardware accompanied by enthusiastic descriptions echoing the manufacturers' claims for their latest product. For obvious reasons, this gives a highly misleading view of the technology, bathing it in a golden glow which is often quite un-deserved. However, both the military and the weapons makers are keen to preserve this aura, and the fans are happy to go along with them.

In America, technology is king, and there is a long history of looking for technological solutions to military problems. This promotion of technology became most conspicuous in the 1991 Gulf War, which abounded with misleading reports. The idea of the surgical strike was promoted, in which precision munitions took out the target without damaging anything around it. Briefings repeatedly showed bombs and missiles striking with amazing accuracy, in spite of the fact that less than 10 per cent of the bombs used were 'smart'. The other 90 per cent were unguided 'dumb bombs'.

The Patriot missile was deployed to intercept Iraqi Scud missiles. Reports at the time suggested that 90 per cent or even 100 per cent of the Scuds were intercepted. Later the Pentagon downgraded this claim to 80 per cent but an Israeli report on the Patriot batteries protecting Israel downgraded it further to 40 per cent. Some time afterwards it was suggested that the rate of effective interceptions in which a Patriot missile actually brought down a Scud rather than just exploding near it was even lower than this. However, during the war it hardly mattered; morale had been boosted just when it was needed. But for a cool assessment of how effective technology actually is we need to examine the facts more closely.

The British have their own version of the military myth. According to this viewpoint, the British soldier is the finest in the world, and wars are only won because of his sterling qualities. British equipment is traditionally seen as inferior to other nations, if not actually defective. The British media seize on every negative report of equipment, such as the now infamous SA-80 rifle with its tendency to jam, the Apache helicopters which were not cleared to fire missiles, and the failure to modify the Challenger tank for desert conditions. Everything from boots to radios is routinely reported as being defective in the press.

Every equipment failure is recorded. As I write this the newspapers are reporting that eighty Royal Marines have returned from a training exercise with frostbite; the story is that the sleeping bags they had been issued with were substandard. Whether this turns out to be the case or not is immaterial, what matters is that the British public perceive it as a kit failure. (Meanwhile in America, the Army are working on clothing with built-in heating and cooling for the combat soldier).

Bigger items, like the Eurofighter Typhoon, are attacked for being too expensive or delivered too late, whereas similar projects in the US are hailed as being world-beating triumphs of American technology which secure the future. British defence procurement is criticized as an outrageous expense, US procurement is a way of creating employment and securing votes in the right areas.

To some extent this vision is shared by the armed forces themselves. The British military are not such great believers in technology as their American counterparts, preferring to rely on more human factors such as good leadership and good soldiering. Technology has an alarming tendency to break down or behave unexpectedly under battlefield conditions, but the British fighting man (or woman) will always be there, and, in British eyes anyway, will always be the best in the world.

This popular myth which the media subscribe to and augment gives a rather deceptive picture of the British armed forces. British kit is among the best in the world, and there are very few nations which can boast the type of hardware possessed by the British military. However, this does not fit with the accepted view of 'our brave boys' let down by duff kit supplied by incompetent bureaucrats and uncaring politicians. It does, however, put a great deal of pressure on defence spending and discourages the kind of blue-sky approach seen in the US, where billions put into research and development are seen as a national asset rather than as a black hole. Ambitious schemes are more for the US, which has much deeper pockets and a more optimistic attitude towards the military benefits of high-tech.

Defence spending has been far greater in the US than Europe since the Second World War, which accounts for why so much development comes from the other side of the Atlantic. In crude terms, the US spends over $40 billion a year on defence research, while the UK spends less than $4 billion. As a result, one might expect to see ten times as much coming from the US. In fact, the situation favours the big spenders and has a multiplying effect: more money on basic research means more technology to take forward; a huge internal defence market means that any project has a much better chance of attracting funding. In the UK, even where there are good ideas they may wither for lack of support from basic research ('we need a new material with this conductivity and strength') or failure to find sponsorship ('it looks great, but even if you can develop it, there's no money to buy the end product').

This does not answer the question of why the USSR, America's great rival in the same period, produced comparatively few civilian spin-offs from military technology during the Cold War. To understand this, we need to look at the contrasting approaches of the two nations.

## America v Russia: Different Types of Rocket Science

During the Cold War years, while the US was going for high-tech, Russia adopted a different tack. Russian rocket scientists vied with the Americans for the honours of being first in space flight, and chalked up a notable series of firsts – first satellite, first man in orbit, first probe to the moon – but the rest of the Russian military establishment was otherwise engaged.

In military terms, Russia had huge reserves of untrained manpower and plenty of raw materials. The Second World War had been one of attrition for them, wearing down the Germans and overwhelming them with large numbers of tanks and aircraft. A Third World War was expected to follow a similar pattern. It was all very well to have tanks and jets with the latest technological gadgetry on them, but such things are complex. They cannot be easily produced in large numbers, and raw troops cannot be trained to use them quickly.

Instead of the West's emphasis on small, professional, highly-trained forces with the most advanced weaponry available, the Soviets followed a policy of developing less advanced systems which were as cheap and robust as possible and could be used by uneducated teenage conscripts.

This resulted in products like the AK-47 Kalashnikov assault rifle, RPG-7 rocket launcher and SAM-7 anti-aircraft missile. All are cheap and relatively easy to make. In terms of quality they may be markedly inferior to their Western counterparts, but they do not require the same level of maintenance or training, and they will keep working.

These weapons designed in the Soviet Union are used by guerrilla and military forces all over the world. A Kalashnikov is no weapon for a sniper, being crude and inaccurate, but it will keep working when another gun would break down. And in the hands of a teenage soldier, a sniper rifle would be completely wasted. What he needs is something that will fire lots of lead in the general direction of the enemy.

'The fact is,' writes Viktor Suvorov, a high-ranking Soviet officer who defected to the West, 'that Soviet designers realised, decades ago, that only uncomplicated and reliable equipment can be successful in war.'

A Russian saying puts it more bluntly: 'A stupid weapon which works is not a stupid weapon.'

The Russians placed less emphasis on electronics and high-tech than the West, preferring equipment which would withstand a long hard campaign. NATO troops in Western Europe only had enough missiles for a month-long war; the Russians expected a much longer conflict and did not place the same reliance on missiles which could not be quickly replaced.

The multiple artillery rockets used by the US and Russia are a clear example of the different approaches. Both armies have these systems, which perform similar roles of providing heavy fire at long-range targets. They appear superficially similar, but a closer look reveals profound differences.

The Americans have the Multiple Launch Rocket System or MLRS. It is a twelve-tube rocket launcher mounted on an armoured, tracked vehicle. It is highly automated, with a mechanical self-loading system and a fire control computer giving a high level of accuracy. The automation means that it requires a crew of just two skilled technicians.

The rockets are carried in two pods of six; to reload, a mechanical loader simply replaces the pods. Each rocket contains a load of over seven hundred bomblets which it scatters over a wide area. The bomblets are of an advanced design, having a sophisticated fusing mechanism and a warhead that can punch through armour as well as producing anti-personnel fragments. A twelve-rocket salvo can be fired in thirty seconds, raining down over 8,000 bomblets over an area the size of several football pitches. The US Army call MLRS 'Steel Rain' for its power to cover a wide area with lethal fragments.[1]

Future plans for the MLRS call for guided rockets to give even greater accuracy, and warheads containing 'smart bomb' submunitions. These will separate from the rocket at high altitude and glide down, seeking out enemy tanks and targeting them.

This level of sophistication comes with a hefty price tag attached. The tracked MLRS vehicles produced through the 1990s cost over $2.3 million dollars apiece. Each rocket with its bomblets cost $20,000 – almost a quarter of a million dollars for each salvo.

The Russian approach is different. The Russians pioneered the use of massed rocket launchers known as 'Stalin Organs' during the Second World War and found them highly effective. The modern version has changed little. The BM-21 Grad ('Hail') is simply a Ural truck with a rack of forty unguided rockets on the back. These rockets have a range of 20 km they are not very accurate and have simple high-explosive fragmentation warheads. Massed fire rather than precision is the Russian approach: great accuracy is not necessary when blanketing an area several hundred metres across.

The Grad requires a crew of five, who are needed mainly for manually reloading the rockets; weighing 60 kg each, the process of lugging forty of these around by hand is labour intensive. It is not a complex system, and conscripts can be trained to use it fairly quickly.

The simplicity of the Grad makes it cheap and easy to produce compared to the MLRS. By the end of the Cold War, the US Army had ordered some 600 MLRS, many of which had not been delivered (the British Army has just fifty-four). In contrast, the USSR had well over 6,000 Grads already in service.

In a short war, there is little doubt that the Americans would have the advantage. The MLRS performed well in both Gulf wars. But when it comes to protracted conflict, it would take a long time to build new systems to replace losses. In Russia, where the automotive industry is a reserve for the military, the truck factories could turn out Grads by the thousand. Grad rockets could be manufactured in the most basic facilities, but the same cannot be said of the complex and expensive MLRS rounds.

In addition, the MLRS requires much more logistic support with specialized vehicles to transport the pods of rockets – each loaded pod weighs over two tons and so cannot be moved manually – whereas the Grad rockets can be shipped like crates of corned beef or pickled cabbage.

The same approach can be seen across the Russian armed forces, from tanks and missiles to aircraft and submarines. The Russian kit is built simply and solidly, and does not have the refinements of technology seen in Western systems. Individually, Western tanks and aircraft are superior to their Russian counterparts; the big question is whether one high-tech tank is better than three low-tech ones. Fortunately for the world, this was never put to the test.

This approach has meant that Russia has not been at the leading edge in many technologies. Starting from a lower base, the Russians knew that they could not win this sort of race, and did not attempt to. However, it meant that there were few of the obvious spin-offs associated with Western military technology.

Although there are some of areas where the Soviets stayed competitive or even ahead in terms of military technology (such as ramjets), this was not translated into civilian products. This was partly for economic reasons – the centrally planned economies of the Eastern bloc did not put a high priority on consumer goods – and partly for cultural ones. The policy of secrecy meant that new discoveries could not always be shared and freely distributed among the scientific community.

For example, Soviet computing benefited from the input of some of the best mathematicians around. But the lack of good quality electronic components and free exchange of information always held back progress. Although the Russians possessed supercomputers, notably the Elbrus range which were used by the defence industry, there was little danger of them overtaking the West in the commercial sphere.

This underwent a sharp change at the end of the Cold War when there was an urgent need for hard currency. Russian military devices like night-vision scopes started turning up in the West. They were crude compared to the Western models, but, according to the standard doctrine, rugged and far cheaper. It took some years for Western (and Far East) suppliers to match the Russians for price. In this instance there was a ready-made product being manufactured for the Army that could be marketed to civilians; in cases where an entirely new product has to be created the Russians' lack of experience and equal lack of essential knowledge in sales and marketing has been a major handicap.

Russian scientific expertise may be of a high quality, but translating high technology into a finished product – military or civilian – was always a problem. What a Pentagon report described as 'lingering quality control problems' were always an issue. This is partly a reflection of a system where firing staff was very difficult and many workers effectively had a job for life.

Without financial incentives on offer, motivating workers to do their best rather than do the least possible was always difficult, and so it was always easier to produce rough and approximate products rather than highly finished ones. While this may work well enough for tractors, it was not as successful for electronics. Russia suffered from a plague of exploding television sets as a result of those quality control problems; Pravda estimated that over 2,000 people were killed as a result of television fires or explosions during the 1980s. It is understandable why it proved preferable to import components and products from the sophisticated Far East producers than build from scratch.

Given this situation, it is easy to see why the Soviet Union, and Russia after it, has rarely been in the forefront of producing high-tech goods. In the new post-Soviet world, the military laboratories are facing the challenge of surviving in a commercial environment. While their science may be at the leading edge, a lack of commercial know-how and foreign investment has left Russia lagging behind the market leaders in moving military technology into the civilian sector.

This contrasts strongly with the US, where many of the companies involved in military programs also have a civilian side. This includes many household names like Boeing and McDonnell Douglas, Texas Instruments and Hughes Electronics (a division of General Motors), whose staff will be well aware of the commercial value of their work.

In America, scientists and engineers are highly motivated to take out patents and share them with their employers. The Soviet Union did not offer this path to riches, and given the difficulties of getting a product to the marketplace it is easy to see why this rarely occurred.

The US also has an active program of technology transfer which helps ensure that anything developed under a federal program for military use can find civilian customers. The National Technology Transfer Centre has a slogan of 'making commercialisation deals happen', and deals with thousands of different projects. They maintain a vast database of available technologies, from high-temperature alloys developed for use in rocket exhausts to new ways of measuring the quality of a video image, which industry can call on.

The massive military spending in the US has a significant effect on the economy. Boeing, biggest of the US defence contractors and one of the largest companies, is something of a landmark. For the first time in 2003, they fell behind their European rival – Airbus – in the sale of airliners, and this situation is expected to continue. At the same time, for the first time Boeing's military sales outstripped their civil ones, and the company is becoming increasingly dependent on the Pentagon. However, after a series of scandals including one involving the supply of tanker aircraft to the USAF, their relationship is precarious.

Social commentator Noam Chomsky has gone so far as to suggest that the main function of the Pentagon's research is to provide a gigantic and continuing government subsidy to industry which allows high technology to be developed outside the constraints of the free market. He also sees it as a way for companies to get a big cash injection without having to produce anything useful for the consumer and without having to label it as a state subsidy. Perhaps, ironically, this would make the US defence industry a close cousin of the giant state-funded corporations on the other side.

## Not Technology Alone

It is not just technology that travels from the military to the civilian world, as fashion has an influence as well. When the military adopt or redesign something that is particularly useful or convenient, this may lead to it being taken up by the rest of the world.

Thomas Burberry designed a raincoat for the British Army in 1901. This was very popular during the First World War, and was taken up by several other armies, becoming known as a 'trench coat'. The civilian version still has the same styling as the military original, including epaulettes (used for holding gloves or caps), a flap to access a holster, and a belt with D-rings which originally secured grenades.

In 1942, the US Navy submitted specifications to its suppliers for a new type of undershirt called the T-type. Cotton undershirts already existed, but were not widely worn; the Navy's T-shirt changed all that. Sailors, and later soldiers, found they made comfortable work shirts – and that the white cotton vests looked good, creating a semi-undressed look. After the war the T-shirt went back to civilian life with the returning veterans, and has gone from strength to strength ever since.

The trench coat and the T-shirt were both eminently practical designs; neither embodies novel technology as such, but we shall encounter other cases where its design combined with technology has produced a success in the civilian world.

There are also developments which are more a matter of concept than technology per se. The first ambulances were invented by Baron Dominique-Jean Larrey, chief surgeon of the French Army in 1792. It may seem obvious now that having dedicated vehicles (in this case wagons) to convey the injured would improve their chances of getting prompt medical care, but at the time the idea was revolutionary. It caught on in the civilian world, as did the later idea of the casualty evacuation helicopter, also pioneered by the French, which dates from the Indo-China conflict of 1949. The technology was not a great advance, but it proved that the idea was an effective one.

## Military/Civilian

Is there a fundamental difference between military and civilian technology? And is it ever truly possible, as the book of Isaiah exhorts, to beat the sword into a ploughshare? There is no single, straightforward answer to these questions. It depends very much on the particular technology involved. Some manage to become part of the civilian world, and like retired soldiers they can become productive members of a peaceful society. Others have a status which is more questionable, and they continue to lurk about the fringes, never completely escaping from their violent past. The following chapters will look into how and why this happens.

Looking forwards is more difficult. As we will see, the types of spin-off which occur are sometimes obvious (from military jet to civilian airliner) but sometimes completely unexpected (like the invention of the microwave oven). We can easily enough guess what a finished product like a new type of supersonic aircraft might lead to but it is impossible to guess what might come from more basic research.

In the course of this exploration we will see how the jumbo jet got its hump, how weapons research contributed to eye surgery and what IBM's very first computer was designed for. We will also meet the man who would move mountains, the deadliest aircraft of the Second World War, and find out why commuters are not flying around with rocket belts yet. In later chapters which look to the future, we will investigate the prospects for intelligent machines, anti-gravity propulsion, hyper-sonic airliners and sensors that can see though walls and clothes, as well as many other new and often surprising developments in the pipeline.

PART 1

# Looking Backward

# CHAPTER 1

# Rocket Science: V-2 to E4

*We have invaded space with our rocket and for the first time – mark this well – have used space as a bridge between two points on the earth; we have proved rocket propulsion practicable for space travel. This third day of October, 1942, is the first of a new era of transportation, that of space travel.*

General Walter Dornberger,
commander of the V-2 rocket program

Satellite technology is so common that it is invisible. We talk to a friend in America, watch a football match live from Asia, and visit an Australian website without ever thinking about how these miracles are achieved. Satellite television dishes are part of the urban landscape; satellite pictures are just another part of the weather forecast. If news reporters beam back their stories from the latest troublespot using satellite videophones, we only notice because the images are still a bit jerky. When this unsteadiness disappears we will stop noticing another piece of satellite technology altogether.

All of these developments can be traced back to the rockets that rained down on London during the Second World War. Without the impetus given to space research by the German military, orbital space would be an empty desert instead of an increasingly crowded metropolis. Rocket science was never really about putting man on the moon: it is about changing life on earth.

## The Rocket Blitz

In September 1944 London became the first target of a new secret weapon. Without warning, a row of houses in Staveley Road, Chiswick was demolished by a massive explosion. Nobody had seen or heard anything before the explosion; there was just the lingering trace of a vapour trail pointing up into the sky.

The authorities were worried that people would panic, so a story was put out that a faulty gas oven had caused the explosion. London had been under attack from 'doodlebugs', the V-1 flying bombs, since the previous year, but the explosion in Chiswick was something different. Others followed it. The official story about gas explosions began to wear thin, and Londoners joked about Hitler dropping gas ovens.

What the Defence Committee knew about the new weapon was too terrifying to reveal. Intelligence sources had found that it was a type of rocket called the V-2. Travelling at more than four times the speed of sound, it was too fast to intercept and too fast to hear – the sound of the rocket only arrived after impact.

Their speed was such that they were very rarely spotted. Charles Ostyn saw one just before it struck Antwerp: 'It was definitely not a contrail, but it was like a streak from a comet – as fast as a shooting star. It was a long, thin, white streak, more like a flash coming down to the earth.'

The Defence Committee estimated that the V-2 might have a warhead of as much as 10 to 15 tons of high explosive. A report by the Ministry of Home Security estimated that every strike of a rocket with a 10-ton warhead on London would kill 600 people, seriously injure 1,200 and cause over 4,000 casualties in total.

They estimated that the Germans would be able to launch one rocket an hour. After one month, the bombardment would have fired 700 rockets, killing 400,000 people and destroying almost half the buildings in London.

No defence was possible. The only way to stop the V-2 bombardment was to destroy the launch sites and manufacturing facilities.

As with most new weapons the V-2 turned out to be less deadly than feared by one side or hoped by the other. The warhead was only one ton, not ten or fifteen. A total of a thousand of the rockets hit Britain, and caused a total of 10,000 casualties including almost 3,000 killed. Though terrible, the effects of the V-2 bombardment were a fraction of what was predicted. Single raids by the Allies delivered more explosive and killed many more people than the entire V-2 effort; the raid on Dresden destroyed the city and killed more than 35,000 in a single night.

In spite of its lack of effectiveness in material terms, the V-2 had a definite effect on civilian morale. Like the first bomber raids of 1917 it was a leap forward in military technology, seemingly unstoppable. The Germans were the first to fully realize the potential of rocket science, and though it may not have helped their war effort, the impact on the post-war world was considerable.

## Rocket Science

'Rocket science' may be shorthand for intellectual challenge, but in reality making basic rockets is easy. Its simplest form is the plastic bottle rocket. All you need to do is take a plastic drink bottle half-full of water and pump it up so that the air is under high pressure. Turn the bottle upside down so that the opening is underneath and release it; the water shoots out at high speed, and the bottle flies up into the air.

This is science for seven year olds, but it shows the principle that drives everything from fireworks to space shuttles. The distinguishing feature of a rocket motor is that it has an exhaust which goes in one direction, pushing the rocket in the other. As dictated by Newton's Third Law, every action (the exhaust going one way) has an equal and opposite reaction (the rocket going the other).

In the case of the bottle rocket, it is easy to see the exhaust jet and appreciate its mass. With most rockets, however, the exhaust is in the form of a gas, but it is still the weight multiplied by the speed of the exhaust that determines its pushing power. To make a simple rocket powered by chemical energy rather than water and air pressure, part-fill a cigar tube with gunpowder and leave it open at one end. Ignite the gunpowder and the sudden expansion of gases produced by the burning gunpowder rushes out of the open end, propelling the tube in the other direction.

Given the importance on the speed and mass of the exhaust, both of these need to be as great as possible. In the case of rockets going into space this will make up the bulk of the craft. An Apollo rocket is really one huge fuel tank with a small spacecraft on top, with a ratio of fuel to payload of fifty to one.

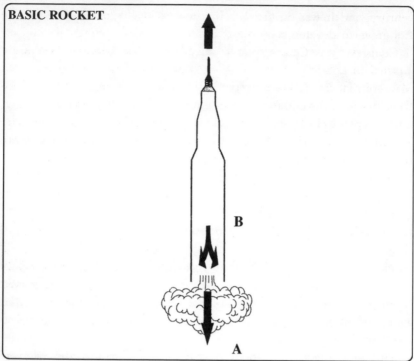

**BASIC ROCKET**

B

A

The basic principle of the rocket: exhaust ejected backwards (A) pushes the rocket (B) forwards.

The fuel-to-payload ratio was the inspiration for multi-stage rockets. Having burned the fuel from a tank, why keep carrying the dead weight of the fuel tank? When a big rocket fires, the stages burn in sequence, and having burned out they are little more than empty metal cylinders.

In order to make the most of the fuel, it needs to be ejected at as high a speed as possible. This has led to a whole area of science to find a mixture that can be burned in a controllable fashion to produce the maximum amount of energy; but initially, rockets relied on the best fuel available – gunpowder.

## Rocket History

The earliest known rockets came from China. Consisting of a wooden tube filled with gunpowder, they were very much like modern fireworks. By the sixteenth century they were being used as weapons in Asia. British troops in India faced rocket batteries in the eighteenth century, and it was captured Indian rockets that inspired Sir William Congreve to develop something similar.

Congreve was Comptroller of the Royal Laboratory at Woolwich Arsenal in London, with the splendid additional title of 'Inspector of Military Machines'. He was later to become an MP and Fellow of the Royal Society; throughout his life he was a prolific inventor, working on a new type of clock, gas meters, stereoscopic images, colour printing, and that old favourite, perpetual motion. But it is as a pioneer in the field of rocketry that he is best remembered.

Artillery in the era of the Napoleonic wars was heavy and cumbersome. Congreve realized that it would be possible to make a rocket launcher which was lighter and more mobile than any artillery piece. Congreve's War Rocket had a long wooden stick to stabilize it, and could be fitted with an explosive or incendiary warhead. The 3 kg rocket had a range of 2 km. Floating batteries of rockets on specially constructed boats were first used against Boulogne in 1806 and proved highly effective. British rockets were later to be used in the war of 1812 against the US, giving rise to the line about 'the rocket's red glare' in the 'Star Spangled Banner'.

The stick made Congreve's design awkward to handle. Stickless rockets were developed, such as the Hale Rocket which was stabilized by spin instead, but neither variety was very accurate. Crosswinds were a great problem. Rockets were much slower than artillery shells and so were pushed much further off course.

Conventional artillery grew steadily more powerful through the nineteenth century. Guns increased in range, accuracy and rate of fire, and their warheads were more lethal. But no solution was found to the problem of keeping rockets on course and they steadily lost ground to the guns. By the end of the century rockets had fallen out of use except in occasional colonial conflicts. Against light opposition, it was still useful to have a piece of artillery that could be transported through mountainous rough terrain by pack mule, but it was clearly obsolete on the battlefields of Europe.[1]

However, the same era also saw the birth of a new type of adventure literature, marked by a spirit of scientific speculation. This new literature came to be called Science Fiction, and authors like Jules Verne and H.G. Wells were hugely popular. Space flight and the exploration of other worlds was a popular topic for speculation. Although Verne settled on a giant cannon to propel his characters from the Earth to the Moon, even he knew that this was highly impractical. Rockets offered a much gentler ride than the bone-crushing acceleration involved in being fired from a cannon.

Konstantin Tsiolkovsky, a Russian schoolteacher, made the earliest known study of rockets for space flight. Another teacher, Hermann Oberth, published a book, *The Rocket Into Planetary Space*, which was a great success and became a handbook for the growing band of enthusiasts. During the 1920s new societies were founded in several countries, all with the object of furthering the exploration of space using rockets.

This led to a great interest in rocket technology. In 1930, a group of rocket enthusiasts formed the American Interplanetary Society in New York, while the British Interplanetary Society was founded three years later in Liverpool.

Gunpowder rockets could only go so far, because gunpowder only releases a relatively modest amount of energy. More powerful fuel would be needed to reach into space. American inventor Robert Goddard calculated that a rocket using liquid fuel such as kerosene combined with liquid oxygen would have much greater potential. Goddard launched his first liquid-fuelled rocket in 1926. It travelled a total of 80 metres and landed in a cabbage patch. The flight was a landmark success and marked the beginning of the new rocket age.

The energy value of the kerosene-oxygen mix was much greater than that of gunpowder. In principle, a large enough liquid-fuelled rocket could break free from the Earth's gravitational field and reach out into space.

Goddard's work was taken up in Germany by the Verein fur Raumschiffahrt (Society for Spacetravel), who clubbed together to fund amateur rocketry projects. The first liquid-fuelled rockets in Europe were tested by Max Valier, a member of the VfR. Valier was a flamboyant character and a great proponent of rocketry who did much to popularize it. He experimented with rocket-propelled sleds, sledges and gliders. He even built a rocket-propelled car – basically a publicity stunt for Opel, the vehicle could reach 145 m.p.h. The VfR relied on publicity to attract funding and their first rocket launch was scheduled to coincide with the opening of a science fiction film about space travel.

Valier wrote a best-selling popular guide to space travel, putting Oberth's ideas into language that ordinary people could understand, and many magazine and newspaper articles about travelling to Mars and similar topics. Valier was killed in a lab explosion in 1930, underlining just how dangerous such rockets can be. Combining liquid oxygen and alcohol is basically a controlled explosion, one that can very easily get out of hand.

The idea of space travel might have remained the province of slightly eccentric enthusiasts, but for the return of militarism in Germany. According to the Treaty of Versailles which ended the First World War, Germany was only permitted a limited amount of artillery. However, the treaty said nothing about rockets. The old war rockets might be obsolete, but new technology promised something much more effective and in 1930 the German Army started its rocket program.[2]

There were many types of small rocket, including some for scattering propaganda leaflets over enemy lines, and others fired in salvoes against attacking aircraft. There were anti-tank rockets and the multi-barrelled 'fog-thrower' which could blanket a wide area with smoke rockets in seconds. But more important than any of these was a new type of heavy guided rocket artillery which would surpass any gun ever built.

The army forcibly recruited staff from the Verein fur Raumschiffahrt, in some cases not having to use compulsion. A young man called Wernher von Braun agreed to do work for the Army in exchange for funding for his research.

'We needed money for our experiments, and since the army was willing to give us help, we didn't worry overmuch about the consequences in the distant future . . . We were interested in one thing – the exploration of space.'

The Wehrmacht's program focused on the liquid-fuelled rocket. In 1936 the researchers set about building the Aggregate-4, or A-4, a giant rocket capable of carrying a one-ton payload a distance of 250 km.

The challenge was to deliver the fuel and oxygen together into a combustion chamber fast enough and evenly enough to ensure that the combustion was stable. Any variation could result in unevenness in the thrust resulting in the rocket veering off course – or could produce an explosion like the one that killed Valier. The engineers developed a pump powered by a steam turbine running at high speed. The pump was itself fuelled by a combination of hydrogen peroxide and sodium permanganate, two secondary fuels that were also extremely hazardous to work with and contributed to the high accident rate.

By 1942 the team had built their rocket and carried out a successful test launch. Hitler was impressed and ordered that the A-4 should immediately be put into full-scale production under the name of V-2 (Vergeltungswaffe 2, the first weapon being the V-1 flying bomb), and declared that here was the weapon that would win the war.

As usual, the Führer was unduly enthusiastic. The technology was in its infancy, and it took another two years and some 3,000 test firings before the V-2 was ready. By the time the first operational rocket could be launched against London, Allied troops had already landed in Normandy and the Russians were sweeping inexorably towards the West.

Although the V-2 attacks on London are well known, it was Antwerp which was the main target. The Belgian port was to be crucial in bringing Allied troops and equipment into Europe, and more than 1,600 V-2s were launched at it. Antwerp was nicknamed 'the city of sudden death' – when a packed cinema was hit, killing over 500 people, a new regulation was introduced banning gatherings of more than fifty.

The V-2 was not the decisive weapon Hitler had hoped for. The V-2s carried a total of less than 2,000 tons of explosive to their targets over six months, during a period when Allied bombers were dropping over 100,000 tons of bombs on German targets every month. The accuracy of aerial bombing was on average within about half a mile of the target; missing by an average of five miles the V-2s were ten times less accurate and frequently failed even to hit the city they were aimed at.

The V-2 manufacturing facilities eventually were destroyed, and the launch sites were overrun, bringing the rocket campaign to an early conclusion. Of the 12,000 rockets ordered by Hitler, only a quarter had been launched.

General Dornberger, in charge of the V-2 program, summed up its effects in two words: 'Too late.'

But even delivered years earlier it would have made little difference. Given a more destructive warhead – nerve gas, a biological agent or a nuclear warhead – the V-2 might have been apocalyptic. With just one ton of explosive it was too weak to have much impact.

The V-2s range meant it could only be fired at Britain from the closest points on the continent, but much more ambitious plans were in the pipeline. If Nazi Germany had somehow survived into 1946, they might have been able to field an even larger rocket, the A9/A10, a two-stage weapon codenamed Projekt Amerika. As the name suggests, this would not just have been able to hit targets anywhere in Britain from Germany, its range of over 5,000 km would allow the Nazis to strike the US. The rocket would have weighed 80 tons, three times as much as a V-2, although it still had the same one-ton warhead. The A9/A10 might have caused alarm as well as civilian casualties in the US – if it managed to hit a populated area – but the effect on the US war effort would have been negligible.

The A9/A10 would have faced other problems. The size of the V2 had been dictated by the maximum that could be transported using the existing road and rail infrastructure. A bigger rocket would have been immobile and the construction/launch facility would be a prime target.

It might have been much better for the Allies if Hitler had diverted even more resources into the rocket program. The V-2 took up a large amount of raw materials and scarce fuel. The total cost in materials and labour involved in building a single V-2 rocket was equivalent to what was required for a fighter aircraft or a medium tank. Without the V-2 the Allies would have had to face thousands more German fighters or tanks.

The overall cost of the V-2 program was staggering. It is difficult to make cost comparisons under war conditions, and estimates have ranged from a total of half a billion to three billion dollars (between $14 and $90 billion at today's rates). This compares with $2 billion for the Manhattan Project, with the important difference that the Allies had far more money and resources to spare.

The technical resources that went into the V-2 meant that other rocket projects received little attention. By the end of the war the Germans had guided anti-tank missiles, surface-to-air missiles and even air-to-air missiles, but none of them were in a fully effective operational state. If the full weight of scientific expertise had been put behind developing these weapons in 1930, instead of the ambitious but ultimately ineffectual V-2, the Germans might have had more usable weapons. Whether this could have affected the ultimate outcome is a matter of debate.

In addition to the cost in resources there was a terrible human toll from the program, not paid by the Nazis but by their forced labourers who had been drafted in from concentration camps. Thousands of these slave workers died of dysentery, typhus and starvation during work on the V-2, victims of appalling conditions and lack of food. Others were simply executed on the spot for refusing to comply with orders. The administrators kept accurate records, so we know that of the 70,000 workers employed, 26,000 did not survive the experience. The program thus killed more than ten times as many slave labourers as it did British civilians.

## After the War

Both the Americans and the Russians realized that rocketry had a tremendous military potential. In particular, it could be combined with the new atomic bomb to create an unstoppable weapon capable of destroying a city from long range. Moral considerations took second place to getting hold of as much advanced German technology as possible.

The German rocket scientists were based at Peenemunde in eastern Germany. In late 1945 they were given orders to join the local militia to defend themselves against the oncoming Red Army. Instead, they moved west and surrendered to the Americans en masse. Under a secret plan called Operation Paperclip, about a hundred and fifty German scientists were given five-year contracts to travel to the US with their families and work for the US Army. Perhaps most controversially, one of the scientists was von Braun, whom some considered to be ultimately responsible for the deaths of thousands of the slave workers.

Relations were strained at first, as the war with Germany was a recent memory and the scientists were not universally liked or trusted. However, relations thawed over time and many of the scientists stayed in the US and eventually became naturalized citizens.

As much hardware as possible was salvaged, and more than a hundred complete V-2 rockets were transported to the US where they formed the basis of the military rocket program. Afterwards, what was left at Peenemunde was taken over by the Russians, who also recruited many of the remaining scientists.

As rocket development continued in the US and USSR, many assumed that the US was well in the lead. Having the advantage of a large number of V-2 scientists and the benefits of an open society which encourages the sharing of academic information, the space race should have been a foregone conclusion.

However, geography and policy intervened. Early nuclear warheads were heavy and bulky, although news reports at the time suggested that they weighed less than five hundred pounds and used a lump of uranium the size of a golf ball. The truth was different: the first atomic bombs were Fat Man weighing 4 tons and Little Boy (4.5 tons). Later and more powerful weapons would be even heavier, as they comprised an atom bomb acting as a trigger surrounded by a shell of material forming a hydrogen bomb. A nuclear warhead would need a rocket many times bigger than a V-2 just to carry it a few hundred miles, and intercontinental distances called for gigantic rockets.

The US had plenty of heavy bombers capable of delivering atomic bombs, and no shortages of air bases in Europe within easy striking range of the Soviet Union, so there was little need for a giant strategic rocket. The situation was different for the Soviets, as Russia did not possess such advanced aircraft, and America was thousands of miles away from their nearest airbase. So while the US concentrated on building short-ranged missiles for tactical applications until smaller nuclear warheads were available, the Russians were pressing ahead with large strategic rockets.

In 1947, a Russian team led by Sergei Korolev built their first rocket, the R-1. This was very similar to the V-2 with only minor changes, but proved that Russia could successfully build rockets using materials produced locally. Several further rockets followed of increasing maturity, and by 1953 the Kremlin was confident enough to order the construction of a true intercontinental ballistic missile. This was the R-7, which became known to NATO by the reporting name of SS-6 or Sapwood.

The R-7 was a two-stage rocket, designed to carry a 3-ton warhead 8,000 miles, enough to strike the US homeland from Russian soil. It was fuelled by liquid oxygen and kerosene and weighed a massive 280 tons, making it fourteen times the size of the V-2.

The program took four years to deliver the first rockets for testing. There were three unsuccessful trials in May, June and July 1957, but on 21 August 1957 an R-7 was fired and successfully reached its test target.

Now that the R-7 was working it could be used for another purpose, and two months later a version was fired which carried something other than a test warhead.

On 4 October 1957, the Earth had two moons: the natural one, and a new artificial satellite by the name of Sputnik (Russian for 'travelling companion'). The first Russian announcements were fairly low key; it was seen as a technical achievement but not an earth-shaking event. However, when the American public reacted with shocked awe at this unexpected development, the Soviets started to make the maximum amount of political capital out of their techno-logical triumph.

On 4 November a second Sputnik was launched. Like the first, it orbited the planet, emitting a faint radio signal that could be picked up by radio hams around the world.

The reaction in the US was one of bewilderment, surprise and fear. If the Russians could put up a satellite, then nuclear warheads soon be drifting over America. Much of the furore that followed was directed at the government, which had failed on two counts: they had failed to let the American people know of this new threat, and, more seriously, they had let the Russians get there first. America had slipped into second place, complained the politicians, and President Eisenhower was accused of failing the country.

Former President Truman blamed the McCarthy era for having deprived the missile and satellite programs of some of their best scientists, and many agreed. There was humour too, with bars advertising a 'Sputnik Cocktail': one third vodka, two-thirds sour grapes.

The Russian domination of orbital space could not go unchallenged. The US had started a project to launch a satellite under the name Project Vanguard in 1955, but this had fallen behind the Russian effort.

The Soviets had a single, centralized rocket development organiza-tion for its strategic rocket forces, and hence for its space program. In the US, the Army, Navy and Air Force each had their own separate rocket development organizations. The Army and the Navy had both put in bids to be the first to launch a satellite. The Navy won the privilege of being chosen for this task, so on 6 December they launched Vanguard 1.

The launch vehicle was based on the Redstone medium-range ballistic missile, which failed immediately after launch and blew up in front of the assembled world's media; the satellite survived, and was seen rolling away, radio transmitter sending out its call sign.

The Navy team had failed ignominiously, and the job of restoring America's lead was given to von Braun and his US Army rocket team. Their rocket was the Jupiter C, a modified version of the Redstone.

Explorer 1, the first US satellite, was successfully launched on 31 January 1958.

Von Braun was hailed as an American hero, and as far as the public was concerned he was rehabilitated. He cemented his position by collaboration with the Disney organization, and acted as a consultant for Disney space-related television programs. He appeared in front of the camera in productions like *Man In Space* and *Man on the Moon*, and advised on the design of Tomorrowland, the futuristic part of the Disneyland theme park.

However, the leap from mass destruction to mass entertainment was not entirely without difficulty and von Braun's questionable past was not overlooked by everyone. Satirist Tom Lehrer wrote a song about the rocket scientist with the memorable couplet:

Once the rockets are up, who cares where they come down?
'That's not my department', says Wernher von Braun.

As a result of the Sputnik crisis, a new organization was formed to unite space research within the US in October 1958: National Aeronautics and Space Administration. This took over from the earlier National Advisory Committee for Aeronautics (NACA) and other bodies and ensured that US space launches would be handled by a civilian agency, albeit one which worked closely with the military. With a staff of 8,000, NASA was dedicated to seeing that the US stayed ahead in space.

Space launches became a vital propaganda weapon in the Cold War, with both sides striving to assert their technical supremacy. Sputnik carried instruments to measure the temperature of the upper atmosphere and Explorer helped to identify the Van Allen radiation belts, but nobody pretended that these were anything other than secondary to the main aim of boosting national pride.

But success in the space race had practical consequences as well. 'Take the high ground' has long been a military motto, and space was the new high ground. Larger and more powerful ballistic missiles were built, but as well as warheads there were rockets designed specifically for satellite launches. Some of these were experimental, like Sputnik, others were for the benefit of the military but would also have some importance in the civilian world.

## Meteorology

In April 1960 the US launched its first weather satellite. This was TIROS I, the first Television Infrared Observation Satellite. It was a joint project between NASA, the US Army, US Navy and the Weather Bureau, and it was seen as having strategic importance.

Weather has been important for the military since the days of sailing ships, and in the modern era meteorological data was needed for air operations as well as estimating the direction of fallout from nuclear weapons.

It was not clear whether a satellite would be able to provide any useful information about weather on earth, or even if the proposed instruments were likely to work. This was completely new ground – until you try pointing a television camera back at the Earth from space it is hard to predict what you will be able to see.

The first TIROS was equipped with two television cameras, one for high-resolution images and the other for a wide view. A transmitter allowed it to send pictures, and it even had a magnetic tape recorder so that images could be stored when it was not within range of the receiving station. The whole thing weighed 122 kg, a far cry from the simple Vanguard payload.

Although it only orbited for eleven weeks, TIROS I was a huge success. It successfully beamed back images of the clouds from above, for the first time giving meteorologists the big picture rather than having to rely on reports from scattered weather stations. A whole series of TIROS satellites quickly followed, with cameras working in both visible and infrared wavelengths.

Satellite pictures show cloud formations whose tone (from white to dark grey) indicates altitude and whose movement allows weather fronts to be followed. IR photography gives an indication of temperature and temperature change. The effect was a swift transformation in the range and accuracy of weather forecasting.

Weather satellites also give warning of imminent dangers. Hurricanes and severe storms cause massive damage across the southern US states, and satellites are the only means of tracking them in real time so that warnings can be issued. In 1961 TIROS-3 spotted Hurricane Esther – the first time a hurricane had been located from space before anyone on the surface was aware of it.

New lenses and new cameras were introduced, and the image quality improved to the point where the US Weather Bureau could offer an international fax service to share cloud pictures with other weather services around the world. The satellites longevity improved as well, and by the mid-1960s the Weather Bureau had continuous coverage of the world's weather from space. Each individual photograph only covered a limited area, but by fitting 400 of them all together into a mosaic it was possible to create a single image of the entire planet seen from space. For the first time meteorologists could see all the world's weather all at once; and mankind had a completely new view of the world.

As well as cameras, the later TIROS satellites were equipped with a radiometer, a device that works in deep infrared and gives precise measurements of temperatures at the surface. An additional instrument, an atmospheric sounder, is capable of giving a picture of the humidity profile from the ground up to the top of the atmosphere.

There are still dedicated military weather satellites as well as civilian ones. In addition to other weather tasks, they have instruments to measure charged particles and electromagnetic fields to assess their effects on early warning radar, satellites and missiles. However, discussions are under way with NASA and the Department of Commerce for a single satellite program that will meet both military and civilian needs. Whether the military will give up this particular strategic asset remains to be seen.

## 'Live by satellite'

Earth-based radio transmitters have problems with long distances, a situation that can be compared to children signalling to each other with torches at night. When there is a line of sight between the transmitter and receiver, everything is easy. If there is no line of sight, communication is possible if there is some object that both can see. If both can see the same building, for example, they can signal to each other by shining the torch beam on to it.

With powerful enough torches, the children could even signal to each other from many miles away over the horizon by shining their torches on to the bases of clouds. And in theory – given ridiculously powerful torches – on a cloudless night they could signal to each other by shining their torches at the moon.

Radio works in a similar fashion. Distant receivers pick up the signal after it has been bounced off the upper atmosphere, like bouncing a torch beam off clouds. At night the atmosphere cools and its properties change, which is why radio reception changes at night. The reflections of the signal will also cause interference, especially if the same signal is received via two separate reflections slightly out of phase with each other. If you wanted truly clear long-distance communication the answer would be to bounce the signal off something very high in the sky – like the moon, or a man-made substitute.

A satellite gets rid of many of the difficulties of earth-bound transmission. So long as you have a line of sight to the satellite, you should be able to get a clear signal. Rain and cloud may reduce the signal somewhat, but not by more than a half.

Even before the first artificial satellites were launched, the communications possibilities were being investigated. In 1946 the US Army Signal Corps succeeded in bouncing a radar beam off the moon. Code-named Project Diana after the ancient moon goddess, this showed that it was possible to send a signal by reflecting it from the moon. The US Navy continued this work and by 1955 succeeded in establishing radio by this means. The first message from outer space came from Dr Robert Page, the associate director of research at Naval Research Laboratory, to his colleague Dr Franz Kurie: 'Lift up your eyes and behold a new horizon.'

The moon was unsatisfactory though, being a bad reflector and given to rising and setting at inconvenient times. The science fiction writer Arthur C. Clarke is hailed as the originator of the modern communications satellite. In 1945, while an RAF electronics officer, he wrote an article in *Wireless World* magazine about using satellites to transmit television around the world. He realized that the length of time a satellite stays visible from the ground depends on its orbit. A very low satellite will circle the Earth every ninety minutes; one at the same distance as the Moon will take twenty-eight days. Somewhere in between – at an altitude of 22,300 miles – the satellite will orbit in exactly twenty-four hours. If it is orbiting over the equator, from the point of view of an observer on the ground the satellite's orbit is exactly cancelled out by the rotation of the Earth. A satellite in such an orbit appears stationary, hanging at a fixed point in the heavens.

These 'geostationary' orbits are the perfect place for communication satellites, which can remain in permanent communication with any point underneath them on the Earth. Three such satellites spaced equally above the equator could provide global coverage.

In 1955, John R. Pierce of Bell Telephone Labs estimated that a man-made communications satellite would be able to carry the equivalent of a thousand simultaneous telephone lines. This compared to the TAT-1, the transatlantic telephone cable that started operation in 1956 and could carry thirty-six telephone lines. TAT-1 cost over $30 million, so in theory a communications satellite costing a billion dollars would still be good value.

Early communications satellites were 'mirrors' that reflected a signal broadcast from the ground, soon to be superseded by repeater satellites that received a signal and rebroadcast it at a different frequency. For satellites thousands of miles away the transmitting station needed to be powerful and, more importantly, to be able to focus its signal into a narrow beam so that the satellite could receive it. The satellite itself would only be able to broadcast a very weak signal because of its limited power supply, but so long as the ground station had a large enough receiving dish it would be able to pick up the faint broadcast.

In fact, how strong the satellite signal is when it returns to Earth depends on how it is transmitted. If you bounce a signal off the moon the returned signal goes in all directions, but a satellite that transmits rather than reflecting can concentrate its broadcast on a small area. If the signal is focused on the Earth rather than all surrounding space, it is a hundred times more powerful. Concentrating it on a fraction of the Earth increases the power density accordingly; you can transmit a weak signal over a wide area, or a strong one over a more limited area. This is why communications satellites have a 'footprint' or area over which their signal can be received.

The principle was proved in 1958 by the military Project SCORE (Signal Communication by Orbiting Relay Equipment), which put a repeater satellite into orbit, and progress was swift. However, the military program ran into trouble soon afterwards. Their new satellite, called Advent, was intended to be state of the art and had new features including a sophisticated stabilization system. But technical problems and cost overruns led to its being cancelled, while civil programs raced ahead.

In July 1962, just four years after the first US satellite launch, Bell Telephone got their wish for a civilian communications platform. It was called Telstar. The launch vehicle was military (a Thor intermediate-range ballistic missile with an additional upper stage), but Telstar was promoted as a messenger of world peace. Less than a metre across and weighing 80 kg, it had a power of just 15 watts, less than a local radio station. Telstar's lifespan was only seven months, but it was a powerful weapon in the propaganda war: the first satellite transmission, broadcast to the US, Britain and France, showed the Stars and Stripes waving in the breeze to the tune of the 'Star Spangled Banner'.

Telstar went on to broadcast baseball games, the World Fair in Seattle and Presidential news conferences. The capacity was half what Pierce had estimated, but the cost was a fraction as much at only $6 million for the satellite itself and a further $6 million to NASA for the launch. The total Telstar network of satellites and ground stations would cost far more to complete, but it was a knock-down price: crucially, the government was not attempting to recoup any development costs. The billions spent on building and perfecting nuclear missiles would provide a benefit to the taxpayer.

Telstar was not in geostationary orbit but a lower orbit with an altitude of 3,000 miles, so a whole series of satellites would be needed to provide continuous coverage. More communications satellites soon followed, including NASA's own communication satellite, Syncom, launched in 1964 which was the first geostationary communications satellite. A debate had started over whether the military should be using commercial communications satellites or whether they needed their own network.

The military pushed on after the failure of Advent, and a replacement system called the Initial Defense Communication Satellite Program (IDCSP) was created. In contrast to Advent it was designed to be as simple as possible, with a very limited capability equal to about ten telephone lines per satellite. IDCSP became the US's first strategic satellite communication system, allowing direct contact between Washington and military units around the world.

By 1967 the demands of the Vietnam conflict led to IDCSP being used for digital data transmission, and a profusion of other communications satellites were launched with higher capacity for various strategic and tactical requirements.

Meanwhile international co-operation resulted in the creation of the International Telecommunications Satellite Organization (IN-TELSAT) so that satellite receiving stations could be set up across the globe. The cost of a single phone line using satellites was almost

$100,000, but prices fell as more satellites with greater capacity joined in.

In the 1970s a second organization was created specifically for mobile users. This was the International Maritime Satellite Organization (INMARSAT), which was originally intended for communication with ships but came to be the standard for land-based users as well. It first came to public prominence during the 1991 Gulf War, when journalists used their INMARSAT satellite phones to file reports back from the front line. By the time of the 2003 Gulf War the journalists were sending back reports by videophone. These were just jerky images of low quality (a result of the need to compress the signal to fit into the bandwidth available), but they gave the media independence; instead of having to rely on the military for communications, they could send back images of whatever they wanted, whenever they wanted.

Further improvements in bandwidth usage and image compression will improve the images to the point where the technology becomes invisible and the satellite connection becomes as everyday as the mobile phone.

Satellites were also an obvious choice for broadcasting; their altitude means that it is possible to cover almost half the planet with a single transmitter, though practical considerations of power mean that the footprint satellite broadcast is limited to a much smaller area. By the 1970s the idea of satellite television had come to fruition. Initially users would require a huge dish antenna, but this was reduced by satellites with greater power and smaller footprints combined with better antenna technology. By the end of the 1990s satellite dishes had become ubiquitous.

Like the Internet, the global reach of satellite communications has had some unexpected side effects. While the original idea was that the Pentagon would be able to stay in touch with air bases in Guam and warships in the Pacific, the end result was that viewers in Britain could pick up hard-core pornography from Europe. Because the footprint of the satellite channel Red Hot Dutch overlaps the UK, in 1995 the British government attempted to ban the sale of decoder cards for the channel because its content was deemed to be unacceptable under British obscenity laws. The decision was challenged in the courts, but Red Hot Dutch went bankrupt before it could be resolved.

Even if they had won the court case, attempting to stop satellite channels would be like King Canute trying to stem the flow of electromagnetic waves. The flood cannot be forced back because satellite communications mean that it is now comparatively easy for an organization to broadcast to any point in the world without having an expensive infrastructure. This makes it difficult for regimes which wish to control their people's access to outside information. The effects have been particularly noticeable in countries like Iran, where camouflaged dishes are a regular feature of middle-class households, and the Arab countries where international satellite channel al-Jazeera is now by far the most influential source of news. State television cannot maintain a monopoly on the truth when other sources are available to anyone with a satellite dish. Al-Jazeera's version of events may be challenged, but the very existence of multiple channels changes the situation permanently.

This may have the effect of putting the TV stations in the front line themselves. The US destroyed the al-Jazeera offices in Kabul during the fighting in Afghanistan, and it is likely that similar action may be taken in future where enemy propaganda is seen to be a danger.

## You Are Here: Global Positioning Systems

When Sputnik I first orbited in 1957, US scientists realized that the exact location of the satellite could be calculated by listening to its radio beacon by taking advantage of an effect called Doppler shift. This effect is encountered in everyday life when a car races past: as it approaches the engine sounds high pitched; after it has passed the note drops, creating a distinctive 'EEEEYOWWW' sound as it passes. This is because the relative movement of the car as it is coming toward you has the effect of piling up the sound waves, producing a shorter wavelength and a higher note. As it goes away, the sound is 'stretched out' and the note is lower.

The Doppler shift of radio signals works in a similar fashion, and it meant that Sputnik could be tracked by several different ground stations, comparing the shift of the signal they received and then calculating the satellite's location. The US military developed a system called Minitrack using this principle so that Sputnik and other Soviet satellites could be monitored. Russian satellites might be spies or even carry nuclear warheads, so they were of considerable interest.

As an offshoot of this, it became apparent that the same process would work in reverse: given signals from a satellite of known location, the observer could calculate their own position. Frank McClure, a scientist at John Hopkins Laboratory, was the first to propose this as a means for a submarine to establish its exact location.

During the Second World War, the navigation of bombers at night was critical; while an error of half a mile may matter little in civil aviation where getting within visual range of the airport is all that matters, bombing an invisible target in darkness requires pinpoint accuracy. This applies to an even greater degree with missiles; as we have seen, one of the great failings of the V-2 was its inability to hit closer than a few miles to a specific target.

The same problem also applied at sea. The US was building a fleet of submarines carrying nuclear missiles which they would not be able to aim accurately unless they could first identify their own location on the surface of the earth. Inertial navigation systems were good, but not quite good enough. Although they started off accurately, INS gets steadily less and less accurate over time and distance. An aircraft is only in the air for a few hours, but a submarine might be at sea for months, building up unacceptable errors. McClure believed that satellites could do much better.

The first satellite navigation system was NAVSAT, also called Transit, built by the John Hopkins Laboratory. Testing started in 1960 and it became operational in 1964 when it was used for ballistic missile submarines. It was not shut down until 1996. NAVSAT could give a fix to an accuracy of about two hundred metres – more than enough for a nuclear missile, as a hundred metres either way makes little difference to its effect.

The Russians followed suit in 1967 with their own satellite navigation system called Tsyklon.

NAVSAT used ten satellites in a low orbit with an altitude of about six hundred miles, but for greater accuracy and better coverage something bigger was needed. Research began in 1973 for a new system known as the NAVSTAR Global Positioning System which would involve a constellation of twenty-four satellites. Users would need to be able to see three of them to get a fix, but with this arrangement there would be at least six visible at any time from any point on Earth, giving a safe margin of redundancy in case of accident or attack.

NAVSTAR finally went live with limited capability in 1989 and was declared fully operational in 1995. However, between the start of the project and its completion, things had changed. Instead of being a purely military system, the Department of Defense and the Department of Transport jointly recommended that it should be made available for civilian use as well. This was partly inspired by an incident in 1983 when a Korean 747 airliner was shot down because it had strayed into Russian airspace. President Ronald Reagan wanted airliners to benefit from technology which could prevent this kind of accident from ever happening again.

The Russian counterpart of NAVSTAR is GLONASS, which was operational in 1993. It originally had twenty-four satellites, but it was in a poor state of repair and by early 2002 the number in operation had dropped to eight, limiting its usefulness. However, three more satellites were launched in late 2002 to restore coverage, and three more followed in 2003. Over the next few years enhanced satellites will be launched – GLONASS-M followed by GLONASS-K – giving improved accuracy. The current GLONASS is accurate to within about 70 metres and the improved system will halve this.

Once a purely military operation, GLONASS is still run by the Russian Ministry of Defence but now has to justify its existence by its benefits to the civilian user community. GLONASS receivers are freely available on the commercial market, as well as receivers which use both GPS and GLONASS signals to give a more accurate fix than either would produce on its own.

Uncertainties over funding have led Western observers to wonder how long the system will continue operating, but GLONASS still appears to be struggling on.

GPS usage has far surpassed the original estimates. The initial plan was for 27,000 military units, but by 2002 there were already 40,000 in use with US forces alone.[3] However, the numbers used by the military is dwarfed by the number sold in the civilian market. Almost three quarters of a million GPS sets are sold every year, with prices dropping to a few hundred dollars. GPS is now used for everything from airliner navigation, to helping the blind to find their way, to in-car navigation. London buses will use GPS to record their position as part of a system which tracks buses and provides estimated arrival time indicators at bus stops. A scheme was recently proposed in the UK to have all cars fitted with GPS so they could be monitored for the purpose of levying taxes on driving in urban areas at certain times.

The number of GPS users may be set to increase far beyond these applications as next-generation mobile phones start to incorporate GPS chips. This opens the door for all sorts of new functions: a mobile with GPS and Internet access could, for example, locate the user and give directions to the nearest pizza parlour or pub. More seriously it could summon emergency services to the caller's exact location if they press a panic button. A similar function could call the nearest available taxi. Linked to the power of computer navigation, a GPS phone user might never need to have a map or directory, but could simply dial up and get directions for wherever they needed to go.

The first applications for 'location-enabled' phones are likely to be games, the first of which are already being used in Sweden. Called BotFighters, it is described as being a type of 'virtual paintball' game in which players used their mobile phones to 'fire' at each other; if the phones are close enough, a hit is scored. The industry is hoping that these games will become a major attraction for phone users.

Driven by on-line gaming and other attractions, location-enabled phones could take off very quickly. With around 400 million mobile phones sold every year, GPS would become truly global.

However, the Pentagon still control GPS and are well aware how effectively it could be used to guide missiles and aircraft from enemy nations. Any third-world nation could get the benefit of the multi-billion-dollar satellite for a few hundred dollars. To counter this, GPS has a feature known as 'Selective Availability' (SA), in which the GPS signal is deliberately degraded in certain parts of the world. In 2000, SA was turned off, but the military reserve the right to turn it on again at any time. US military GPS receivers are capable of picking up a special encoded signal so SA does not affect them.

As one government scientist put it: 'As an illustration, consider a football stadium. With SA activated, you really only know if you are on the field or in the stands at that football stadium; with SA switched off, you know which yard marker you are standing on.'

This is something of an exaggeration (and therefore quite representative of statements by scientists working on military projects). Without SA, GPS receivers are capable of accuracy to about 15 metres; when it is turned on this is degraded to 100 metres or so.

One way of getting around the reduced service is known as 'differential GPS', which uses a land-based radio beacon as well as the satellites. Systems using this can work with high precision, down to a matter of millimetres, and are widely used in surveying.

The crossover between civilian and military technology has raised a political controversy around Galileo, the 2.5 billion Euro satellite navigation system being developed by the European Space Agency. The initial impetus for Galileo came from the fact that NAVSTAR is run by the United States, which reserves the right to turn it off during any future conflict so that potential enemies cannot use the system to target the US or UK forces. While this makes military sense, it causes problems for civilian users: a navigation set-up that may stop working at any time is essentially useless.

NAVSTAR is also less than ideal because it does not provide good coverage at northerly latitudes like Northern Europe, and it can fail in 'urban canyons' where buildings block the view of much of the sky.

The EU decided that a new European navigation system under EU control was the answer. This was strongly opposed by the US, because it would provide potential enemies with an accurate guidance system which could not be blocked. The US insistence only strengthened European resolve, as it indicated that without an independent system the US would hold a monopoly on this technology.

Galileo will involve a constellation of some thirty satellites, due to be put in orbit between 2006 and 2008. This will give fixes to an accuracy of five metres or less – much better than even the un-degraded GPS signal. The arrangements of satellite will give much better coverage in northern areas, and the greater number of satellites will greatly improve the chances of getting a fix in urban canyons. Galileo is also more powerful than NAVSTAR, which will make it easier to receive through foliage – anyone who has used a GPS system will know how difficult it is to get a signal while in thick woodland.

As a civil system, Galileo will have other features not shared by NAVSTAR, one of which is a receiving system compatible with existing distress beacons; this would give a 98 per cent chance of detecting and precisely locating a distress beacon within ten minutes.

Galileo is a huge program. By making some functions such as basic location available freely and charging users for more advanced features such as high-precision fixes, it will potentially tap into a global market worth billions. It has been estimated that it will create up to 100,000 jobs across Europe and will bring great improvements in the reliability of satellite navigation so that it can be widely adopted in safety-critical situations.

The program puts the EU on a collision course with America. The US has announced an intention not to allow hostile activity in space, so there is the possibility that, in the event of a war, Galileo may be interpreted as a threat. There have been extensive negotiations to ensure that Galileo and NAVSTAR are compatible and the signals will not interfere with each other – this was initially an issue because the wavelengths used are quite close. This does not resolve the more fundamental question of a navigational system outside of US control.

'Navigational warfare' is already an active area of research, as the Pentagon seeks for ways to selectively jam Galileo signals if they can be used by a potential enemy in a war zone. This would be the soft option; a more robust approach would be to knock out sections of the Galileo system, which would (from the Pentagon's perspective) solve the problem entirely.

## Rockets for War and Peace

There would be no satellites today without the V-2 rocket. No government would be likely to sink half a billion dollars into such an ambitious and completely unproven scheme in peacetime. Without the work that the German military had done in proving that rockets could reach orbit, and developing the technology to do it, even the Russians would probably have ignored rockets in favour of jet-powered bombers. By the end of the war, man was already halfway towards space, and although with the worst possible motives the Nazis had broken important new ground.

Satellites may have evolved beyond their original military purpose, but space technology remains closely linked with warfare. The major space powers are the US and Russia plus a European consortium including Britain and France – the original nuclear powers. China, Israel and India have also launched satellites, and it is no coincidence that all three possess nuclear weapons.

When North Korea launched a satellite in 1998 it was seen as a thinly-veiled show of force: the ability to put a satellite into orbit implied an ability to send a nuclear warhead over intercontinental distances.

As we have seen, areas like satellite telephone, satellite television broadcasting and global positioning systems still have strong military significance. Rocket science is not simply about mankind's quest to conquer space; it is about projecting force on earth. Those who have the rockets do not just have the power to reach for the stars, they can also dominate neighbours who lack the technology.

# CHAPTER 2

# Air Raids To Airliners

*Life expands in an aeroplane. The traveller is a mere slave in a train, and, should he manage to escape from this particular yoke, the car and the ship present him with only limited horizons. Air travel, on the other hand, makes it possible for him to enjoy the 'solitary delights of infinite space'. The earth speeds below him, with nothing hidden, yet full of surprises.*

Early French advertisement for airline service

Between the 1930s and the 1960s a remarkable transformation took place in civil aviation. When you look at airliners of the two eras side by side, the transformation is obvious: the 1930s aircraft is driven by propellers, the wings stick out at right angles, and it is unpressurized with a fabric covering. Although the main fuselage is built of metal, some parts of the aircraft are still wooden.

The aeroplane beside it is like a spacecraft next to a sailing ship. The airliner of the 1960s has a silvery metallic skin, and the shape has been streamlined with wings swept back at an angle. Wood has been replaced with steel and aluminium alloy; the propellers have been replaced with jet engines. There is also a dramatic difference in size, with the earlier aircraft dwarfed by its gigantic successor.

If you add a 1990s airliner to the collection, it is remarkably similar to the 1960s version. Outwardly it is almost indistinguishable. The 1990s model has the same shape, the same layout, same type of engines as the earlier version. There have been improvements, but they are mainly internal; in general terms the two are the same. The big change in civil aviation took place with the technological leap forward after the Second World War.

**AIRLINER COMPARISON**

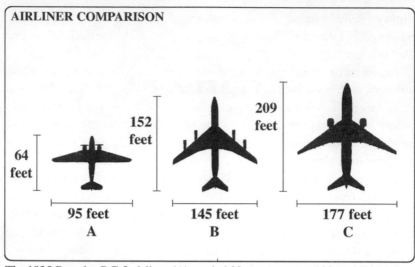

The 1935 Douglas DC-3 airliner (A) carried 28 passengers at 180 mph/288 kph; the 1958 Boeing 707 (B) carried 147 passengers at 560 mph/900 kph, while the 1995 Boeing 777 (C) carries 380 passengers at the same speed.

The change has been so profound that it is hard to appreciate how different things once were. We take air travel for granted these days, but back before the war flight was a novelty and air travel was an adventure reserved for the few.

A 1930s advertisement for air travel suggests that passengers should introduce themselves to their pilot: 'He is always a man of the world as well as a flying ace.'

Even in the early days there was a network of air routes spanning the world, but flying was not easy, quick or cheap. Travelling from London in 1935 you could fly from London's main airport at Croydon to Karachi, which was then part of the British Empire. The journey comprised several legs taking seven days in all. One of the legs, from Basle to Genoa, had to be taken by train, because of a prohibition on flying through Italian airspace. The ultimate long haul was a trip to Australia: London to Brisbane took a total of twelve days.

For such a journey you would be travelling on an airliner like the twin-engined De Havilland Dragon Rapide, cruising at a stately 132 m.p.h, which would give you plenty of time to get to know the other seven passengers on the plane. A few years later you might be one of the lucky few travelling on Short's magnificent Empire flying boats. These boasted the luxuries of an ocean liner, including seven-course dinners, and carried a complement of no less than twenty-four passengers. Cruising speed was 164 m.p.h, at an altitude of around 10,000 feet.

At the time, these speeds seemed amazing. Only a few years before airliners had been biplanes cruising at 100 m.p.h. or less. But even travelling half as fast again, long distance travel was only for those with plenty of leisure time, and the prices were daunting enough to restrict air travel to the seriously affluent. The ticket to Karachi cost £130 in 1935 money, almost £7,000 at modern prices. The Australia trip would have been over £10,000 – and these prices were for a one-way ticket. Only the truly rich could consider flying in both directions.

In the 1950s the first jet airliners entered service. The British Comet was the first true jet airliner, which started with BOAC in 1952. Unfortunately it was grounded after a series of tragic accidents, and it was the Boeing 707 which became the true success story. It flew at more than twice the height and almost three times the speed of propeller-driven airliners; most importantly it carried 140 passengers and economies of scale were a big factor in its success. More than a thousand 707s were built, and by 1960 air travel had been transformed.

The cost of air travel plunged. In 1962 the price of a flight to India was £130 – about £1,700 in modern terms – for a return flight. This represents a decrease by a factor of eight from thirty years before. Perhaps as importantly, the speed and range of jet airliners meant that the journey could be completed in a single day. Long-haul travel was no longer for the leisured classes.

Of course, you still had to have money to be part of the emerging jet set, but the bar had been lowered. For short-haul flights this made the difference between a tiny, exclusive service and a mass-market commodity. The holiday resorts of Europe were opened up to millions of British holidaymakers as a whole new industry was born. Mass travel was already a feature of modern life, but the new jet airliners meant that a week at the seaside could mean somewhere on the Mediterranean rather than in Britain. The Spanish coast and the Greek islands were ripe for development that is still continuing.

Forty years after the first days of popular air travel, the changes in airliner technology have been relatively minor. Prices have continued to fall, and the cost of a flight to India is about a third of what it was in the 1960s, but the aircraft themselves have changed little in terms of externals and performance. Even the latest Boeing 777 is only marginally faster than the old 707, and although the number of passengers on board has doubled again to 300, the changes have been much less noticeable than the transformation from the old propeller-driven airliners to today's streamlined silver jets.

The transformation happened because of a series of technological innovations, mainly from the Second World War, and entirely from the military side. Peel back the skin of a modern airliner and you find the body and the heart of a war bird – although one which has been very successfully domesticated.

## Early Military Aviation: Balloons to Bombers

Generals have always understood the value of the high ground and eleven years after the Montgolfier brothers' first flight, French military observers used a tethered balloon to observe the battlefield during the Battle of Fleurus. By the nineteenth century tethered balloons were an essential adjunct to artillery. Just before an artillery barrage a balloon was raised with an observer whose job it was to report on the fall of shot and direct the gunners. By the time of the First World War the phrase 'when the balloon goes up' became a synonym for an artillery barrage heralding the start of an attack.

Height also has a practical value: the defenders in a siege could throw rocks and other projectiles down on their unprotected opponents.[1] The only counter was to build a siege tower higher than the walls being besieged, so that the defenders could in turn be attacked from above.

Balloons were the first bombers, starting with the siege of Venice in 1849 in which small unmanned balloons were used to carry bombs over the city. Their effectiveness was marginal and they posed a hazard to friendly troops, but the idea had taken hold. Powered airships which could fly anywhere at will opened up new vistas, and early science-fiction writers predicted that fleets of airships would conquer the world as no defences could stand against them. While happy to use such craft for observation, the military were slow to take on the idea of the bomber, until the advent of the German Zeppelin airships of the First World War.

The Hague Convention of 1907 had outlawed the bombing of civilian targets, but few expected it to hold. In January 1915, two airships of the German Navy dropped bombs on, or at any rate near, Great Yarmouth and King's Lynn. By May the Germans were getting better, and airship LZ38 dropped 30 explosive and 90 incendiary bombs across London, from Stepney to Leytonstone, killing seven people.

Faced with the shocking fact that London, the very heart of the British Empire, was coming under aerial bombardment, the War Office came up with a variety of plans for defence.

'It appears probable that, for the immediate present, the surest method of attacking airships would be by a superior force of airships,' concluded a report by the Committee of Imperial Defence, and plans were drawn up for the R Series of airships which would tackle both Zeppelins and U-boats.

In the end the Zeppelins were stopped by a new type of aircraft: the heavier-than-air machine. The Wright flyer was not immediately recognized as the weapon which would transform warfare in the twentieth century, and early craft were slow, flimsy and unreliable compared to the lumbering airships.

'It is by no means proved that aeroplanes are likely to have the better of an encounter,' noted the Committee. The US government was similarly sceptical, having had its fingers burned by spending $50,000 on a failed steam-powered aircraft project called the Aerodrome in 1898. Although the US Army was involved with the Wrights on observation planes from an early stage, there was no large-scale investment at first.

But during the course of the First World War the heavier-than-air aircraft developed rapidly. More powerful engines were built, and they gained the performance needed to intercept airships. The BE2 of the early war years, with its top speed of 75 m.p.h. and a ceiling of 12,000 feet gave way to the Sopwith Camel which could achieve 120 m.p.h. and 19,000 feet.

On the Western Front, aircraft graduated from observation and reconnaissance to attack. The counter to an aircraft was another aircraft, and while the infantry laboured in the muddy trenches, the fighter pilots above them became the new knights of the sky. For the first time, factories across the US and Europe turned out aircraft by the thousand.

The high altitude and speed required warm clothing, and pilots adopted a type of short leather jacket. This was the first garment to make use of the new zip fastener which had been patented in 1913. After the war the design was taken up by police motorcyclists and others. According to Mick Farren's book, *The Black Leather Jacket*, the jacket we know today can be directly traced back to German First World War pilots. (Bomber jackets, predictably enough, are derived from the pattern worn by US bomber crews in the Second World War).

The lumbering Zeppelins could not escape the new fighters which could intercept them at their maximum altitude. The Germans responded by building a series of huge bombers with a 23-metre wingspan to carry out attacks on London. With a cruising speed of 80 m.p.h. and a load of half a ton of bombs, they terrified whole cities and pointed the way to the future of warfare. The raids caused little real damage, but killed dozens of civilians and gave a warning of the new age of air power. After the war, the evolution of bigger and more powerful bombers continued, spurring the development of faster fighters to intercept them.

## The Need for Jets

For their first four decades, aircraft were driven by propellers, a logical evolution of the propellers that drove ships. The maximum speed of such aircraft is limited by how fast the propeller can push air. Though slow by today's standards, early aircraft were much faster than any other form of travel.

There was a constant competition for greater speed and a glamorous international air-racing scene in which competitors flew ever-faster planes. The biggest drive for speed came from the military; in the words of the fighter pilot's saying: 'Speed is life'.

Throughout the 1930s, fighters and bombers were designed with ever greater speeds and altitude, with the war applying even more pressure. Speed was the trump card in air-to-air combat. A faster bomber could not be intercepted by a slower enemy, and the pilot with the faster machine could always put his foot down and break off the fight if it was going against him.

This led to larger and larger engines, which meant more and more weight. By 1938, the Mark 1 Spitfire had a speed of 350 m.p.h., leaving the Sopwith Camel trailing. By 1944 the Mark XIV Spitfire could manage 450 m.p.h., although this required a doubling of the engine power, and an increase of 50 per cent in the weight of the aircraft.

More powerful engines needed bigger aircraft to carry them, and the limits were being approached. A better power-to-weight ratio would improve matters, but piston engines were already reaching the theoretical limits. One engineer calculated that the engine would have to weigh more than the entire aircraft to drive it past the speed of sound.

There were other fundamental problems. Air resistance increases with speed. This can be overcome by flying higher, where the air is thinner and there is less resistance – but the efficiency of the internal combustion engine driving the propeller is reduced in the thinner air, so the speed falls off again.

Propellers are powered by an internal combustion engine similar to a car engine, referred to in aviation circles as piston engines. The limits of the piston engine led engineers to look at alternative ways of driving an aircraft, including one which had previously only been used for power generation: the gas turbine.

A water turbine harnesses the energy from a stream of water. The water flows through the turbine, causing the blades to revolve, driving a shaft. Water turbines, an evolution of water wheels, have been around for centuries.

Replace the water with wind, and you have a wind turbine. Wind is a fickle and unreliable source of power, but burning fuel produces a steady stream of gas which can drive a gas turbine. An intake sucks in air, compresses it and mixes it with fuel.[2] The resulting mixture is ignited, burning efficiently at high temperature and producing a stream of high-speed exhaust gas to drive the turbine.

By using some of the power of the turbine to drive the compressor, the gas turbine becomes self-sustaining. Once started it takes in air, burns fuel and produces power for the compressor by turning a driveshaft. This could potentially be much more powerful than a piston engine, though it would still be limited by the propeller as mentioned above.

Early gas turbines were tested in 1908; in the 1930s they were used for electricity generation, and many are still used today in this role.

A faster propeller can increase the speed, but only up to a certain point. As the speed of the propeller tips approach the speed of sound, they produce shockwaves, making the propeller less efficient at shifting air. The shockwaves also cause vibrations which threaten to destroy the propeller, putting a practical limit on the speed a propeller can achieve.

**TURBOJET**

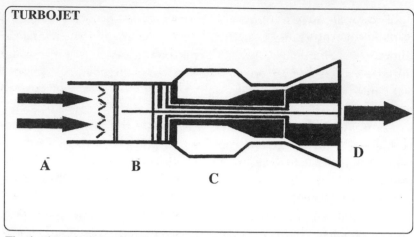

The basic principle of a turbojet: the inlet (A) sucks in air, which is then compressed in the compressor (B). Fuel is injected and the fuel-air mix is burned in the combustor (C) producing exhaust (D), which generates thrust.

In 1928, a student at the RAF College, Cranwell submitted a thesis proposing a new type of aircraft engine which did away with both the piston engine and the propeller altogether. The student was Frank Whittle, who had been an aircraft apprentice at the RAF for several years when he was selected for officer and pilot training. His final thesis for the RAF College included the basic idea of the jet engine. By 1929 he had formulated the idea of using a gas turbine to power it, and in 1930 he applied for a patent for 'a reaction motor suitable for aircraft propulsion', which was granted two years later.

For electricity generation, the stream of exhaust gas is purely used to drive the turbine blades. For an aircraft engine, it is the stream of gas which provides the propulsion; the turbine blades simply draw power for the compressor part of the engine and do not provide thrust.

This is the jet engine in its simplest form.

Whittle recognized that the efficiency of the jet engine would increase in thinner air, making it ideal for high-altitude flying, the power-to-weight ratio being much higher than an internal combustion engine. He also knew that the speed of the exhaust from the jet – which would ultimately dictate the speed of the aircraft it was driving – was potentially far greater than anything which could be achieved with a propeller.

Whittle was not the first in this field; others had tried before and failed.

'I was well aware of the many prior failures in the gas-turbine field early in the century,' said Whittle, 'but was convinced that the cause of these – low compressor and turbine efficiency, and lack of suitable materials – could be overcome in due course, especially in the case of aircraft, where the very low high-altitude air temperatures substantially offset the adverse factors.'

However, 1930 was not a good year to be selling a new invention. The global economy was depressed, air travel was a luxury for the rich and there was no market for new civilian aircraft engines. The Air Ministry was stretched for funding, and their own analysis of the available compressors suggested that Whittle's idea was not practical – the idea, coming from a junior officer just twenty-two years old with no experience in the industry was not looked upon favourably. The view was that Whittle was hardly likely to succeed where others had failed before; Whittle himself has commented that if he had been older and more senior he might have been taken more seriously. He was a small and slight man, and perhaps his great technical skill needed a more imposing presence to back it up.

But he was not deterred, and along with two ex-RAF pilots he set up a company, Power Jets Ltd, to develop his ideas. It was not a large organization, taking more after the pioneering spirit of the early backyard inventors. One senior RAF officer referred to their site as 'Whittle's rabbit hutch'.

Still an RAF officer, Whittle pushed ahead with jets while working on other projects. He tested seaplanes and catapult systems for launching aircraft from ships, did the engineering course at RAF Henlow (scoring a remarkable 98 per cent) and wrote papers on engineering.

Power Jets were in a difficult position, because the Air Ministry had classified the details of their invention as secret, but the Ministry itself would not fund them. They scraped together what investors they could, including £200 from an old lady running a corner shop near Whittle's parents.

By 1937 Whittle had successfully bench-tested a jet engine, finally proving his theory.

'The invention was nothing,' he wrote later. 'The real achievement was making the thing work.'

The RAF was supportive, but in spite of his progress the Ministry remained sceptical. They noted that the jet turbine required materials of a strength and heat-resistance at the limit of what could then be manufactured. The core of the engine would generate gas at a temperature approaching 2,000°C, and although the gas would be cooler when it reached the turbine blades, it was difficult to see what material these could be made of to survive for any length of time. It was not that they doubted that a prototype could be made to work, but they were sceptical that it could be flown day after day.

It was not until 1939 that the Air Ministry's Director of Scientific Research accepted that Whittle's ideas might be practical. War with Germany was looming, and Whittle finally received government backing.

Other nations were also developing jets. In Russia, Arkhip Mihailovich Lyul'ka produced plans for a turbojet at Kharkov in 1936. He encountered less resistance than Whittle, and by 1938 development was underway, but the German invasion changed priorities and Lyul'ka was reassigned, though he later returned to jets and by 1944 his team had produced a working prototype.

The British may have had a head start, but it was the Germans who reached the finishing line first. Hans von Ohain invented a jet engine independently, publishing theoretical work in 1933, some three years after Whittle. It was his good fortune that the aircraft maker Ernst Heinkel was actively looking for new types of high-speed propulsion when he received von Ohain's proposal. The inventor was given a team of engineers selected from the best in the company, and by 1937 Heinkel had a working laboratory test rig. This was purely a company development as the German Air Ministry was not involved until later.[3]

The German government were quicker to appreciate the potential of the invention and gave it full support. The first jet-propelled aircraft to fly in August 1939 was a Heinkel 178. It was followed by the Messerschmitt 262, the first operational jet fighter. The Me-262 first saw action in July 1944, and was the only jet aircraft to play a significant role in the war. The first experimental British jet did not fly until 1941, and although the Gloster Meteor entered service with the RAF in 1944 its only combat operations were against V-1 flying bombs.

By the standards of the day, the Me-262 was an awesome aircraft. Its top speed of 540 m.p.h. far surpassed any Allied plane (compare that with 450 m.p.h. for the Spitfire), and its high rate of climb made it ideal as an interceptor. In the early days there were cases recorded of Allied pilots baling out rather than facing a jet in combat.

But the jet engines had serious drawbacks. They consumed fuel quickly, limiting the range and duration of flights and wartime fuel shortages meant that the jet project was put on hold for three months. They behaved differently to propeller-driven aircraft, and getting pilots sufficiently trained to fly the aircraft in combat proved difficult. The accident rate was predictably high.

The Me-262 required a long runway to get airborne, leading one Allied pilot to comment that this was why Hitler invaded Russia – simply to get enough room for the jets to take off. The long runways were plainly visible to Allied reconnaissance, marking out the locations of jet bases so they could be attacked.

Worst of all, the new jet engines were unreliable. The 'mean time between failures', the length of time the engine ran before breaking down on average, was very low. As the British Air Ministry had feared with Whittle's design, the steel alloys of the turbine blades were not rugged enough. Running at high temperature (700°C), the centrifugal force on the turbine blades caused 'creep' in which the metal gradually deformed and the blades lengthened. The engines had to be changed before the creep was dangerous, and the early engines could only work for ten hours before they needed replacing.

Improvements in the turbine blades increased the engine life progressively, but after six months of development they still only lasted twenty-five hours. It was the problem of producing a reliable engine that slowed down the introduction of the jet fighter rather than Hitler's demand for fighter-bombers. Even with the improved turbine blades, at any given time at least 30 per cent of the jets were grounded waiting for engine changes.

It has often been claimed that Hitler's intervention slowed down the German jet effort because he was interested in bombers rather than fighters, and that without his interference Germany would have had fighters which would have gained complete dominance of the air. It is certainly true that the Führer was interested in strategic bombers capable of devastating an enemy's cities, not short-range fighters which he saw as a purely defensive weapon. But his edict that the Me-262 should be a fighter-bomber rather than a pure fighter was only issued in May 1944 when the jet engine was just starting to enter mass production, and it is dubious whether it affected the wider entry of the plane into the war.

When the Me-262 jet fighter took to the skies, it was not invincible.

Allied pilots found that the jets were vulnerable when they were at low speed, after take-off and just before landing. Allied aircraft patrolled over German airfields, ready to ambush the jets. Lt Urban Drew, a US pilot, shot down two Me-262s in one action by ambushing them as they rose from the runway.

In other engagements the jets showed that high speed is not itself decisive in dogfights. This is because when your only weapons are machine-guns and cannon with a range of a few hundred metres, it is impossible to engage a target if you are flying past it at several hundred miles an hour. To fight someone you need to be travelling at similar speeds.

The dogfights themselves were conducted at relatively slow speed, and although it gave the German pilots the option of breaking off combat, it did not mean they could win every time. Much air-to-air combat was essentially ambush: in something like half of all dogfights the loser was not aware of the enemy until they started firing. Under these conditions the faster jets could be just as vulnerable as the slower aircraft.

On other occasions the Me-262s came back to find their long runways damaged by Allied bombing, and some were lost while trying to land on cratered runways.

Most Me-262s went down in air-to-air combat or in accidents. Although more than 1,200 were delivered to the Luftwaffe, only about 300 saw action and they failed to make much impact. Part of this was simply because of the sheer number of Allied aircraft. In one action, thirty-seven Me-262s were sent to intercept a raid of over a thousand Allied bombers – with a protective screen of more than 600 fighters. The Messerschmitts were simply swamped by the Allied forces, unable to do more than token damage.

An attempt to build a small, cheap jet interceptor using as few resources as possible was undertaken. This could have been built in much greater numbers than the Me-262, but the project did not progress far before the end of the war.

Although it may have had little practical impact, the jet engine had shown what it could do. Once the technology was mature and reliable engines could be produced, jets would leave piston-engined planes standing. In future the only thing that would be able to catch a jet was another jet.

## The Pulse Jet: Doodlebugs over London

The Germans also introduced a second type of jet engine, the pulse jet. This is a much simpler type of engine. It does not have the compressor and the turbine of a true jet engine; it is really little more than a metal tube with a set of shutters at one end.

As the name suggests, the pulse jet does not work continuously like other jets, but in pulses. The shutters close off the front of the jet tube, and fuel is injected and ignited, propelling the jet forwards. As the ignition gases expand, the pressure inside the jet drops, opening the shutters to allow more air in, so the cycle can be repeated. It is mechanically a fairly simple device, without the complexity of the gas turbine and its need for high-performance materials.

The pulse jet causes severe vibration and has a lower maximum speed than the turbojet, though it can give propeller engines a good run for their money. It also has high fuel consumption and is extremely noisy, making it a poor choice for a manned aircraft. But it is considerably simpler than a propeller engine, and correspondingly reliable. There was an attempt to use two pulse jets to power an aircraft (the Me-328), but the vibrations were too much, causing parts of the tailplane to fall off or damage to the airframe.

While it was unsuitable for manned aircraft, the pulse jet was ideal for an unmanned aircraft designed for one-way missions: the V-1 flying bomb.

At the same time as the German Army was developing their V-2 rocket, the Luftwaffe were working on the V-1 to carry a one-ton high-explosive warhead. The distinctive sound of the jet pulsing forty times a second gave it the name of 'buzz bomb', and it had a range of over 200 miles before the fuel ran out. An ominous silence indicated that the bomb was plunging to earth, and was followed by a devastating explosion.

Launching was something of a problem. The pulse jet only developed thrust when it is already moving, because it needs forward movement to keep opening the shutters after each pulse. To give the V-1 a kick-start, a type of steam catapult launching system was devised to fire it into the air where the pulse jet could start working. The Luftwaffe also launched some V-1s from bombers.

An unmanned aircraft is very vulnerable, unable to defend itself. But the makers counted on the V-1's good speed (345 m.p.h.) and low altitude (3,000 feet or less) to make it difficult to intercept. It was a cheap and simple weapon, intended to be mass-produced in large numbers. It was wildly inaccurate, not being launched at a target so much as in the general direction; the average miss distance from the aim point was more than 8 km. As a means of attacking specific targets like factories or airfields it was useless, but as a terror weapon to bring fear to a civilian population it was more than adequate.

The V-1s could be shot down by an alert anti-aircraft gunner, but more were intercepted by fighters. Interception was a hazardous business. The pilot had to spot the small, fast-moving V-1 and position himself a few hundred metres behind it for a shot, always aware that when the V-1 was hit there was every chance that it would explode with full force.

The defenders quickly learned to take advantages of the V-1's weaknesses. Anti-aircraft guns with radar guidance shot them down as they came over the coast, and the faster fighters could intercept them. Skilled pilots found that by flying beside a V-1 and nudging it with a wingtip they could bring it down or send it off course, flying back out to sea.

There was also a massive Allied campaign to attack the V-1 production facilities and launch sites. The launch catapult proved to be vulnerable; because it took up to a week to set up, bombers could destroy the launch site before it was used.

It is worth noting that the Allies also investigated the idea of a flying bomb and produced them in large numbers. The JB-2 (Jet Bomb 2) was a copy of the German V-1, made by Republic Aviation Corporation, with an engine from the Ford Motor Company, based on an unexploded V-1 from Europe.

Known as the 'Loon' or 'Thunderbug', about a thousand of these US flying bombs were produced, with plans for a production run of 75,000. They were to be used for bombarding Japan preparatory to an invasion, launched from carrier-based aircraft. Production was stopped at the end of the war, though the test launches continued until 1947. Ballistic missile development meant that the flying bomb was left behind, though the idea of the comparatively slow cruise missile was later resurrected.

The pulse jet found a curious new life after the war in a completely different field: spraying crops. Crop sprays blast a mixture of water and insecticide into a cloud of tiny droplets and blow it towards the crops, and a modified version of the pulse jet proved to be just the thing.

The pulse jet is still favoured by model aircraft builders for its cheapness and simplicity, although its basic limitations have still prevented it from becoming more widespread. But there may be ways of overcoming these limitations, and a more sophisticated form of the pulse jet may yet revolutionize air travel in the twenty-first century, as we will see in a later chapter. But for the meantime, turbines have dominated the jet scene and will do so for the foreseeable future.

## Turbofans: The Jet Age

A licence fee of $800,000 bought the US full details of Whittle's work and permission to use it for both military and civil purposes. A test engine was also included in the deal, and by October 1942 the Bell XP-59A Airacomet was flying, powered by two jet engines from General Electric. GE subsequently became one of the world's biggest jet engine manufacturers.

The turbojet changed air warfare. A jet bomber could fly too high and too fast to be intercepted by piston-engined fighters, so jets became essential for all combat aircraft. Propellers were confined to transport aircraft and others which did not need to hold their own in combat.

Turbojets were adapted to civilian use, although their noise and fuel consumption made them less than ideal. Military aircraft engines were optimized for maximum thrust for the size of engine: the all-important power-to-weight ratio for achieving quick manoeuvres and rapid climb rates.

Military jets also have afterburners, allowing jet fuel to be injected into the jet's exhaust to produce added boost. This sends out an immense jet of flame and can burn up the plane's entire fuel supply in a matter of minutes. The added thrust is helpful when intercepting or breaking off combat or during take-off when fully loaded.

What the civil sector wants is reliability, low maintenance costs and good fuel economy. The fighter pilot's steed may have a large team of technicians to cosset it, but civil airlines want something that can be run with a minimum of effort.

Noise was not a major consideration for military aircraft but it became all important in civil aviation as strict new regulations came into force about allowable noise levels from new jet engines. A loud engine, however powerful, was simply unacceptable, and afterburners were right out.

In spite of the disadvantages of turbojets, they soon took over the civilian market. The first jet airliners like the Comet and the Boeing 707 had military powerplants taken directly from bombers. These mature jet engines were reliable and their 'work rate' was very high: instead of one slow flight, a jet plane could carry out two fast flights over the same route, doubling the turnover while providing a better service. Fuel consumption was a problem though, with the early aircraft barely able to make the transatlantic routes without refuelling.

One solution to the problems of the turbojet was the turboprop, which uses the rotation of the jet to drive a propeller. It has the benefits of the power and reliability of the jet combined with the additional fuel efficiency given by the propeller. Unfortunately it suffers from the same limits as other propellers so that very high speeds are not possible. Turboprops could only fly at about 350 m.p.h., compared with 550 m.p.h. or so for the early turbojets, putting them at a significant disadvantage, and they were never popular.

By 1961 there was a new type of jet engine: the turbofan. The turbofan differs from the turbojet in having the addition of a fan inside the engine. This is powered by the same rotary motion that drives the compressors. In addition to the thrust provided by the jet exhaust, the fan acts like a propeller and the exhaust is a mixture of hot air from the combustor and cold air driven by the fan.

The ratio of cold air to exhaust is known as the bypass ratio. According to the laws of thermodynamics, the temperature of the exhaust determines its efficiency at low speeds.

The concept was developed by Whittle as far back as 1936, who saw it as the ideal engine for a long-range bomber. But the turbofan was not pursued, and after the war there was not seen to be a requirement for such a bomber.

The Convair 990 Corona was the first of the turbofan airliners, but it was not a great success and the story might have ended there had it not been for the B-52 bomber. The USAF's new bomber for nuclear strike was going to be the biggest jet bomber with the longest range possible. Turbojets would not do so a fuel-efficient turbofan was called for. This prompted the two US engine makers, General Electric and Pratt & Whitney, into turbofan development. The efficient new engine helped the B-52 become the first aircraft to fly around the world non-stop (though air-to-air refuellings were still needed). Rolls Royce, the British engine manufacturer, followed afterwards.

Pratt & Whitney led in the civilian market. Their JT8D was a turbofan based on a turbojet they had built for the US Navy. This was the engine the market had been looking for – quiet and efficient, providing a long range on one tank of fuel. It was adopted for the Boeing 727, the 737, the Caravelle and the DC-9 and became the world's leading jet engine with more than 12,000 seeing service.

The oil shock of the early 1970s, when fuel prices were virtually quadrupled, made fuel efficiency vital if airlines were to operate profitably. By increasing the bypass ratio to 5:1, modern turbofans can achieve fuel consumption several times better than their turbojet equivalents. As a result, the civil market is now entirely dominated by turbofans, but they could never have existed without having gone through the evolutionary stage provided by the military turbojet.

## Swept Wings

Compare any aircraft of the 1930s with its modern equivalent and you will see that the angle of the wings has changed. Early aircraft are almost cruciform, with wings sticking out from the fuselage at right angles; later ones have their wings swept back into a V-shape.

The importance of swept wings for high-speed flight was first pointed out by a German scientist, Adolf Busemann, in the 1930s. He showed that the nose of an aircraft generates a shockwave, like the bow wave of a boat, which spreads out at the speed of sound. If an aircraft travelling at high speed does not have swept wings, this shockwave will impinge on the wingtips, causing vibration and other problems. If the wings are swept backwards, the shockwave misses the wings entirely, and the aircraft can travel faster. Although there were some experiments using swept wings for other purposes – they moved the aircraft's centre of gravity back slightly – Busemann's work was largely ignored.

The swept wing was first used on the Me-262 where it proved highly successful. Postwar tests showed that it could outfly the straight-winged American Lockheed P-80 jet fighter designed several years later.

The Allies were slow to catch on to the advantage of swept wings, but serious research started after the war with the help of German scientists. The British De Havilland company modified a Vampire fighter-bomber which became the first British high-speed swept-wing aircraft. The knowledge gained from this project was used to help design De Havilland's Comet 1, the first civilian jet airliner. In the US, Boeing first adopted the swept wing for its B-47 jet bomber.

The advantages of the swept wing soon became apparent as the B-47 set new records for cross-country speeds, outpacing the jet fighters sent to escort it. The wing soon became the standard for all new jet aircraft.

Other improvements to wing design are less apparent to the naked eye, but are still important. When an aircraft is flying at high subsonic speed, some of the airflow over the wing may be supersonic. This is related to the thickness of the wing, because the air flowing over the wing has to travel further as it is pushed out of the way. The supersonic airflow causes compression and drag, slowing the aircraft down. It can also have more serious effects – the compressed airflow behaves differently to a subsonic airstream. When Spitfire pilots found themselves in a very high-speed dive, the supersonic airflow over the wings meant that attempting to pull the stick back simply increased the dive.

Careful engineering of the thickness of the wing in combination with wing sweep ensures that this supersonic airflow will not occur. The work on this was carried out by Albert Betz – yet another German wartime aerodynamicist who is little remembered but who played an important role in making the modern airliner possible.

## 'Fly by wire'

Flight control is a key issue for aircraft. In 1901 Wilbur Wright said that if the control problem could be solved, flight would be a reality 'for all the other difficulties are of minor importance'. The difficulty is finding a balance between a design that is stable, and therefore difficult to manoeuvre, with one that is unstable and so hard to keep in a straight line. Post-war jet fighters were fast, but without agility they were no better than slower aircraft in dogfights. The solution was active control, in which the instabilities of the aircraft are managed by automatic systems.

Direct control means that the pilot has to work the individual flaps, rudders and ailerons by hand to steer the aircraft, remembering to take into account how they will affect each other. With active control, the pilot just pushes the stick to the left and the control system calculates exactly what to do to move the aircraft in that direction. A pilot may take a fifth of a second to tell whether a rudder movement has had the desired effect; an automatic system can sense within a millisecond whether the rudder needs to be moved more or less and adjust it accordingly.

The origins of active control also date back to the Second World War. Forbidden by the Treaty of Versailles from developing powered aircraft, Germany had become a centre for gliding, and gliding clubs were popular at universities. As with other aircraft, gliders had an instrument for measuring airspeed, although it was impossible to tell how fast the glider was moving with respect to the ground.

A German student and glider enthusiast called Helmut Hoelzer set out to devise a way of calculating ground speed. Hoelzer envisaged an arrangement of weights and springs that would translate the force of acceleration into an electric current. The greater the acceleration, the stronger the voltage. An extra circuit would multiply voltage by time to derive an exact ground speed.

The project was too ambitious for a student, but Hoelzer returned to it after graduation, and when he was assigned to join the rocket team at Peenemunde he had a shrewd idea of what was needed, and his design proved successful in practice.

A device that would give the ground speed of an aircraft or rocket could be combined with a compass to determine its direction. Putting the two together meant that the rocket's distance and direction from its starting position could be calculated; more importantly, the distance and direction from the target could also be calculated, and this information could be used to adjust the direction of the rocket. This was the basis for an automatic guidance system.

Hoelzer's V-2 guidance system was the first true fly-by-wire control system. In effect it completely automated the process whereby the rocket determines 'I am here and going in this direction, so in order to get to the target I have to change direction this way.' The key factor is that the guidance system operated continuously, constantly making minor corrections until the desired result is obtained. It is the direct ancestor of modern control systems which have spread from rockets to warplanes to civilian airliners.

The technology was taken over by the US after the war. After some decades of military use it was taken up by the civilian sector, notably in Concorde. The latest refinement allows inherently unstable aircraft such as the latest Airbus to be piloted with ease; such aircraft need fewer control surfaces than stable designs, making them smaller, lighter and more economical.

The Airbus A320 was the first digital fly-by-wire airliner in 1987, and it has become the new standard. There was considerable controversy over the crash of an A320 at a Paris air show in 1988; the pilot claimed that the fly-by-wire system was responsible for the crash, but the official report has put it down to pilot error.

The lines used for the control signals can also be used to power electrical actuators that work the controls; this is known as 'power by wire', and it replaces the heavy hydraulic and pneumatic systems used in earlier aircraft. Already used in military unmanned aircraft, it should be safer than systems which can suffer from mechanical breakdowns and leaks, and it is easier to build in back-up systems.

## Finding the Way:
## Compass and Watch (and Sunglasses)

Navigation has also followed the path of military technology. Night flying was rare until the First World War, and the first aircraft night navigation systems were used during German raids on Britain.

The gyrocompass is a device which finds 'true North' rather than magnetic North, by interacting with the earth's rotation, and thus avoids the problems associated with using a magnetic compass in a metal ship. An early gyrocompass was developed by a German, Herman Anschütz-Kaempfe in 1904 and quickly adopted by the world's navies for ships and torpedoes.

An American, Elmer Sperry developed a more advanced design in 1908. When Sperry attempted to sell his design to the German Navy in 1914 his patent was contested by Anschütz-Kaempfe. The German won the case, with the help of a physicist who had previously worked in the patent office – a certain Albert Einstein.

Over land pilots could use roads and other landmarks, but seaplanes needed better guidance and the US Navy took up the gyrocompass, which Sperry developed into a basic autopilot in 1929. Sperry went on to develop more advanced gyrocompasses and gyro-based guidance systems for the US Navy and Air Force, and his company (now part of Northrop Grumman) went on to become one of the giants of the US defence industry.

Radio beacons were first used in the 1930s but these were supplemented by the Inertial Navigation System or INS. The basic gyroscopic navigation system designed by Hoelzer used three mechanical gyroscopes set at right angles to each other to measure acceleration along each of the three dimensions. The gyros were comparatively big and suffered the limitations of any delicate machinery. They were still extremely useful though: a military Boeing 707 flew over the North Pole, navigating with the aid of INS in a flight path for which a magnetic compass would have been useless. This was not just a stunt; the shortest route for ballistic missiles between the US and USSR was over the pole, and it showed that accurate guidance was possible in the area where compasses would not function.

During the Cold War, huge amounts were spent on INS developments for missiles, and also for submariners who needed more accurate navigation tools than ever before. It was not enough to know their location to within a few miles. If they were firing ballistic missiles, any error in the position of the submarine would translate into an error in targeting.

In 1967, Britain's Ministry of Defence sponsored a project to develop an INS using lasers, known as a 'Ring Laser Gyro' or RLG. The RLG uses three lasers, each going around a loop. If there is any acceleration in the plane of the loop, it will affect the time taken for light to complete the circuit. The minute change in the time taken can be measured and the acceleration calculated.

RLGs could be made much smaller than mechanical systems, and are far more accurate. They are also not susceptible to influences like temperature. Unfortunately, having proved that it worked, the MoD lost interest, but within five years the rest of the industry had discovered RLGs and they were adopted as a standard for military and civilian aircraft.

This simplified the task of navigation hugely. If the INS is programmed with the location at take-off it can keep track of where the aircraft is, with an accuracy of about one mile per hour of flight. So after a ten-hour flight the location of the plane is known to within 10 miles, easily enough for it to find the radio beacons and radar of the destination airport if not actual visual contact. INS and similar technology meant that there was no longer a need for specialist navigators in civil airliners.

Navigation also requires exact timing. To know your location, you need to know not just your speed and bearing, but exactly how long you have been flying. A deviation of miles may be acceptable for civilian uses, but when it is a matter of dropping bombs at night, or targeting warheads, maximum precision is called for. Mechanical clocks which had been in use for centuries were no longer good enough.

As far back as 1857, Jules Lissajous showed that a tuning fork can be kept vibrating indefinitely with an electric current. A tuning fork sounds a single note with a fixed number or cycles or pulses per second. If the sound has a frequency of 10 kilohertz, then every 10,000 pulses equals precisely one second. This was used in the early twentieth century for scientific measurement of time, but it was not until 1927 that Warren Morrison of Bell Telephone Labs developed a 'quartz clock'. This was based on a piece of quartz crystal which could resonate at an exact frequency (32, 768 hertz) in response to an electric current. The crystal was robust and reliable. Such devices were important for applications like radar which required very precise timing.

Early quartz clocks were expensive, and outside the lab they were mainly confined to military applications including navigation and radar. The Second World War brought a series of advances in the US, particularly in the key area of reducing wastage by developing techniques of checking the crystals before cutting (previously much of the output had to be discarded) and making them stable so that the frequency of the crystal did not 'drift' as it aged.

In the 1950s and 1960s further developments allowed military electronic company Motorola to start mass-producing quartz oscillators, producing near-identical crystals with a high degree of precision. This enables missile guidance systems to be built more cheaply, but more importantly it revolutionized the watch-making industry. Centuries of painstaking craftsmanship were about to be set aside: accurate time-keeping no longer depended on careful construction of gears by a skilled watchmaker when it could all be done by a chip of quartz.

In 1967, the Centre Electronique Horloger (CEH) in Neuchâtel developed the world's first quartz wristwatch, the Beta 21. But it was the Japanese company Seiko that captured the mass market with their Astron in 1969.

Early quartz watches were quite chunky – the large Motorola crystals were designed for missiles, not wristwatches. But others copied and refined their crystal-cutting techniques, and soon the market was awash with cheap crystals from Asian companies. Suddenly, even cheap watches were accurate to within seconds. Today more than 90 per cent of watches use quartz crystals thanks to techniques developed by Motorola.

Incidentally, the wearing of wristwatches is largely a military by-product. Pocket watches were the rule until the First World War, but in 1917 the US Army commissioned Waterbury to make a wrist-carried version of their Yankee pocket watch, as this would be more practical in the trenches. Hundreds of thousands were produced and the wristwatch spread from returning veterans to the public at large.

Another fashion comes directly from the military. In 1930, the US Army Air Force (precursor of the USAF) commissioned the optical instrument makers Bausch & Lomb to develop glasses to protect aviators' eyes from the bright glare they were encountering at high altitude. B&L responded with new 'Anti-Glare Aviator Sunglasses' with green lenses optimized to reduce the type of glare encountered by pilots. The green was the result of tests to find the best UV and infrared absorption to protect the eyes without affecting colour vision. They were an immediate success, and became standard issue for pilots. The commercial version was launched in 1936; the name 'Anti-Glare' was not popular, so instead 'Ray Ban' was adopted as a brand name for the new glare-blocking lens, and advertised using the images of famous flyers like Amelia Earhart. 'Champions and Leaders wear Ray Ban,' said the adverts.

They also became popular with people who wanted to look like the glamorous aviators, and, crucially in California, movie stars took up the look. Wide-brimmed hats which had been used to shield eyes from the sun for centuries were no longer quite the thing, and sunglasses became an essential accessory.

The military did not invent sunglasses – they can be traced back centuries – but Ray Ban Aviators broke the mass market for the first time and millions were sold. They were followed by Ray Ban Wayfarers, the bestsellers ever; Ray Ban also went on to develop the first mirror sunglasses in association with the USAF (they were not aiming at a cool new look so much as a technology that reflected 30 per cent more glare away from the eyes). Foster Miller had been selling sunglasses since 1919, but it was only with Ray Ban Aviators in the 1930s that they really became popular – a small but highly visible sign of military technology.

## A Wall of Instruments

The Wright Flyer had only three instruments on board: a revolution counter to show how fast the engine was running, an air speed indicator and a stopwatch to measure flight duration. A few years later and the number of instruments had doubled to the dials known as the 'primary six': an airspeed indicator, a vertical speed indicator showing the rate of climb or descent, turn-and-bank co-ordinator, a heading indicator showing the magnetic compass course, altimeter and attitude indicator.

By the jet era the amount of information had increased to the point where the cockpit began to look like a clockmaker's workshop, with dials and displays everywhere. As well as the systems for flying aircraft, there were new sets of controls for radar and communication. More extras proliferated, dealing with electrics, hydraulics, pneumatics, pressurization and auxiliary power units. The cockpit sprouted rows of circuit breakers for all vital systems.

Greater altitude and night flying meant that visual landmarks could not be used and navigation became more demanding. Radio communication with the ground was vital as air traffic multiplied.

The extra complexity demanded more crew, and in the postwar period an airliner might require not just a pilot and co-pilot but also a navigator, a flight engineer and long-range communications operator. Some routes that went close to the North Pole might require an extra navigator because the closeness of the magnetic pole meant compasses were of limited use, so a total of six aircrew might be necessary.

The proliferation of instruments was causing problems of its own. At any given time only about 10 per cent of the instruments were relevant, and the others were a potentially dangerous distraction. When warning lights started flashing and buzzers sounded from a completely irrelevant instrument, it could draw the crew's attention away from real problems.

This problem was familiar to the military, where the situation was even more acute. Pilots had to deal with enemy aircraft, anti-aircraft weapons and their own mission, in addition to simply getting from A to B. The mass of information had to be brought under control.[4]

Communications were automated with the introduction of radios which tuned themselves to pre-set frequencies, allowing the equipment to be simplified so that a separate radio operator was no longer required.

The military were also dealing with the profusion of instruments. The banks of displays were replaced with a few computer screens, resulting in what is known as the 'glass cockpit'. The screens are programmed to show the information which is most relevant in any given situation, with different displays for take-off, navigation, landing, and, in the case of the military, ground attack and air-to-air combat. All irrelevant information is excluded and the pilot can concentrate entirely on the task in hand. The computer screens can also be used to display weather radar or other data, and the computer software has automated many of the standard procedures.

The screens themselves are evolving. The original screens were cathode ray tubes, using the same technology as a television set, but these are now being replaced with liquid crystal displays – familiar from laptop computers. These displays use less power and take up less space, and also have a better graphics capability. Airbus are now fitting LCD screens in some of their older planes, replacing the CRTs with screens that they say will be clearer and more reliable.

Lockheed's next-generation fighter, the F-35 Joint Strike Fighter, has an instrument panel that consists of a single giant display surface which can be programmed to show any combination of instrument information as required.

Another military technology now appearing in the civilian world is the Head-Up Display or HUD, in which flight information is projected on to the windscreen in front of the pilot, allowing him to fly without looking down from the scene in front of him. The military use HUDs for targeting, projecting the image of the aim point of guns or bombs. Guided bombs have a 'basket' which defines the space where a bomb can be released and find its target; this is shown to the pilot as a series of hoops which he must fly through to guarantee a hit on the target.

In the civilian world, HUDs are useful because they can improve the situational awareness of the pilot, displaying vital information in front of him and removing the need to look down at instruments while flying. HUDs are now beginning to be installed in luxury cars, so drivers do not need to take their eyes off the road to check their speed.

As with the military, HUDs are used in airliners to display additional visibility aids. These include infrared cameras which can see through cloud and fog, projecting an image on top of the visible world. Future developments will follow the military pattern, integrating low-light television, radar and other sensors into the same picture.

An alternative is to have the display mounted in a helmet visor. This gives the pilot a complete 'virtual world' overlaid on his field of view, and is used for the pilots of fighters and attack helicopters, using the same kind of liquid-crystal display as the HUD.

The next generation is already on its way: an image projected directly on to the retina. Instead of having to wear bulky, heavy goggles (weight starts to become significant when manoeuvring at several Gs) which can restrict his range of view, the pilot has a headset with a small laser diode which uses a laser to paint an image on his retina. Known as Virtual Retinal Display (VRD) technology, this has been under development by the US Army since the mid-1990s. The US Navy is carrying out a similar project.

According to the makers, the VRD has many advantages over HUDs and helmet displays. 'The VRD is the only display technology that has sufficient luminance to be used as an augmented display over the pilot's real world view in bright sunlight.'

Little has been released about the VRD in recent years, but a close study of military budget documents shows that it is still being pursued.

Given the ability to project television and other imagery to the pilot wherever he is, there is no longer a strict need for the cockpit to be in the front of the aircraft. Cameras and other sensors will give 360° vision wherever the pilot is. He could be in the cockpit, the middle of the aircraft, or the back where he is far less vulnerable.

There is no reason why the pilot could not, in principle, be on the ground a thousand miles away: the 'virtual cockpit' can be wherever the information is. This leads to the question of whether the plane needs a pilot on board at all.

In the past, air crews have had to rely on a large number of standard drills and procedures for handling normal operations and the various possible emergencies, backed up by extensive checklists. Computerization meant that many of these routines could be carried out at the press of a button.

The improvements in flight-deck management mean that the role of flight engineer has become increasingly unnecessary, and the flight crew for modern aircraft has been reduced to two, pilot and co-pilot. Safety considerations are likely to mean that this is the standard for years to come, but this is not because of technological limitations.

Military aircraft already have the ability to operate practically without a pilot. The F-117 stealth fighter can be pre-programmed for a complete mission. The aircraft can take off, go through a complex flight path, release weapons over a specified target and return to base and land. The only input required from the pilot is to press the release button for the bombs; even the USAF is slightly wary of putting the power to kill people entirely into the hands of machines.

## The Whole Plane

As we have seen, all the important technologies used in modern airliners originated in the military sphere. Sometimes the involvement is even greater though, and it is difficult to exaggerate how close airliners are to military aircraft and how they have benefited from military funding.

In the early days of air transport some passenger aircraft were simply modified bombers. One of Boeing's first airliners, the Model 247 which flew in 1933, was a civil version of their B-9 bomber. This was notable for its metal skin – a common feature of military planes, but the first time it had been seen on an airliner.

Boeing has always been closely tied to the military, and the Air Force has played a vital role in the company's most important developments. In 1954 the requirement for a new jet-powered tanker aircraft led Boeing to invest heavily in a prototype called the Model 367–80. This drew heavily on their earlier success with the B-47 bomber which featured swept wings and jet engines in underslung pods. The prototype, called the Dash-80, served a dual purpose, being built to the Air Force's specification but also being adaptable to civil use.

The prototype was aimed firstly at the tanker market and had few windows and no seats. A week after its first flight, the Air Force ordered twenty-nine of them, giving the aircraft the designation KC-135. This gave Boeing a solid financial base to work from. It was needed, as there was tough competition in the civil market from another new jet, the Douglas DC-8.

Finally, five months after the Air Force order, Pan American ordered twenty of the airliner version. The production version was given a new name, Boeing 707, and as noted earlier it was the first truly successful jet airliner. Over a thousand were built; but it should be added that over 700 KC-135 were built as well as 50 C-135 transports.

Ten years after the first 707s entered service, the military transport version was still the most effective transport aircraft in the USAF. During one exercise the C-135s flew two complete non-stop missions in the time it took propeller aircraft to complete one mission with a break for refuelling. But even the C-135 could not carry the really heavy items like tanks, and the USAF outlined a requirement for a huge freighter aircraft with unparalleled capacity and range.

Contracts of $6 million were given to Boeing, Lockheed and Douglas to develop plans for this next-generation transport aircraft, then designated C-X; engine makers General Electric and Pratt & Whitney were awarded $11 million each to develop suitable engines.

The C-X would carry more than twice the payload of the C-135, and carry it 50 per cent further. Its wide cargo bay would be able to accommodate 99 per cent of the Army's equipment, including a pair of main battle tanks. The competition was won by Lockheed whose winning entry became the C-5 Galaxy transport.

Boeing's entry lost, but the company was to put their design effort to good use. The experience in airframe design and dealing with the engineering issues involved with large aircraft were vital, and their C-X proposal formed the basis of a new type of airliner. Because it was originally designed to carry vehicles, the airliner was much wider than previous civil aircraft, giving it much greater passenger capacity. The new wide-bodied aircraft was the 747 jumbo jet.

Its most distinctive feature was a hump, originally designed to keep the cockpit and support crew cabin out of the way so that the nose could swing open for loading cargo. This was irrelevant for passenger transport, but would have been vital for carrying heavy vehicles.

The C-5 had proved that the technology was viable, and the progress of the 747 was a matter of economics. The air travel market was uncertain, but once an order was secured – again from Pan American – the aircraft went into full scale development on an extremely tight schedule. The order was placed in 1966 and Boeing had to build the plane and get it flying for delivery by 1970.

Needless to say, this gamble paid off and the 747 went on to become the most successful long-haul airliner in the world, making transatlantic air travel almost routine and opening up new horizons for mass-market travel.

The military can also lay claim to the most successful small jet. The

Learjet, the first business jet to enter mass production, also traces its ancestry back to military roots, though in this case it is descended from a fighter rather than a transport. William P. Lear set out to transform the business aviation world with a small affordable jet aircraft with space for seven passengers. His starting point was a Swiss fighter jet, the P-16 which had an innovative design but never went into production. Lear recruited a team of Swiss designers, and kept the wing shape and airframe of the original, but incorporated a passenger cabin.

The Learjet was an instant success and became a best-seller around the world. The maximum speed of 564 m.p.h. was far greater than any propeller aircraft, and in businesses, where time is money, the Learjet was seen as a money-saver as well as a token of corporate clout. Ironically enough, it was so successful that several air forces (including Argentina and Yugoslavia) adopted modified Learjets for military missions.

## Conclusion

'It wasn't until the jet engine came into being and that engine was coupled with special airplane designs – such as the swept wing – that airplanes finally achieved a high enough work capability, efficiency and comfort level to allow air transportation to really take off.'

This was the conclusion of Joseph F. Sutter, head of Boeing Commercial Airplanes. The figures bear him out. In 1939, the world's airlines carried three million passengers; by 1959 the number had gone up to almost a hundred million. The 707 had a work rate – in terms of passenger-miles per year – sixty times greater than the pre-war airliners.

The turbojet and later the turbofan were crucial in building and sustaining this growth. Swept wings allowed airliners to fly faster, and a series of other developments improved efficiency and reduced the crew requirements, transforming the economics of air travel. All of these developments, including the jet engines and swept wing mentioned by Sutter, came from the military.

Not all military technology has been taken up by the civilian world. Innovations like air-to-air refuelling, vertical take-off jets, ejector seats and swing wings could have been adopted in airliner manufacture, but it has never been economical to do so. If the situation ever changes and, for example, there is a requirement for an airliner that can take off vertically, then the hardware is already on the shelf and ready to be adapted to civil use.

The move from romantic propeller-driven flying boat to the packed transatlantic 747 flying its routine services across the Atlantic has been a transition from a glorious adventure to an everyday reality. Air travel is now more universally available than ever before, and it has all been a matter of the application of military technology.

# CHAPTER 3

# Computing:
# Codebreaking to Netsurfing

*With the advent of everyday use of elaborate calculations, speed has become paramount to such a high degree that there is no machine on the market today capable of satisfying the full demand of modern computational methods . . . the present invention is intended to reduce to seconds such lengthy computations.*
Patent for ENIAC, the first electronic digital computer, 1947

It is hard to overstate the importance of the military in the development of the computer. Everything from the digital electronic computer, the silicon chip, to interactive computing and most recently the Internet has been the result of military requirements. As with rocket science and aviation, this is a technology that has moved from camouflage to business suit without being noticed.

## Prehistory: Human Computers

Computers are part of the modern age, and it is not obvious why anyone would need a computer in the eighteenth century. But in the age of sail, navigational computing was a matter of life and death.

Navigation beyond the sight of land involved complex techniques of calculating position based on the angles of the Sun and the stars. The navigator relied on an almanac with tables giving the locations of the stars and Moon, and mathematical functions (logarithms, sines, cosines etc.). Used correctly, these would permit the navigator to plot accurately the location of the ship and deduce the course to set. If things went wrong the results could be catastrophic.

In 1707 a fleet of five British warships commanded by Sir Clowdisley Shovell was sailing from Britain when it ran into bad weather. Poor visibility meant there was some risk of running aground, but the fleet's navigators agreed that they were safely west of Brittany.

A sailor reportedly approached Shovell to say that the navigators had calculated wrongly, and according to his own reckoning the ships were in severe danger. Shovell promptly had the man hanged for mutiny, as 'subversive navigation by an inferior' was expressly forbidden by Navy regulations.

The next night, all five ships ran into rocks off the Scilly Isles, sinking four of them. Almost 2,000 men were lost. Shovell himself survived the wreck, and was washed up on shore but his luck did not hold and he was murdered by a local woman for an emerald ring he was wearing. In spite of his role in the disaster, Sir Clowdisley Shovell is commemorated by his own marble monument in Westminster Abbey.

The error did not always originate with the navigator. Accurate timepieces were needed to determine longitude, and the British government offered an enormous prize to anyone who could build an accurate clock. The prize was claimed by John Harrison in 1759, a story recounted in Dava Sobel's excellent book *Longitude*. But this did not solve the problem entirely: even if he took his bearings with complete accuracy and possessed an accurate chronometer, the navigator was still at the mercy of the tables he used to calculate his position. The *British Nautical Almanac* of 1818 had a good reputation, but was discovered to contain fifty-eight errors, any of which could sink a ship. The quality of the annual almanac declined further after the death of Nevil Maskelyne, the Astronomer Royal of the time, and subsequent editions contained increasing numbers of errors to the point where they were considered dangerously unreliable.

Mathematical tasks such as updating the tables in the almanac were carried out by computers, the 'computer' being man (this was a male occupation) employed to carry out calculations. Teams of them were needed to carry out the annual updates.

As long as there was the risk of human error, every entry in the almanac was a potential time-bomb waiting to wreck an unwary navigator.

This was the situation which inspired Charles Babbage's obsession with mechanical calculation. A mathematician, he was Lucasian Professor of Mathematics at Cambridge – a post held earlier by Isaac Newton and later by Stephen Hawking – but he never gave a lecture. His energies were entirely taken up in the development of a mechanical computing device to transform the practice of mathematics.

Babbage was impressed by the new mechanical looms which wove to a pre-determined pattern without error, their actions controlled by punched cards. It struck him that this offered a way for long and tedious calculations to be carried out by machinery as tireless as a steam engine without human labour. Even better, the element of human unreliability could be removed.

It was an era of grand schemes spawned by an enthusiasm for all things scientific. Swift had lampooned the new men of science in *Gulliver's Travels*, in the inhabitants of the flying island of Laputa. The Laputians pored over plans for releasing sunbeams from cucumbers and similar projects. However great their scientific skill, Gulliver remarks, 'In the common Actions and Behaviour of Life, I have not seen a more clumsy, awkward, and unhandy People, nor so slow and perplexed in their Conceptions upon all other Subjects.'

The interest in science was a fad, and there were impractical schemes afoot from constructing mechanical ducks to electrical healing machines and 'scientific' astrology. Unlike many others, Babbage managed to attract government funding for his project.

The government realized that navigation tables for ships, and artillery tables for guns, were important for naval power.

The 'mathematical art of artillery' had been appreciated since Elizabethan times. For every type of cannon a table could be drawn up to show what angle of fire was necessary to propel a cannonball of a given weight to a given distance. By consulting these tables a gunner could ensure that his shot was true.

Of course it was perfectly possible to plot a route or fire a gun without the best tables, but doing so would put you at a disadvantage if the enemy had better technology. Accurate tables would be an advantage for every British man-of-war, and the struggle against Napoleon was a recent memory. The security of the British Empire was at stake, and the government financed Babbage to the tune of £17,000, a fabulous sum in those days.

The Difference Engine, commissioned in 1822, was to be a huge mechanical device. It used gears and levers, and was quite different from any previous calculating machine in the way it stored a representation of a number. An abacus requires a skilled operator; in effect the wooden beads act as an aid to human memory. But the Difference Engine did all the work itself, and only needed an operator to put in numbers; it even printed its own output to prevent copying errors.

The Difference Engine did resemble an abacus in that both are digital rather than analogue devices. In a digital machine, numbers are represented as discrete units or digits, whereas in an analogue machine they are stored as a continuous physical value. In the Difference Engine, or an abacus, or an adding machine, the number 23 might be stored as a 2 in the tens column and 3 in the units column. A simply analogue device might store 23g of sand as 23. For this analogue device, numbers could be added by combining different weights of sand and weighing the result: unlike a digital machine, it can add 23.4534 and 17.9432 as easily as two plus two – but it runs into difficulty when dealing with very large numbers.

Analogue devices such as slide rules would prove to be useful aids to calculation, while astrolabes and other such machines had been used to predict the movement of the planets for centuries. But the Difference Engine would be much more flexible; an astrolabe, once built, could not be changed, but Babbage's machine could be used to calculate anything from the tides to the stars to mathematical tables.

The project did not go well. Babbage complained of his workmen's failure to construct gears with the fineness and accuracy that the Engine demanded. Others said that Babbage's lack of management skills, his constant criticism of everyone around him, and general bad attitude had a lot to do with the problems. Swift's description of the man of science being 'slow and perplexed' in worldly matters was sadly accurate. Babbage's wife and children died during this period, which can only have made matters worse.

Babbage was forced to commit increasing amounts of his own money to the project, which was taking longer and costing more than originally estimated (giving it much in common with most other computer projects of the last two centuries). By 1834 the Difference Engine was still not complete.

'We got nothing for our £17,000 but Mr. Babbage's grumblings,' wrote the Reverend Richard Sheepshanks, one of Babbage's critics and a secretary of the Royal Astronomical Society. 'We should at least have had a clever toy for our money.'

The government did not decide to cease funding, but rather it failed to decide to continue. No more money was supplied, and digital computing technology came to a halt.

Improvements to navigation were to come from the human computer rather than the mechanical one. The big leap forward was the widespread introduction of logarithms, which simplified multiplication and division (with logarithms these are turned into the much easier tasks of adding and subtracting). Calculations, which would have taken a great deal of time and effort, and with the risk of error, could be carried out with ease.

This required accurate logarithms in the first place, a project undertaken by the French government. They employed six mathematicians and seventy computers for two years, and when completed, the new *Tables de Cadastres* comprised seventeen manuscript volumes. It was not considered possible to print the work, as this would have introduced too many errors. Anyone wishing to look up the exact figures would have to consult the original in Paris.

Reliable tables of logarithms were a great assistance to calculation. In 1834 when the Difference Engine ground to a halt, logarithms were used to calculate the tables in the *British Nautical Almanac* for the first time, and the errors were largely eliminated.

The Navy had their accurate tables, but the Difference Engine was never completed, and the opportunity for the computer to make its mark had passed. However, his experiences on the Difference Engine had inspired Babbage on to a much grander project, the Analytical Engine, the true forerunner of the modern computer. As well as an input and output, it had a store, corresponding to the memory of a computer, a mill which was the processing unit, and a transfer system for moving data around.

The Analytical Engine would have been the first recognizable, modern computer if it had ever been built. Of course the plan was far too ambitious, and in the light of the failure of the earlier project it is not surprising that Babbage failed to win backing. It is hard to gauge what effect Babbage's Engines might have had, but it is perhaps significant that later computing developments went down similar routes.

## The Second World War: Germany

Nazi Germany, often seen as a great technological innovator, missed its chance with the digital computer, although both analogue and digital computers did play a part in the war effort.

We have seen how Helmut Hoelzer's analogue computer was used in the V-2 program. Having shown what could be done with electronic calculation machines Hoelzer set out to build something better. The V-2 rocket program involved a seemingly endless series of test launches (over 3,000 of them), many ending in failure as the rocket plunged into the sea. Some V-2s even ended up in Sweden, providing the Allies with valuable intelligence on the rocket program.

The source of the problem was often impossible to trace. Scientists could not tell whether the problem lay with the rocket itself or the guidance system. The only way of testing the guidance system was to try it in a live rocket test, but if the test failed the problem might or might not be the guidance. Hoelzer set out to build an analogue computer that would model the behaviour of a rocket in flight, so the responses of guidance systems could be tested in a situation that would exactly mimic a live test, but without the expense of a rocket launch.

Although initially discouraged by his superiors (even von Braun advised him to 'stop playing with electronic toys'), Hoelzer persisted and eventually constructed a working machine. Weighing over a ton and powered by a vacuum cleaner motor, the analogue computer was successfully used to test new guidance systems.

After the war, both Hoelzer's invention and the man himself were drafted by the US army at the Redstone Arsenal in Alabama. Hoelzer wrote a doctoral dissertation on analogue computers based on his work, and in 1946 he became the first recipient of a doctorate for research in computer science.

Germany also had its digital computing pioneer, who was virtually ignored. In 1938 Konrad Zuse, an engineer at the Henschel Aircraft Company, built his own computer to help with the calculations of the stresses on wings. Called the Z1, the device was entirely mechanical, made of metal bars between glass. This was a completely independent development and Zuse said later that he had not even heard of Babbage at the time.

When Zuse was conscripted into the German armed forces, he was given the resources to build an improved machine, the Z2, in which the mechanical switches were replaced by electrical ones. This worked better than the Z2, and was followed in 1941 by the Z3 which boasted over 2,000 relays.

Like the Z1, the Z3 was put to use by the aircraft industry, again mainly for calculating stress. The next stage would be an even more powerful machine which used electronic valves rather than electric switches. But Zuse's proposal was turned down, the reason given being that Germany was so close to winning the war that the machine would not be required.

In spite of the lack of support for an electronic version, Zuse continued work on the Z4, which was also known as Versuchsmodell 4 or V4. He was still working on it in 1945 when Germany was overrun by the Allies. British and American troops arrived at the house in Bavaria where the machine was stored. The troops were looking for a new German secret weapon called the V4; expecting some kind of weapon like the V-1 and V-2 they were surprised to find a collection of electronic components without a warhead to be seen.

Zuse is credited with developing the first computer programming language, Plankalkül, which ran on the Z3, and used it to create a chess-playing program. However this was not published until 1972, so Zuse is not usually recognized as a pioneer in this area.

After the war Zuse founded a company which was later taken over by Siemens, and did important work on second-generation computer languages. Considering the uses which the Allies found for computers, it is fortunate that Zuse's talent was not recognized earlier.


## The Second World War: Britain's Bletchley Boffins[1]

If the Germans failed to capitalize on the possibilities of the electronic computer, the British might be said to have won the computer war but lost the peace.

The story of the code breakers of Bletchley Park is a classic tale of scientific endeavour in difficult circumstances. One of the tasks facing the team was breaking codes created by German teleprinter ciphering systems. These teleprinters were used by the German High Command to send reports, orders and information covering everything from troop deployments against Russia to the disposition of U-boats in the

Atlantic. The signals could be intercepted, but the Germans were not greatly concerned as they were convinced that their cipher was unbreakable.

The cipher was fiendishly complex. In standard teleprinter transmissions, each letter is represented by five elements, each of which can be one or zero. For example, 00001 represents the letter E, 11100 is X. The ciphering system took each element of a letter and used a key to change it, transforming 00001 to 11100, for example. The new letter would then be enciphered again using a second key, so 11100 would be changed to 01011.

The output gave no clue as to what the original was. Doubly encrypted, the message could be sent without risk of being understood; without the keys, it was impossible to decipher the message. With the keys, it was a simple process of reversing the second encryption and then the first one.

The cipher was difficult to break because a different key was used for every letter of the message, the keys being automatically generated by ciphering machines like the famous Enigma. Even when the Allies managed to get hold of an Enigma machine, they could not decipher a message without knowing the exact key settings being used for that particular message.

The limitation of the Enigma was that it was a mechanical device and the keys were generated by toothed gear wheels. The first wheel of the Enigma machine had forty-one teeth in it, and each time it operated the wheel turned around one notch. This meant that the cipher would repeat every forty-first revolution. If there was only one wheel, this would have limited the machine to forty-one different ciphers, and the decoders could simply have tried each possible cipher in turn. However, Enigma possessed ten such gears, each with a similar number of teeth, so the total number of ciphers was $41 \times 41 \times 41 \times 41$ etc., creating more than ten million billion possible ciphers, a truly astronomical number. No wonder the Germans thought the cipher was secure.

This ciphering process did still leave discernible patterns in the encrypted data, allowing the Bletchley Park team a number of approaches. One of these was by applying statistical analysis to encrypted messages. Knowing that certain letters appear more often than others, this information could be used to deduce the settings on the encryption machine and work backwards to the original message.

Statistical analysis was a slow process, often taking weeks, so Max

Newman, one of the code breakers, decided to build a machine to do the job automatically. He was building on the work of Alan Turing, a brilliant mathematician who had already revolutionized the Bletchley Park operation by designing electrical calculating machines for simple deciphering. These could be used to quickly test deciphering solutions rather than doing the job by hand.

Newman approached the Telecommunications Research Establishment in Malvern to build his machine. It was called Robinson, after Heath Robinson, a cartoonist famous for his pictures of improbable inventions. The encrypted message was punched into paper tape and fed through the machine. Driven at high speed by spiked wheels, the paper ripped continually. Worse, the mechanical imperfections meant that Robinson was inconsistent and could give different results with the same data on different runs.

Tommy Flowers, an engineer called in to help with Robinson, concluded that this was not the best approach. Flowers was a telephone engineer with some knowledge of switches, and he suggested electronic switches would work better than the electro-mechanical ones. The tape could be read by a photoelectric cell and driven by smooth wheels which would not break it.

The new electronic version of Robinson was called Colossus, and it started work in 1943. The first practical large-scale computer controlled by programming, it was a forerunner of all subsequent digital computers, was far more reliable than Robinson and proved highly successful. An improved version (with 2,500 valves compared to the original 1,500) followed.

Colossus was not a complete solution, and did not translate the Enigma cipher into plain text. But it did strip out the effect of the first wheel on the ciphering process, reducing the deciphering task to a job that could be carried out by humans.

Although it was successful, the extreme secrecy of the entire project meant that even the existence of Colossus was not revealed until many years afterwards. Even then its significance was only appreciated by a small number of people in the computing field.

The cracking of the Enigma Code was itself highly secret. If the Germans had known that their ciphering had been compromised they would have stopped using it, so it was vital that this was not discovered. Even after the war, the work from Bletchley Park was kept secret, as their task had moved on to cracking Russian codes and it was important that the Russians should not realize how advanced the Allies were in this area.[2]

Enigma itself had a curious postwar life as well. The British government helpfully provided ex-German Enigma machines to the emerging former colonial nations in Africa for their security services. The cipher was still thought to be unbreakable, and right up until the 1970s some countries were still using these machines for top secret communications, unaware that British intelligence could tap into them.

The secrecy that surrounded Colossus meant that it remained in the shadows and was never developed further into a true general-purpose computer; those who worked on it were not able to share their knowledge. To find the true origins of modern computing we remain in the Second World War era, but we must look to America.

## The Second World War: American Artillery

The need for gunnery tables persisted into the Second World War. Every type of shell fired by an artillery piece has its own characteristics – the same gun might fire high-explosive, shrapnel, armour-piercing and smoke rounds, each of which had a different weight and shape. Each required individual calculations for every possible trajectory and weather condition.

At sea the situation was even harder, as the gunner had to calculate the movement of the ship and the target vessel, and mechanical analogue computers were developed to help with these calculations during the First World War.

What had been a major task was becoming an impossible one, as the rate of technological change during the war meant that a constant stream of new guns was being produced. Most nations started the war with 37mm tank guns like the British two-pounder and successively upgraded through 50mm and 75mm to even larger calibres. A proliferation of new types of ammunition made the situation even more challenging as each had slightly different characteristics. Producing tables for all of the new weapons was creating a huge workload. The US Army needed a new machine for the job.

The state of the art in calculation was the Bush Differential Analyzer (sic), the invention of American scientist, engineer and all-round genius, Vannevar Bush. Like Hoelzer's creation, this was a mechanical, analogue device rather than a digital one. It used a series of drive belts, shafts and gears to carry out calculations rather like a sophisticated mechanical calculating machine, although it might be more accurately described as an automated slide rule. The Analyzer was a sizeable piece of machinery incorporating 150 electric motors. This was about the limit for analogue computing; something more sophisticated was needed for ballistic calculations.

The International Business Machine Company, better known as IBM, was already a big player in the field of office automation. Founded as the Tabulating Machine Company in 1896, IBM made mechanical calculating devices based on punch cards which could process data much faster than manual methods. By the 1940s their machines were everywhere; the government used punch cards to record wage data about every working American.

IBM built an electro-mechanical computer for the US Navy using electrical switches, based on some of Babbage's ideas. The man behind it was Howard Aiken of Harvard University. Known as the Mark I it weighed 5 tons and with a 50-foot drive shaft, it made a noise like a textile mill when it ran. Although usually referred to as the 'Harvard Mark 1', the machine was not made for the academics but was specifically funded by the US Navy. It did some work calculating navigational tables just as the Difference Engine was supposed to have done, but it was not fast enough to meet the requirements of the Army's ballistics tables calculations.[3] IBM heralded the Mark I as Babbage's dream come true, although it was actually missing some of the functionality that Babbage designed, but it was shortly to become obsolete in the face of a new development: electronics.

Earlier machines relied on comparatively slow mechanical or electric switches, but as Konrad Zuse concluded, computing would be far faster with electronic switching. This was based on vacuum tubes, sometimes called thermionic valves, or just valves. As the name suggests, the vacuum tube is a small glass bulb containing a vacuum with two or more electrical junctions. The most basic type is the diode, which has high resistance in one direction and low resistance in the other. More important for computing are triodes, which are in effect electronic switches with no moving parts – the essential building blocks of a computer, the physical counterparts of the ones and zeros in the computer memory. With no moving parts, the switches can flick over at literally lightning speed.

In June 1943, the US Army Ordnance Corps contracted with the University of Pennsylvania's Moore School of Electrical Engineering for research and development of the Electronic Numerical Integrator And Computer or ENIAC. The initial investment was $61,700, with the specific requirement for a machine capable of calculating shell trajectories faster than existing methods. It was, quite incidentally, to be the world's first digital electronic computer.

There were a vast number of technical hurdles to be overcome. One of the most fundamental was the limitation of the vacuum tube. Tubes blew all the time; the most complex device at the time was a counter with 200 tubes, whereas ENIAC would have several thousand. The chances of all of them working at the same time were minute, and made worse by the wartime conditions that took skilled staff away from factories. Wartime valves were of lower quality than earlier models.

Presper Eckert, one of the builders of ENIAC, spent a great deal of effort finding the highest quality tubes available, but he eventually settled on some which had been designed for a transatlantic cable and so were made to be as reliable as possible. He discovered that by running them at lower than their intended voltage they would last longer – in the end ENIAC ran the tubes at a tenth of their design voltage, and the necessary reliability was achieved.

ENIAC started operating in 1944. It was a giant, even compared to the Mark I, the completed machine weighing in at 27 tons and occupying 139 square metres of floor space. It used 200 kilowatts of power when it was working, and even when it was inactive it needed a large amount of power just to keep the tubes warm.[4] By 1946 the cost had reached just under half a million dollars, eight times the original contract.

What the US Army got for their money was the fastest computer in the world. Calculating a sixty-second trajectory of a shell would take a human with a slide rule twenty hours. With the aid of the Bush Differential Analyzer, this could be cut to fifteen minutes. ENIAC could do the job in just thirty seconds.

ENIAC was not the easiest machine to work with. It was the most complex electronic device built up until that time and suffered from frequent breakdowns, but gradually became more reliable as problems were resolved. Programming was a manual task: ENIAC had to be reconfigured manually for each type of calculating job, with wires physically connected in the required way. It took weeks to reconfigure

ENIAC for a new problem, such as moving it from ballistic calculations to weather prediction, but once set up it was far more efficient than earlier machines.

By the time ENIAC was fully functional the war ballistics tables were no longer so urgent, for there were now even more important tasks to be done. The atomic bomb had ended the war, and now theoreticians at Los Alamos were considering an even more powerful weapon, the hydrogen bomb. Edward Teller believed the new bomb would work, but his colleague Stanislaw Ulam raised objections. Neither would give way and the project hung in the balance.

Whether or not the bomb would function depended on what was happening during nuclear reactions taking fractions of a millionth of a second. The scientists could only roughly calculate how this would work with slide rules. The number-crunching for a more exact model would take far more computing power.

Computations by ENIAC – some of which are still classified – showed that Ulam was right and Teller's proposed design would not work. However, the results enabled Teller and Ulam to collaborate on an improved design which would lead to the first hydrogen bomb three years later.

ENIAC gradually took over the workload of US military calculators, working out all the ballistics tables for the US Ordnance Corps in the evocatively named Project Chore.

By 1949 ENIAC had become the US's main national resource for all forms of scientific computation, carrying out work in the fields of weather prediction, wind-tunnel studies, atomic weapons design, rocket ignition and cosmic ray analysis. By the time it was retired in 1955, ENIAC had been superseded by other machines based on the same principle (EDVAC and ORDVAC) and it was clear that digital computers were here to stay.

And somewhere in there, in 1951, the first digital computer was used for commercial purposes. A purely commercial version of ENIAC was built incorporating some improvements. Called UNIVAC, it was first used for the 1951 US census. In 1952 it won fame when CBS news featured the first computer prediction based on early returns in the Presidential elections. The television presenters intended it to be more as a piece of entertainment than a serious scientific endeavour, but the machine predicted the results with 89 per cent accuracy, which was far better than the human pundits who were amazed at Eisenhower's landslide victory. The media loved it, and the 'Giant Electronic Brain' entered public awareness.

After their first entry into computing on behalf of the US Navy, IBM were well aware of the threat to their business from electronic calculating machines. They were already starting to use electronic components in punch-card multipliers and other devices, and an IBM electronic computer was the next logical step.

IBM appointed Wallace Eckert (no relation to Presper Eckert) as Director of their new Pure Science Department. Wallace Eckert had worked at the Naval Observatory in Washington and was Director of the US Nautical Almanac Office. In this role he had played an important role in computerizing the calculation of nautical almanacs, using machines to calculate and print tables – exactly the same aim that inspired Babbage.

Eckert's wartime role was important because the previous tables only listed the locations of celestial bodies at one or two times in the day, but Eckert's team produced tables with much more detailed listings. This cut down the time taken to get an accurate location from half an hour to a few minutes, vital when directing aircraft to the location of an enemy submarine, as the delay was crucial if the sub was to be caught before it could get away. The team also produced triangulation tables for locating U-boats when their transmissions were intercepted.

The problem was always the lack of computing power: working continuously on the almanac, Eckert's team could never get more than a week ahead. The 'submarine book' was created using the same machines during the night shift and took three months. The combined effect of these aids is credited with helping reduce shipping losses from 30 per cent to just 6 per cent.

Eckert had a keen appreciation of the importance of more powerful machines, and had helped build some of the fastest ones in the world, including the NORC (Naval Ordnance Research Calculator), started in 1950 and reckoned to be the first supercomputer. NORC was used for the Navy's ballistics calculations – gunnery tables again.

With Eckert's help, by 1952 IBM had its first all-electronic computer, the Model 701. In 1953 they produced the Model 650. Unlike earlier machines, this was not designed to push the limits and provide more power, but was intended as a cheap computer for the general business market. At only half a million dollars and fitting into a single room, the 650 was a far more attractive proposition than the

million-dollar UNIVAC. IBM also had another key advantage –
companies already used their punch-card systems, and the 650 was
compatible with existing machines. IBM already had a huge customer
base, a solid business and an effective sales force. By stressing that the
new equipment was compatible with the old, they held on to the
customer base and expanded, and within four years they dominated the
market.

This pattern was to be repeated with the invention of the Personal
Computer in 1981. Other companies had introduced desktop computers before, but IBM's economic clout meant that their design became
the standard. This brought about a growth in personal computing that
would otherwise have been impossible, though in the case of the PC
IBM lost out. This was because it turned out that competitors could
manufacture 'IBM-compatible' PCs of the same standard more cheaply
than IBM itself.

The military still relies on computers for ballistics calculations, and
these have become increasingly advanced. Instead of a book of tables,
everything is calculated on the move. Tanks have fire-control computers
that allow a vehicle travelling at 30 m.p.h. to hit a moving target
several miles away with lethal accuracy.

A modern ballistics system can also calculate trajectories so that
several shots can be timed to arrive simultaneously. One such system
claims to be able to fire five shells over the course of a minute and have
them all land on target within the space of three seconds. This type of
one-gun barrage gives the targets no chance to run for cover. Smaller
versions of the ballistics computer are now integrated with weapons as
small as grenade launchers; the next stage will be to have them built
into rifles. These devices have a laser rangefinder and can automatically
compensate for the distance and target movement, so that all the solder
has to do is hold the sight on the target and pull the trigger.

## Integrated Circuitry: Chips with Everything

ENIAC, UNIVAC and their contemporaries were all based on vacuum
tubes. They worked, but vacuum tubes had significant disadvantages
when put together in large numbers. They were big and bulky,
consumed a lot of power and gave out a corresponding amount of
heat, and they had a short lifespan. As ENIAC showed, putting a large

number of tubes together meant that a regular stream of spares would be needed.

The replacement for the vacuum tube was a device called a transistor. The transistor is a small piece of solid semiconductor that can do the same task as a vacuum tube, but using less power. It is much more robust, not needing a fragile glass bulb. Although its roots go back much further (the theory dates back to the nineteenth century), the modern transistor was a creation of Bell Labs in 1947.

Later on, the transistor would turn out to be a milestone of the twentieth century and the inventors (Bill Shockley, Walter Brattain, and John Bardeen) were awarded a Nobel Prize in 1956. The transistor made all sorts of cheap electronics possible from radios and televisions to hi-fi. However, it is only a matter of luck that the transistor was not classified a military secret. In June 1949, Ralph Bown of Bell Labs gave a presentation of the new invention to the US Army, showing how the transistor could amplify a signal more efficiently than a vacuum tube. After some days of deliberation, they decided to allow Bell to release information to the public, and a press launch went ahead a week later.

Transistors made little headway at first; vacuum tubes were a mature technology and were good enough for what they did. When a transistor radio was demonstrated to the media, the reaction was muted: it might be a fabulous technological breakthrough, but it did not behave any differently to the normal valve radios. In addition, the new transistors were expensive compared to vacuum tubes.

But transistors made new applications possible. In 1958, two Texas Instruments silicon transistors were used to make the first miniature cardiac pacemaker.[5] Pacemakers which artificially controlled the beating of the heart already existed, but they were big and bulky; a pacemaker small enough to be surgically implantable was far more effective and has saved thousands of lives.

In a few years transistors would become cheap and easy to massproduce. The first Japanese transistor was produced in 1955 under licence from Bell, leading to the start of the Asian electronics manufacturing boom. The transistor radio made music cheap and portable, and it arrived just in time for rock 'n' roll.

However, if transistors were to be useful for computing there was another problem to overcome. Transistors themselves were small, but the individual transistors still needed to be attached to each other with wires. Even the most skilful technician had difficulty putting together something when the mass of connecting wires was bigger than the transistors.

The need was already apparent. In 1947, the creators of ENIAC were contracted by Northrop Grumman to build a computer to control the flight of their new Snark missile. The idea was that this computer could be transported and operated in an aircraft; in addition to the problems of making the computer compact enough, there was the added difficulty of the fragility of the vacuum tubes. The system was not a success.

What Northrop really wanted was a computer small enough to fit inside the limited space of the rocket itself. The computer would have to be as small and reliable as possible. Small, rugged transistors were the answer, but only if they could be joined together in an efficient way.

The US Army Signal Corps sponsored the Micro-Module Program as a possible solution to the problem of connecting transistors. All components were to be made of a uniform size and shape with built-in connectors. Equipment could then be made by snapping these building blocks together, removing the need to solder connections between them.

Jack Kilby of Texas Instruments was working on the Micro-Module Program when he devised a better solution. The idea occurred at the same time to Robert Noyce of Fairchild Semiconductor, and both filed patents for methods of miniaturizing electronic circuits in 1959. The idea had been around since 1952, having been put forward by Geoffrey Dummer, a British scientist working for the MoD's Royal Radar Establishment. However, Dummer was not successful in his attempt to fabricate the device; Noyce and Kilby made the idea work.

Printed circuits had been around since the 1940s. These start with a sheet of insulator which is covered with a thin layer of copper, which in turn is covered with a light-sensitive film. Light is shone on to the board through a mask, which has the pattern of an electrical circuit on it. The sensitive film hardens where it is exposed to the light. When the board is washed and treated with acid, the surplus copper is dissolved away, leaving only the pattern of the circuit. The components can then be soldered on to the circuit board. The intricate pattern of connections does not need to be wired up, but is printed on to the board.

A similar idea could be applied so that minute transistors could be incorporated into a circuit without having to solder them on.

Dummer reasoned that instead of starting with a number of transistors and trying to solder them on to a circuit board, the circuit board could be built up with the transistors already on it. The printed circuit showed that you could create copper wiring by painting it on to the surface. Why stop with copper?

Kilby and Noyce started with a plain sheet of silicon and painted the components on to it, building up diodes and other components from layers of various materials. This might involve many stages of deposition of material, baking impurities into it, 'metallization' or adding metal connections, and etching, but all of these were technically possible.

The result, a small piece of silicon with several different components joined together on it, became known in the industry as the Integrated Circuit or IC, and to the rest of the world as the silicon chip. The chip was the basis of the first electronic calculators and it transformed computing. It has become an essential component for virtually every household appliance we have today, from mobile telephones to fridges to washing machines and cars.

Texas Instruments developed the chip, and although the US space program is often quoted as the driving force for the development of the integrated circuit, this is not entirely true. Chips were first used in the Solid Circuit Network Computer in 1961. Described as 'a micro-miniature digital computer utilizing semiconductor networks', this marvel was a 300 gm computer which did the same job as a vacuum-tube system weighing 13 kg. It was designed for the US Air Force's missile program, to show that chip-based computers would be useful for guidance and telemetry, and was soon taken up on a massive scale for the new Minuteman missile.

NASA was an early customer for chips and needed them for much the same reason as the Air Force – missile guidance. NASA launched just a handful of spacecraft (only seventeen Apollo launches for example); compare this to the Minuteman intercontinental ballistic missile program, for which a thousand missiles were built during the first phase alone. Minuteman was calling for more than 4,000 chips a month.

At one stage it was envisaged that the chip would be entirely confined to military applications and there were doubts over whether it would be declassified for civilian use, but as with the transistor the military did not put up objections to the silicon chip being used commercially.

Early chips had just a handful of components on them, and were described as Small Scale Integration. The next generation, with more than a hundred components on each chip, were known as Medium Scale Integration. Large and Very Large have succeeded them as the number of elements went into thousands, tens of thousands and more. These days chips can contain millions of microscopic transistors, and there is no sign of the progress slowing down.

In 1968 Robert Noyce left Fairchild to found a new company called Intel. Intel's main business was to be the microprocessor – a chip containing the processing elements for a computer. In 1971 the first microprocessor, the Intel 4004, was smaller than a fingernail and packed in as much processing power as ENIAC had boasted twenty-five years earlier. In subsequent years processors have steadily increased in speed and power, and a whole industry has developed around the manufacture of silicon chips. Intel processors went on to power the IBM Personal Computer, the ubiquitous PC, while Noyce himself acquired the soubriquet 'Mayor of Silicon Valley'.

It is understandable why NASA, America's favourite high-tech institution, should be given the credit for developing the silicon chip, but this cannot be entirely justified. Without the USAF's need for better missile guidance, the huge growth in the electronic industry might never have happened.

## Military Software

So far we have looked at the machinery of computing, known as hardware. There is another side, the software used for controlling the computer. The earliest machines like ENIAC were programmed by physically rearranging the hardware, changing connections by pulling out cables and reconnecting them. Later programming could be carried out by sending commands which would open and close switches to achieve the same thing. Initially these control commands took the form of codes which were only meaningful to the computer itself, forming a language known as 'machine code'.

Machine code is difficult for humans to master. A way of translating a set of commands was needed that would turn something more like human language (such as 'add this number to that number and divide the result by a third number') into machine code that the computer could understand.

In 1951, Grace Hopper of the US Navy started work on a piece of software called a compiler. She was later to become Rear Admiral Hopper, nicknamed 'Amazing Grace', and the discoverer of the first computer 'bug'. This happened when she traced a computer problem to a moth trapped in the relays (the moth was duly pasted in the logbook and is now a museum exhibit). The compiler was the first attempt to produce machine code from a more human-like language, and started the process of transforming computers into machines which can be easily programmed by normal people rather than computer scientists.

In the early days of computing there were a series of computer languages such as FORTRAN (short for FORmula TRANslation) and LISP (LISt Processing) which were mainly used for academic and engineering purposes. Business computing did not really take off until the advent of COBOL (COmmon Business Language), a military development.

There were many companies jumping on to the computing band-wagon in the US, building hardware to different standards and producing their own programming languages. A program written for one manufacturer's machine would be useless on any other machine, and a computer programmer could only work with one type. The Pentagon, guided by Hopper, formed the committee to set the standards for a common language which could be used across all businesses. As one of the biggest customers for the computing industry, the military were in a strong position to set standards: whatever they decided on was likely to be adopted as a standard by the rest of industry.

COBOL brought programming into the world of commerce. A single piece of software, with minor changes, could be sold to users of all sorts of different machines. Before COBOL, programming was a job for technicians and scientists, but COBOL with its commands that resembled English could be used much more widely. (Though speaking as an ex-COBOL programmer, the author found this language to be distinctly unfriendly, with the simplest programs taking many pages of coding and producing reams of error messages.)

The first computers were used for batch processing: you fed the machine a program and a pile of data, and came back a few hours later when it had finished processing. This was acceptable for projects like trajectory calculations and nuclear simulations, but a computer which could be controlled in real time would be much more useful. The first steps towards creating an interactive digital computer came from

Project Whirlwind, which started as a US Navy flight simulation project.

The flight simulators at the time were analogue devices, with a different one being built for each new aircraft type. Jay Forrester, the leader of the project, realized that the power of the new digital computer would allow one computer to imitate many different types of aircraft, responding to signals sent in real time. Instead of sending off a batch of commands and waiting for them to be processed, the computer could send back a response every time a key was pressed.

By 1946 Forrester had expanded his aim to that of building a computer capable of carrying out all sorts of processes in real time. The project kept growing, until at one point it was absorbing 10 per cent of the Office of Naval Research's budget. Technical advances were being made, but the budget was threatened until a saviour appeared in 1949, with the emergence of what would now be called a 'killer application'.

SAGE (Semi-Automated Ground Environment) was the control system for the US's air defence network. It had to take information from over a hundred different radar systems and display the output to operators so that they could track and identify aircraft. This was not a system which could wait for overnight processing, it had to give an instant response to inputs. SAGE was vital to the defence of the US against Soviet bombers; money was no longer an object.

SAGE saw the first implementation of shared software operating in real time. This was to revolutionize the computing industry: for the first time, dozens of operators could sit at remote terminals, all with the ability to call on a central computer at a moment's notice.

The technology was quickly adopted in the commercial world. Over 7,000 IBM employees worked on SAGE, and much of the same technology was used in SABRE, the airline reservation system developed by IBM in 1964 which soon became the heart of the airline industry and continues to the present day. The technology from SAGE meant that a customer anywhere in the world could make a telephone call and guarantee themselves on a particular flight weeks or months in advance. SABRE continues to the present day in upgraded form, with thousands of users across dozens of countries.

Real-time computing touches almost everybody's life. Whenever you put your card into an ATM machine you are using technology that originated with Whirlwind.

Without these vital contributions to programming and real-time processing, computers would never have become the force that they are in the world today. Computers might still be machines whose mystery and inaccessibility meant that they were operated solely by a caste of technicians, rather than appearing on every desktop and an increasing number of homes.

## DARPA and the Internet

The origins of the Internet have generated endless argument. The debate even cropped up in the 2001 Presidential election campaign, when George W. Bush mocked his opponent Al Gore for claiming to have invented the Internet.

This stemmed from a March 1999 interview with Wolf Blitzer, in which Gore said, 'During my service in the United States Congress, I took the initiative in creating the Internet.' Although Gore was claiming that he promoted its development, not that he had any technical involvement in the Internet, in many people's minds it became a vain boast.

Gore did actually play a significant role; Internet pioneer Vinton Cerf said that: 'He was certainly among the first if not the first in Congress to realize how powerful the information revolution would be . . . he has been enormously helpful in supporting legislation and programs to help further develop the Internet.'

But the damage had been done. Needless to say, Gore lost the election.

The problem is that the Internet is a complex system made up of many different elements which were developed by different people at different times. It has no single inventor – but there is no doubt that the Pentagon is in the strongest position to claim the credit on the basis of work by DARPA.

## A Network of Networks

As we saw in Chapter 1, in 1957, the US Armed Forces had no centralized research, with separate Army, Navy and Air Force teams working on ballistic missile development. Rivalries and security issues prevented the free sharing of information, slowing down progress.

NASA was formed to unite the space program under a single roof. Another organization was needed to handle technology in general, and so President Eisenhower created ARPA, the Advanced Research Projects Agency. (The word Defense was later added to the title making it DARPA). It was given a budget and ordered to go out into academia and industry and draw on America's intellectual reserves, to find new ideas for military technology and develop them into usable form.

ARPA was a great success. The core organization was small and mercifully free from bureaucracy. It could hand out a research contract on the basis of the briefest of proposals, and consistently picked winners. This was a contrast with the Pentagon's usual way of doing business which was highly conservative and bureaucratic. It was said that the government documents specifying the requirements for a new battleship weighed more than the ship itself. Some have suggested that ARPA's advantage was simply that it was young and had not yet swelled to the Pentagon's bloated proportions.

In 1965, Bob Taylor was appointed as Director of ARPA's Information Processing Techniques Office. Arriving at his new office, he found three separate computer terminals. One connected to a machine at MIT, one to Berkeley, and one to Santa Monica.

'The obvious question that would come to anyone's mind,' said Taylor later, ' "Why don't we just have a network such that we have one terminal and we can go anywhere we want?" '

Computer networks were a familiar enough idea, but the difficult part was to put together a network connecting completely different machines. It would be like putting an Englishman, a Hungarian and a Tibetan together without a shared language. But if it could be achieved it would be invaluable. Workers at any of the three sites would be able to access information and processing power at any of the other sites without having to install new connections and terminals. The machines could exchange information and work together. They could even be programmed to split the workload of a major task between them, combining their power into a single supercomputer.

In February 1966, Taylor went to his boss and outlined the idea of a network to connect together the three disparate computer systems. According to Taylor, it took about twenty minutes for the Director of ARPA to agree with him and hand over a budget of a million dollars for the project.

The original aim was not to ensure that the US computing infrastructure could withstand a nuclear strike and keep functioning. At its inception, it was simply a way for the military establishment to make more efficient use of computing resources.

Rather than laying new cables for the system, the new network would use existing telephone lines. The first trial of a direct computer-to-computer connection over a phone line was a qualified success: data was successfully transferred, but the poor quality of the trans-American phone line showed that communication was going to be difficult.

The first idea of the project involved all the big computers connecting directly to each other using dial-up phone lines. The computers would then have to translate from their own language into the language of the network, as well as coping with the process of routing messages to the other computers. The additional workload made it unpopular with users, who did not want power from 'their' machines taken up with other tasks.

There was also the problem that every new computer would mean a new connection. With three machines, this was simple enough: each computer would be connected to the other two. With a hundred computers, each machine in the network would have ninety-nine phone connections to handle. The number of computers in DARPA was increasing rapidly and the growth was set to continue. What would happen if they ever had a thousand computers?

Wesley Clark from the Lincoln Laboratory proposed a solution. There would be a network of interconnected small computers, each of which would act as a node on the network. These small computers would not all connect directly to each other, but each would be able to communicate via other machines in the network. For example, machine A is attached to machine B, and B is attached to C. Whenever A wants to send a message to C, it is passed to B with an instruction to pass it on.

In this set-up when a new computer is added it only needs to be connected to one other machine on the network. The small computers which do the connecting became known as Interface Message Processors or IMPS, and their task was to handle the routing of messages around the network. There would be no extra load on the users' machines.

In 1967 the term ARPANET was used for the first time to describe the new network layout. The new concept was more popular with the users, who were assured that their computers would not be lumbered with an additional workload and an endless spaghetti of proliferating telephone connections. The way was clear for the network to be constructed.

## Which Switch?

There remained the question of how messages would be passed around the network. Telephone networks are 'circuit-switched': for A to communicate to C, first a connection has to be established between A and B and then a second one between B and C to make a complete circuit. When C wants to send a message to someone else, the circuit is broken and new connections are made. The switches involved are physical ones; a telephone exchange is simply a collection of thousands and thousands of switches.

This is fine for voice communication, in which you need to use the line for the entire conversation. Data communication is different. You might be sitting at your computer for several minutes reading something on the screen, then send an email of several hundred words taking a second or two before lapsing into silence for several more minutes. If the phone line is in use for the entire time, then for the vast majority of the time it is completely wasted.

Ideally, during data communication the circuit would only be established when you need to send a message and then it could be cut off again. That way dozens of users could use the same phone line, and since none of them have it for more than a second or two at a time, for each of them it would feel like they had a permanent connection. Each message carries with it the instructions for when it is to be routed, and these instructions control the switching process, hence the name applied to this type of network: message-switched.

But what happens when one of the users decides to send an encyclopaedia, or, more likely, download a large adult-oriented graphics file? The other users are left unable to connect to the system until MYVIDEO.AVI has finished downloading. This is an inherent problem with message-switched systems; any number of short messages will be held up behind a long one, like frustrated motorists backing up behind a slow tractor.

The man who solved it was a British physicist and computer scientist, Donald Watts Davies, who had earlier worked with Alan Turing. He saw that if every message was broken up into small pieces the system would never have to deal with this sort of traffic jam. The small pieces, termed 'packets', would be of a uniform size and all equally speedy. There would be no delays; a message would take as long to send as the packets that made it up.

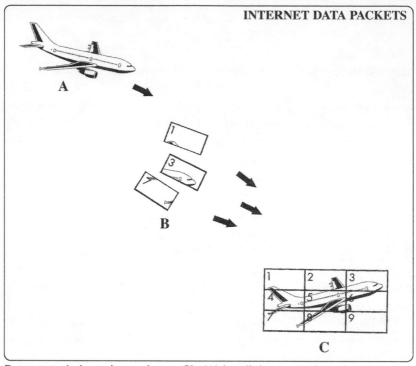

**INTERNET DATA PACKETS**

Data transmission using packets: a file (A) is split into several numbered packets (B), which are sent to the destination via various paths. On arrival, the packets are assembled in order to recreate the file (C).

Davies decided on a size of 128 characters for his packets. As well as the message content, each packet would also have a header which corresponds to the address label on a parcel saying where the packet was to be directed.

Each packet also had a sequence number so the packets could be assembled in the correct order, and, as an additional feature, a check digit. This is a way of determining whether the content of the packet has been received correctly; if the check digit indicates a problem, the receiving computer sends back a request for the packet to be re-sent. The check digit means that messages can be sent without the risk of any error even over noisy telephone lines that would otherwise tend to lose or distort the data, so the poor quality transcontinental lines that dogged the first tests would not be a problem. The small packet size means that there is a good chance that any given packet will make it through intact, and the packets that get damaged can be re-sent. Bad lines only slow down the process of transmission rather than stopping it.

Another advantage in this scheme is that the packets can take different routes through the network. They can take advantage of any spare capacity and arrive at the destination from several different directions. If a connection is broken partway through a message, the rest of the packets take a different route. The header means that all the packets arrive at the right place, and the sequence number ensures that they can be assembled in the right order.

Davies worked for the British National Physical Laboratory, but his idea was rapidly picked up by DARPA. This advanced form of switching became known as packet switching and is the basis for how the Internet sends data.

## A Network is Born

By October 1969 the technical details had been agreed, the hardware had been built, and the first DARPANET connection was ready. The IMPs to connect computers to the network were small, but only by the standards of the day: each was the size of a fridge-freezer and weighed half a ton.

The very first connection was made between machines at UCLA and SRI. An operator from SRI typed 'login' to connect remotely to the UCLA machine. Unfortunately, when he reached the letter 'g', the UCLA machine unaccountably crashed. This was like Neil Armstrong following his 'One Small Step' speech by toppling into a crater.

Things went more smoothly afterwards, and from an initial config- uration of four computers, the network of networks started to grow. By 1972 it had expanded to thirty-seven computers. The first computers were at universities within the US, but it soon spread out to encompass government departments and commercial operations around the world.

In 1973, a group headed by Vinton Cerf from Stanford and Bob Kahn from DARPA started work on a means for different networks to communicate with each other. This was known as the Transmission Control Protocol, a protocol being a formal set of instructions for how different systems can talk to each other. One of the papers describing TCP first used a new term to describe the interconnected network of networks: 'Internet'.

By 1983 the Internet had grown to include over five hundred hosts. Not everybody attached to it was involved in DARPA work the way the original members had been. In this year there was a hit film called *Wargames*, in which a teenage hacker manages to break into the Pentagon's computer system by dialling numbers at random. Thinking that he has found the a games company, he is misled by the computers asking 'Do you want to play nuclear war?' and sets in train a series of events that almost leads to catastrophe.

It might be a complete coincidence, but in the same year as *Wargames* the military moved to a secure network. The military network, MILNET, is not physically attached to the Internet, so hacking into the Pentagon's war computers via the phone lines or the rest of the Internet is no longer a possibility.

## Patterns of Development

The outline of computing history sketched here shows a clear pattern. There are false starts, and even the best line of progress will run into a dead end at some point. Babbage's Engine could have been the wonder of the age, but its mechanical construction meant it would never progress beyond a certain level. The electro-mechanical switches of the Mark I and the Z3 worked, but could never have been used to make fast computers. The vacuum tubes of Colossus, the Z4, and ENIAC had to give way to transistors and integrated circuits before it was possible to build computers whose memory was measured in megabytes.

Each step in development required a sponsor, and in each case this was the military. Once the usefulness of the computer had been established, commercial interests could take over development. When it is a matter of life or death – or in the case of Colossus, victory or defeat – money and resources become less important issues and even a faint chance of success is enough to prompt massive investment in an idea.

The original ideas have tended to come from enthusiastic academics, from Babbage to Turing to Donald Watts Davies; once they gained the support of the government in the form of military funding, their ideas were turned into concrete products. Some were dead ends, but when they were successful the results have been remarkable.

Computers are still evolving, and there is not yet any sign of an end to the growth in their processing power. Moore's Law, which says that the power of computers for a given price doubles every eighteen months, has proven valid for the last twenty years. It is not the increased processing power of the chip at the top end which has made it important in our everyday life, so much as the easy availability at the mass market level. Cheap chips have come to control everything from fridges to cars to mobile phones, quite apart from the proliferation of computers. Digital cameras, camcorders, music players and radios are all part of the computing revolution. 'Convergence' is the next trend: the mobile phone that is also a digital camera and MP3 player, and in time, a TV, video and camcorder.

While computers may seem to be ubiquitous today, there is every sign that they will become even more common in the future, expanding into new areas. As we shall see in a later chapter, the pattern is set to be repeated with influential developments from the military sector.

# CHAPTER 4

# Death Rays to DVDs

*Then slowly the hissing passed into a humming, into a loud long, droning noise. Slowly a humped shape rose out of the pit, and the ghost of a beam of light seemed to flicker out from it.*

*Forthwith flashes of actual flame, a bright glare leaping from one to another, sprang from the scattered group of men. It was as if some invisible jet impinged upon them and flashed into white flame. It was as if each man were suddenly and momentarily turned to fire.*

H. G. Wells, The War Of The Worlds, 1897

Archimedes supposedly invented the original death ray in 212 BC. The Greek mathematician and engineer is said to have used mirrors to concentrate the sun's rays on the attacking Roman fleet: 'A fearful kindling of fire was raised in the ships, and at the distance of a bow-shot he reduced them to ashes.'

It is a good story, but has never had much credibility; there have been occasional attempts to replicate the feat with partial success, but the solar death ray never caught on. Some 2,000 years later, in the late nineteenth century, the discovery of X-rays and other new phenomena reawakened interest in this area. Wells' Martian heat ray caught the popular imagination, and death rays of various types became an established feature of science fiction. By the 1920s Buck Rogers and Flash Gordon found that ray guns were essential props for fighting evil in comic strips and on the radio.

Scientifically speaking it seemed like fantasy; a parabolic reflector could be used to direct heat at a target a few metres away, but it was impossible for the target to be made hotter than the source of the beam.

New radio technology looked as though it might offer the answer. Early radio transmitters were omnidirectional, broadcasting radio waves in all directions just as a light bulb emits light all around. But in 1924 the radio pioneer Guglielmo Marconi developed a directional shortwave transmitter. Rather than wasting energy, Marconi's new invention concentrated the transmission into a tight beam, greatly increasing the range at which the signal could be picked up. Marconi's idea was soon put to use by the Imperial Wireless Chain – a series of radio transmitters and receivers that could pass a message over long distances. Marconi's Wireless Telegraphy Company was a great success, but directional transmission might have other uses as well.

In her autobiography, Rachele Mussolini, widow of the fascist dictator, says that Marconi put the possibility of a radio beam weapon to her husband, but the idea does not seem to have been accepted.

Marconi was not the only one thinking along these lines. The First World War had brought a terrible shock as it revealed the new face of warfare. New aerial weapons, the bomber and the Zeppelin, brought war to civilians for the first time, threatening whole cities with destruction.

'The bomber will always get through,' Prime Minister Stanley Baldwin warned the House of Commons in 1932, giving expression to a popular belief at the time. Air defences, anti-aircraft guns and fighters could not prevent bombers from destroying cities.

What was needed was a new weapon to bring down enemy aircraft before they could drop their bombs. It might also be effective against ground troops making wars short and decisive again – and with an outcome in favour of the power that had the new technology. The race was on to build the radio beam death ray.

Such a weapon would have immense advantages. Operating at the speed of light, it would be able to strike its target instantly, even a fast-moving target like an aircraft. There would be no need to calculate lead, windage and drop associated with physical projectiles, and aiming it would be as easy as pointing a torch. Given enough power, a heat ray would be able to burn or melt its way through anything at any range. But it would not necessarily require high power. A beam of radio energy with relatively low power might be able to interfere with the aircraft's engines and stop them from working.

Marconi does not seem to have progressed the idea beyond his original suggestion, but others did. Among those who claimed to have invented radio beam weapons was Nikola Tesla. Born in Croatia, Tesla emigrated to the US where he became one of the great pioneers of the electrical age. He discovered the advantages of alternating current for electricity transmission and designed the first power generators at Niagara Falls. He also invented the Tesla Coil, still used for generating high voltages. In later life Tesla became somewhat eccentric and could sometimes be found talking to pigeons. He had difficulty getting sponsors for his work; having sold the patent rights to his system of alternating-current dynamos, transformers and motors to Westinghouse in 1885 he missed out on the big money.

Although the quality of his work declined, Tesla was always a great showman, performing spectacular demonstrations such as producing an artificial 40-metre bolt of lightning, and lighting bulbs from a distance.

'We are enabled to project electrical energy in any amount to any distance and apply it for innumerable purposes, both in war and peace,' Tesla told the *New York Times*.

In 1940, at the age of eighty-four, Tesla offered the secret of his Teleforce Projector to the US government. He said that this could bring down aircraft at a range of over 200 miles, protecting the US with an invisible wall. However his price was $2 million, and nobody was inclined to take him up on the offer. Conspiracy theorists claim that Tesla's secret papers were plundered after his death by shadowy government agencies, but there is no evidence that Tesla's ray existed outside his imagination.

In Britain we had Harry 'Death Ray' Matthews. He too was a genuine inventor, having worked on an early wireless telephone and talking pictures. In 1923 he gave journalists a demonstration of an electric ray, which could stop a motorcycle engine from a distance.

'I am confident,' he announced, 'that if I have facilities for developing it I can stop aeroplanes in flight – indeed, I believe the ray is sufficiently powerful to destroy the air [sic], to explode powder magazines and to destroy anything on which it rests.'

He claimed that his apparatus could kill mice and shrivel plants at a distance, as well as igniting explosives, and that an operational weapon would have a range of up to 8 miles.

The media lapped it up, and the new death ray was a public

sensation, helped by rumours that the Germans were working on their own version. Public pressure forced the War Office to take the 'diabolical ray' seriously, in spite of allegations that Matthews had been involved in scientific fraud. In 1924 the Air Ministry asked him to give them a demonstration.

Matthews failed to impress the military. They were not allowed a close inspection of the experimental set-up and were dubious of Matthews' claims, suspecting that the whole thing was a conjuring trick with rigged apparatus. Thwarted, Matthews attempted to sell the idea in France and America but without success. By 1930 he had given up on death rays and was working on a way of projecting advertisements on to clouds.

The Air Ministry offered a reward to anyone who could produce a death ray capable of killing a sheep at a range of a hundred yards, but it was never claimed. In the same period the US Army's Aberdeen Proving Ground offered a similar prize to anyone who could kill a tethered goat with a death ray, and this too was never claimed. It might be possible to kill a stationary animal at point blank range with a powerful enough emitter, but doing at a distance was a much greater challenge.

A ministry scientific committee was appointed to look into the matter, and after some research, Dr Robert Watson-Watt reported back on 4 February 1935 that the power needed for an electromagnetic beam that could damage an aircraft was far in excess of what could be achieved with existing technology.

The committee went on to ask Watson-Watt whether, in the absence of death rays, anything useful could be done with radio waves. Post Office engineers had earlier noticed that the radio signal was disturbed when aircraft flew close to the transmitter masts; this suggested that radio waves might be used to detect approaching aircraft.

Watson-Watt drew up a paper entitled 'The Detection and Location of Aircraft by Radio Means', and he was asked to carry out tests to demonstrate the theory. Three weeks after the report on death rays, Watson-Watt organized an experiment in which a Handley Page Heyford bomber was flown past the BBC transmitter at Daventry, while scientists looked at the reflected radio signal using a receiver attached to an oscilloscope. They found that the bomber was reflecting a signal strong enough to be easily detected, exactly as predicted.

The Air Ministry funded further development work for an apparatus for Radio Direction Finding, which later became known as RAdio

Detection And Ranging – or RADAR for short. No less than seven other nations independently developed their own versions of radar, but by 1940 the British had an advantage. The invention of the resonant-cavity magnetron provided a means of generating very high-frequency radio pulses. As a result, radar played a vital role in the Battle of Britain. By detecting incoming Luftwaffe raids at a distance, the RAF could intercept them at will rather than relying on scrambling fighters when they were seen visually.

Radar became steadily more sophisticated as the war progressed. In addition to being used to detect aircraft, an aircraft-mounted version was used by allied bombers to locate targets in darkness and under cloud cover. Later the proximity fuse was developed which used a radio signal to detect when it approached a target. Anti-aircraft proximity fuses gave improved defences against the V-1 flying bombs, while proximity-fused artillery shells could burst in the air, making their shrapnel far more lethal.

Others were still pursuing death rays. In Germany, the Nazi government funded several programs without success. Some of these were over-optimistic, and in one case deliberately so. Professor Richard Gans produced calculations which showed that a type of X-ray generator called a Rheotron could be converted into a death ray.

In this concept the X-ray generator had a cathode that was concave rather than flat, and like a concave mirror it would allow the beam to be focused. At extremely high energies and short wavelengths, the X-rays would be emitted in a tight beam like Marconi's radio.

This beam would be directed at enemy aircraft. A report on the device stated, 'Aeroplane engines could be "pre-ionised" with the bundled, highly penetrating radiation, so that the ignition would fail and the machines could no longer fly.'

However, Gans became aware that in his calculations he had omitted something called the Compton Effect which made the whole scheme impossible. He continued work on it anyway: as a scientist of Jewish origin in Germany, his life depended on being useful to the regime. It is possible that his masters were not concerned about whether the weapon would be viable; the propaganda value of the death ray was enough to justify going ahead with it.

The Japanese also worked on a similar weapon. Built by Norohito Laboratories, the device could kill rabbits at a range of 30 metres. However, it only worked if the target remained motionless for an extended period; tests on monkeys were not successful because the animals simply would not stay still in the beam.

The US Army Air Force Scientific Advisory panel looked at the Japanese designs after the war and decided that, like the German death rays, it would never make a usable weapon. Radar technology had advanced tremendously during the war years, but it was clear that a radar installation powerful enough to be deadly would be too big and clumsy to be practical. (However, radio-beam weapons have not entirely been abandoned – later chapters discuss non-lethal directed-energy weapons and the 'e-bomb'.)

Very short wavelength radio waves became known as microwaves, and military work into radar and allied fields such as microwave communication led to a boom in the production of this equipment and a flourishing industry.

In 1945 the first experimental microwave communications relay was used between New York and Philadelphia. Microwaves could be formed into a highly directional beam, and the short wavelength allowed more signals to be packed into a microwave beam than a radio beam. Microwave relay towers were used increasingly to distribute telephone and television traffic. In 1965, the new Post Office Tower was the heart of British telecommunications, an 189-metre/620-foot tower studded with microwave dishes.

The work on microwaves had an unexpected spin-off. In 1946 Percy Spencer, a self-taught scientist and inventor, was working for the military electronics company Raytheon. Spencer was standing next to a magnetron, a device for generating microwaves, when he discovered that a chocolate bar in his pocket had partially melted. Driven by curiosity, he wondered whether the heating was caused by the magnetron, and experimented by placing a carton of popcorn kernels in the same position. The popcorn duly started crackling and popping.

The third food item tested was an egg, which on heating exploded in one of the experimenters' faces. The principle had been proven – microwaves could be used to heat food, though some foods can be cooked more successfully than others. Microwave heating works by warming water molecules. The water molecule is polar, with one end having a positive charge and the other a negative one. This causes it to twist around to align itself with electromagnetic radiation; radiation of the right (microwave) wavelength will cause it to spin rapidly, absorbing energy.

Within a year Raytheon had produced the world's first microwave oven, a giant 3,000 watt contraption for commercial use. Other defence contractors with experience of radar and magnetrons produced their own versions. The American name for the device – 'RadarRange' – is suggestive of its origins.

The idea took a while to catch on. The first microwave ovens were huge commercial units costing thousands of dollars, but by the 1970s they were firmly established on the US domestic scene as an essential labour-saving device.

Military work on short-wavelength radio had another important spin-off as well. In the 1930s, radio transmitters and receivers were barely portable, weighing 15 kg or more. By miniaturizing the components as far as possible, and using higher frequencies (and so shorter wavelengths), American military radio maker Galvin Manufacturing produced the 'Handie Talkie', a handheld two-way radio weighing just over 2 kg with a range of up to 3 miles. The shorter wavelength led to this type of radio being known as 'short wave', and more than 100,000 were produced during the war. There was also a larger backpack version, the 'Walkie Talkie', which had a range of up to 20 miles.

The new portable communication devices were revolutionary in that they used Frequency Modulation (FM) rather than Amplitude Modulation (AM), making them more resistant to interference and requiring less power. Equally importantly, FM transmissions do not interfere with each other in the same way, so it is possible to have more FM transmitters working over a given set of frequencies – vital when several military units are operating in the same area.

The handheld, short-wave FM radio transceiver would gradually evolve over the following decades into the mobile phone. Introduction was delayed by the requirement for base stations and the need to patch the radio into the phone system – bulky radio-telephones as used by ships can have a powerful transmitter, but a hand-held device has more limited range, so large numbers of mobile phone masts are needed to gain coverage.

Galvin changed its name to Motorola in 1947, and some people date the start of the mobile phone era to a call made by Martin Cooper, the company's Manager of Communications Systems on 3 April 1973. This was the first call from a portable phone that linked directly to the public telephone system. The handset he used weighed about a kilogram, half as much as the Handie Talkie but still rather weighty; mobile phones have progressively shrunk since then but are still basically the same device. Motorola is still a major manufacturer of mobile phone handsets, producing almost one in six of the world's output in 2003.

The radio frequency death ray was eclipsed in the 1960s by an entirely new concept: the laser. The theoretical basis of the laser lies in the work of Albert Einstein. In 1916 Einstein described a new interaction between atoms and light. It was known before then that an atom could absorb a photon (a 'particle of light') which would push it into a higher, 'excited' state. After a while the excited atom would spontaneously drop back into its normal, low-energy state and emit a photon.

Einstein's idea was that the emission of photons was not necessarily spontaneous but could be triggered. If an excited atom were struck by a photon of the right wavelength (which depends on the material), it would emit a second photon of the same wavelength and drop back to its neutral state. This type of reaction was known as stimulated emission. Rather than being emitted at random, the second photon would be going in exactly the same direction as the original.

If other atoms in an excited state surrounded the atom, then the two photons it emitted would also stimulate emission, and there would then be four photons. In any material in which the excited atoms dominated, an incoming photon would produce a shower of photons of the same wavelength. This would be a remarkable effect: on shining a light through a block of the material, you would get more light out than went in. Instead of absorbing some light, such a material would lose energy and amplify the beam.

The effect was understood at a theoretical level, and was given the name 'negative absorption'. It looked like an obscure quirk of theoretical physics without any practical meaning, like the negative-time concepts in relativity theory. Physicists believed it was impossible to produce a group of atoms in which the excited state dominated, and experiments appeared to support this theory. The laws of physics seemed to dictate that there would always be more unexcited than excited atoms, and absorption rather than negative absorption would be the overall effect.

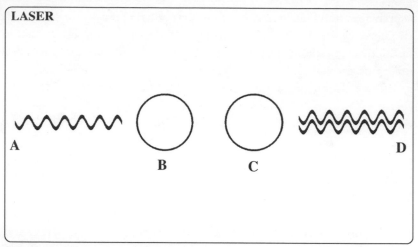

The basic principle of the laser: a photon (A) strikes an excited atom (B). The atom drops into a neutral state (C) emitting two photons (D), which are identical to the first.

Although there was some research in the area in the 1920s and 1930s, no real progress was made in stimulated emission until the boom in radio and microwave research after the war. The microwave oven produces a few hundred watts of energy at most. What scientists wanted were more intense microwave sources, for radar, communications and perhaps weapons. Three separate groups came up with the idea of boosting the output from a microwave using negative absorption. A group of excited atoms would act as an amplifier for the microwave beam; negative absorption meant that the beam coming out of the excited atoms would be stronger than the beam going in.

Charles H. Townes of Columbia University coined a name for his proposed device: 'Microwave Amplification by the Stimulated Emission of Radiation', abbreviated to MASER. He acquired funding from the Department of Defense, but after spending $30,000 over two years on the project Townes still had not produced a result, and cynics suggested that MASER actually stood for 'Means of Acquiring Support for Expensive Research'.

However, in 1954, Townes, along with colleagues Herbert Zeiger and James Gordon made a breakthrough. Microwaves have a wavelength of a few centimetres; by using a cavity the same size as the microwaves they could benefit from a resonance effect. This resonance is the same as the effect that shatters a wine glass when a soprano hits exactly the right note. If the wavelength of the sound matches the size of the glass, it produces a resonance which greatly amplifies the sound. Similarly, a resonant cavity matched to the wavelength of microwaves (a few centimetres) could boost the source.

By directing a beam of excited ammonia molecules into a resonant cavity, Townes' team produced a working maser at last.[1]

Masers soon proliferated in laboratories across the world, and researchers started work on a maser that operated at the frequency of visible light. In spite of the amplification, the power levels for masers were disappointingly low compared to other devices, which is why your microwave oven uses a magnetron rather than a maser to generate microwaves. But the situation would be better with shorter wavelengths – visible light – which should be millions of times as powerful.

The concept was originally called an 'optical maser', a name which was dropped in favour of the less cumbersome LASER – Light Amplification by Stimulated Emission of Radiation.

Visible light has a wavelength millions of times smaller than microwave radiation. A resonant cavity for visible light would be less than a thousandth of a millimetre across; but without resonance it was impossible to produce detectable stimulated emission.

Townes, the maser pioneer, recognized that this could be solved by using two flat mirrors facing each other with the lasing medium in between. A photon might simply pass right through the lasing medium without striking an atom, but adding the mirrors meant that the photon – and any subsequent photons it produced – would bounce back and forth, until it did strike an atom and stimulate more light. A tiny hole in one of the mirrors acted as an output for the beam.

The Pentagon took an immediate interest. Masers would prove useful for communications, but the military had bigger ambitions. Something a million times more powerful was beginning to sound like a weapon. The Department of Defense's Advanced Research Projects Agency realized that a working laser would finally give them the chance to build Wells' Martian heat ray for real.

A small company called TRG Inc employing some of the most promising maser scientists put in a request to ARPA for $300,000 to build a laser. ARPA liked the idea, but instead of offering them less money than requested, as often happened, they gave TRG a contract for a million dollars.

However, the competition was intense, and TRG was working under a handicap. Their best scientist was Gordon Gould, who had shown an interest in Marxism in the 1940s. Although he renounced it in 1947 he was still considered suspect, the laser project attracted a high security rating, and Gould was not given clearance. The other TRG scientists could talk to Gould but he was not given classified details. In spite of having the best funding, TRG were beaten to the punch.

Hughes Research Laboratories was a branch of the Hughes Aircraft Corporation, a company perhaps best known for its flamboyant aviator-industrialist-movie producer founder, Howard Hughes. During the Second World War, the company made aircraft parts and munitions, but Hughes was enthusiastic about more ambitious schemes.[2] Hughes diversified into military electronics and related items, such as radar-guided missiles. In 1960, Hughes was paying 33-year-old Theodore Maiman to work on a ruby laser, in spite of claims by other physicists that ruby was not a suitable material for lasing.

The light source required by a ruby laser would have to operate at a temperature of over 5,000°C to generate the correct wavelength – an unfeasibly high temperature. Nothing could stay that hot.

However, as Maiman realized, it would only be a problem if the source had to maintain the high temperature for any length of time. There were flash lamps that produced a momentary pulse at high temperatures. Xenon flash lamps, for example, reach 8,000°C for a fraction of a second, and a fraction of a second was all that was needed to create a pulse of laser light.

'I didn't see any reason why I had to do this continuous wave,' said Maiman later. 'Pulsed mode was perfectly fine . . . I was just trying to demonstrate that this could be done, not find the ultimate system.'

In May 1960, Maiman demonstrated the very first laser. The apparatus was so small that the entire assembly – lamp, reflector and ruby laser rod – could be held in one hand. The public relations photographer was not impressed and decided that bigger was better, so the official publicity photographs showed Maiman with a much larger laser system. These publicity pictures, still used by Hughes today, were picked up on by other scientists, who copied the larger system.

## The Laser Boom

What followed was a laser boom with other laboratories replicating the ruby laser and developing their own types. In addition to pulsed lasers using different types of crystal, the first gas laser was successfully tested in December 1960. This was a continuous-wave laser rather than a pulsed one, and it was developed by Bell Labs. Best known as a telephone company, Bell also produced radar, communications gear and other military electronics.

The power for gas lasers comes from an electric current, but as with the ruby lasers whose power source was a flash lamp, the energy output was always going to be limited.

The military wanted something much more powerful. There are two considerations at work: one is the efficiency of the laser, and the other is how much energy it can handle. A ruby laser has an efficiency of something like 20 per cent, which means that only one-fifth of the energy input comes out in the laser beam. Clearly, this would require some giant flash lamps at the laser end to generate a powerful beam, and in the process an immense amount of wasted energy would be produced. The other problem is simply how to get enough energy into the laser; stimulation by flash lamps and electrical impulse was not providing sufficient.

In 1962 General Curtis LeMay proposed that a laser system could be used to defend the United States against attack from ballistic missiles. This takes us right back to the situation earlier in the century: like the bomber, the ballistic missile would always get through, and could only be stopped by some dramatic new technology, preferably one working at the speed of light. But as before, nobody could generate a powerful enough beam; the lasers of the time were clearly not up to the job. DARPA sponsored Hughes Research Labs to build a massive laser, 32 feet long and 4 feet wide. Completed in 1969, it had a power output of less than 2 kilowatts – less than a two-bar electric fire. A laser to meet General LeMay's requirement would be gigantic.

'A laser big enough to inflict militarily significant damage wouldn't even have to work,' said one wit. 'Just drop it on the enemy.'

In the 1960s military researchers did find a way of increasing the power of gas lasers enough to turn them into deadly weapons. The technology was promptly classified as top secret. An expanding mixture of hot gases can produce enough excited molecules to make an effective laser. For this to be a continuous process the gas has to keep expanding; rather than simply having a tube full of gas, the hot gas has to pass by in a high-speed stream. It bears some resemblance to a rocket engine, in that both use streams of hot gas passing through nozzles at high speed.

This type of laser is called gas-dynamic because of the dynamic flow of gas, to distinguish it from other types of gas laser. Some details of the gas-dynamic laser work were declassified in 1970, apparently in response to the publication of Soviet work in this area. The US did not want to be seen to be falling behind, and revealed that a laser forty times as powerful as the Hughes device was already in operation.

Even gas-dynamic lasers were not powerful enough for serious weapons applications and the military still wanted something better. The two most important types were chemical lasers and X-ray lasers.

Chemical lasers draw their energy from a chemical reaction, usually between two different gases. This produces a lot of energy very quickly, just as an explosion turns chemical energy into explosive power very quickly. The first chemical laser reacted hydrogen and chlorine together; more recent versions have used deuterium and fluorine. Although this makes the raw materials far more expensive, the deuterium-fluoride laser produces a beam which is absorbed far less by the atmosphere and can travel much greater distances. At low power levels, the inconvenience and expense involved mean that chemical lasers lose out to other types, but if you want something really powerful then chemical lasers are the only choice. Practically the only chemical lasers today are for high-powered military use.

In the quest for ever more powerful sources of energy for lasers, Lawrence Livermore Laboratory were reported to have reached the ultimate in 1981. They had tested a laser powered directly by a nuclear explosion, a power source potentially millions of times stronger than even a chemical reaction. The intense energy source results in a laser output with a very short wavelength, much shorter than visible light – in fact they were X-ray wavelengths.[3]

The bomb-driven laser only produces a very short pulse of light before being destroyed itself by the nuclear blast, but it is incredibly powerful. Edward Teller, known as the father of the hydrogen bomb and a keen advocate of all things nuclear, seized on this new technology.

Contemporary lasers could not achieve General LeMay's goal of shooting down a ballistic missile – but an X-ray laser had power to spare. The best place to put it would be in space, where the explosion would do least damage and where there was no atmosphere to absorb the beam. This led to a grand plan for a series of orbiting X-ray laser platforms. When detonated, each would produce a number of X-ray laser beams, knocking out missiles thousands of miles away.

'A single X-ray laser module the size of an executive desk,' said

Teller, '. . . could potentially shoot down the entire Soviet land-based missile force.' Teller's enthusiasm for the X-ray laser helped persuade Ronald Reagan that ballistic missile defence was possible, leading to the Strategic Defense Initiative. Critics referred to SDI as 'Star Wars', suggesting that it was a technological fantasy, but those working on it adopted the name 'Star Wars' as a badge of honour.[4]

The critics had the last laugh: technical studies revealed that the X-ray laser was flawed. An argument broke out over whether the early tests had really been as successful as claimed, and further experiments cast doubt on its effectiveness. Funding was cut, and the X-ray laser is no longer in the running as an anti-ballistic missile system.

Chemical lasers still show some promise, and there are a number of high-energy chemical laser weapons under development. Perhaps the most important is the Airborne Laser, mounted in a modified 747 airliner and scheduled to be in operation from 2008. The Tactical High Energy Laser or THEL is also very advanced, and prototypes have already been able to shoot down tactical rockets in tests.

For the moment, the dream of military death rays remains unfulfilled. But they have found plenty of uses elsewhere.

## Lasers in Practice

'Do you expect me to talk?' asks James Bond, as the deadly laser beam slices through the slab of metal he is chained to, inching towards him.

'No, Mr Bond,' replies his arch-enemy with a laugh. 'I expect you to die.'

When Ian Fleming wrote *Goldfinger* in 1959, he had the captive Bond being threatened with a buzz saw. In 1963 when scriptwriters were adapting the book, they wanted something more modern. In the film version, Bond is menaced by the very latest in technology, Goldfinger's industrial laser.

Like most Hollywood lasers, it is not a very accurate depiction. The laser beam itself is visible, and the machinery makes a suitably ominous electronic humming noise. Real lasers are as invisible as any other beam of light, only showing up if there is a trace of smoke or mist to scatter them. Most are silent in operation, with the exception of chemical and gas-dynamic lasers which make a noise like a rocket rather than a discreet hum.

The biggest anomaly is that Goldfinger's laser slices through a thick sheet of metal without difficulty. In the real world, lasers have not replaced buzz saws; as we have seen, the power levels are not high enough. Work is progressing on using high-powered lasers which may be used for drilling oil wells, but practical application is still in the future. While industrial lasers are used for cutting plastics and other materials, this is not a major use. Instead they have branched out into dozens of different applications. The brute force of the high-energy death ray is not what makes lasers useful. Low-powered lasers are used for everything from CD players to supermarket bar code readers, from laser printers to eye surgery.

All of these applications depend on properties of the laser which have little to do with the military's original motive for developing them. A laser is not just a powerful light source; it produces a light which is different to anything else in nature.

The special properties of laser light are consequences of the 'stimulated emission' effect where one photon triggers the release of a second photon. One feature is that the photons are all travelling in the same direction. Unlike normal light sources which produce photons in all directions, lasers produce a beam of light which is almost parallel. Another is that the photons are all of exactly the same wavelength, in other words, the same colour. (Laser light is said to be 'monochromatic', i.e. 'one colour').

A laser pointer is the simplest laser device. One of the useful features of lasers is that the beam consists of almost parallel rays and does not spread out like the beam from a torch. The spot from the pointer is practically the same size and brightness at one metre or ten. Some of the earliest laser pointers were 'target designators' used by the military which serve the same function as the lecturer's pointer – drawing people's attention to items of interest, though in this case they were used by ground troops to direct aircraft on to a target.

Moving up from the simple pointer is the laser rangefinder. Again, the laser's ability to project a very narrow beam that is visible at long distances is important. These simply project a laser beam and use a very rapid electronic clock (pulsing at a billion hertz or more) to time how long it takes for the beam to bounce back. This method is enough to measure distances of several kilometres with an error of only a few centimetres. Laser rangefinders were originally developed for tanks and other artillery, but they are now used for everything from surveying to finding the distance to the pin in golf.

The bar code reader is slightly more sophisticated. A laser scans across a bar code, and the reflected light is picked up by a simple light-detecting device called a Charge Coupled Device or CCD. The intelligent part of the scanning process is the software which interprets the series of light-and-dark returns from the scan, working out what the bars on the bar code look like. The information from bar codes is encoded in the relative width of the light and dark bars. This way a bar code can be scanned from a distance, or at an angle, without difficulty, because these will affect the size of each of the bars equally.

The laser is useful for bar-code scanning because it produces a narrow beam which can be easily controlled, and the reflected light is all of the same wavelength, so there is no risk of confusion with other light sources.

A bar code contains perhaps a dozen characters. There is no reason why it could not contain many more by extending the number of bars. It could contain a complete novel given enough bars; and the bars could be made extremely small – bar codes are only made large for our convenience. It only takes a slight extension of this idea to invent the CD player.

A compact disc is made of several layers. The base layer is clear plastic, and instead of black and white bars it has a series of microscopic rectangular 'dimples' in the recording track, like the bars in a bar code, which represent the stored data. A thin reflective layer covers the dimples so that they show up better, and a protective transparent layer goes over this. The CD player shines a very narrow beam of laser light on to the recording track and a sensor detects the reflected returns. Dimple/no dimple can be read as ones and zeroes. These ones and zeroes form a binary code which can be used to store any type of information – music, data or video.

DVDs – digital versatile discs – are virtually identical to CDs, but represent a step forward in that they can store several times as much data. They do this by having a narrower track and smaller dimples. The dimples on a CD are 830 nanometres across, those on a DVD are 400 nanometres across (a million nanometres equal one millimetre). The total track length on a DVD adds up to about 12 km, storing several gigabytes of data. It's an impressive amount, but as far as the consumer is concerned, the technology is still only storing about two hours of video.

One approach is to move into the third dimension. The reflective layer on a DVD is partially transparent, so by focusing the laser in the DVD player it is possible to see through it. By having two separate layers, a DVD can effectively store twice as much information. If the DVD is double sided rather than single sided this can be doubled again, packing four times as much data or video on to the single disc.

Why not make the dimples smaller still and store ever more information? Unfortunately, the size of the dimples is already at the physical limits of what can be read. It is not possible to resolve an object smaller than the wavelength you are seeing with (a later chapter explores this topic with regard to new types of sensor), and the lasers used in current CDs and DVD players have a wavelength of 630–680 nanometres, at the red end of the spectrum.

As described above, the wavelength of laser light depends on how it is produced. The race is on to produce commercially viable short-wavelength laser diodes – a blue laser with a wavelength 400–470 nanometres. Nichia Corporation demonstrated the first such device, and Sony and other big players were not far behind.

A shorter wavelength would mean that the track could be narrower and the dimples smaller, fitting several times as much information on to a single DVD. A DVD can store up to eight hours at present. The new technology could quadruple this again. Entire television series with additional commentaries and extra features will fit on a single DVD with ease. Alternatively, smaller and cheaper DVDs could be used which hold a few hours of video, ideal for the giveaways distributed free with magazines.

Over the next few years, blue laser DVDs will appear in the commercial market, first in Japan and then elsewhere. They will be compatible with old CDs and DVDs, but consumers will be under pressure to upgrade to the new standard. New products will tend to be released in the new format only and older players will become steadily obsolescent.

Of course, the spectrum does not end with blue light, and by the time blue laser DVDs are everywhere, the technology may well be in place for ultra-violet lasers. These will potentially multiply the data storage capacity again leading to discs which are even smaller or hold even more. This could eventually lead to button-sized media – anything smaller would be too difficult to handle – manufactured at negligible cost with massive storage capability.

Personal video players will effectively have unlimited capacity, as well as personal music players, digital cameras and camcorders.

Before long you may find a small free gift in your breakfast cereal, which could turn out to be a whole year of a television soap opera, the collected works of Elvis Presley, or a copy of every book printed before 1900. Whether you are likely to toss it away is another matter.

## Talking Lasers

There are other lasers which we may never be aware of but which we may use every day. In 1970, researchers at Corning Glass managed to create optical fibre, a type of glass that can be spun out into long strands like wire. Because of the optical properties of this glass, if you shine a light down one end of the optical fibre it will come out at the other end – an effect called 'total internal reflection' ensures that the light does not leak out. Some years before it had been suggested that light could be used as a far more efficient way of communicating than electricity and that it would be possible to send much more information down a single line; the problem was that there was no cable capable of carrying light. But that changed with optical fibre.

Bundles of optical fibres have been steadily replacing copper cables for high-volume telecommunications. The first transatlantic fibre-optic cable was laid in 1988, and fibre-optic cables now criss-cross the developed world. Computer networks are built around fibre-optic 'backbones' which link into the copper wiring used for short distances.

Although fibre-optic glass only absorbs a tiny fraction of the light passing through it, this adds up over long distances. Fibre-optic cables need to have boosters to increase the signal, and these are provided by laser. Any long-distance phone conversation is likely to have been transmitted by laser.

## Laser Eyes

While high-energy military lasers are still in the future, there is one area in which lasers have already been used on human flesh. But rather than being a matter of brute power, surgical lasers are useful because of their precision and delicacy.

The parallel rays of a laser beam mean that it can be focused on a very small spot. It did not take any great leap of imagination to try using a laser as a substitute for a scalpel – indeed, Goldfinger's planned bisection of 007 might be seen as a rather radical version of this idea.

The effect of a laser on flesh, or on anything else, for that matter, depends on how the beam is absorbed or reflected. Early lasers were found to be unsatisfactory because they operated in the red wavelengths which are not very well absorbed. However, with the introduction of gas lasers which emit infrared light, a new surgical tool became possible. Soft tissue consists mostly of water, which absorbs infrared wavelengths very well, so an infrared laser is a very effective way of heating it.

The new lasers cut like a scalpel, a scalpel that could be made almost infinitely sharp. As a further advantage, the laser did not cause bleeding. Any blood vessels it cut through were instantly sealed by high-temperature cauterization. A typical surgical laser has a power output of 20 watts, concentrated on to a spot barely 0.1 mm in diameter – a tenth the size of a full stop.

Surgical lasers are not the answer to everything, though. The amount of energy used is low, but still enough to cause destruction of tissue around the incision. They produce smoke – sometimes so much that a smoke extractor is needed – which can be toxic. Although preferred for some surgical procedures, lasers are a specialist tool rather than a general-purpose one.

The amount of tissue damage meant that gas lasers were not suitable for the most delicate eye surgery. There was no laser capable of performing this kind of surgery until 1975 when the Excimer laser was developed.

The Excimer laser was first demonstrated by James J. Ewing at AVCO Everett Laboratory, yet another company carrying out military research. They were looking at ways of making lasers more efficient so that they converted more of the input power into laser energy, and also at expanding the range of wavelengths which could be produced.

The atmosphere absorbs red and infrared lasers so they have a limited range. Although by this time it was clear that a death ray was not practical, the military had found other uses for them in designating targets and range finding.

A laser-guided bomb works by looking for reflected light of a particular unique wavelength and homing in on it; a laser is ideal for this because the light is all of the same wavelength and will not be confused with background lighting. In spite of being called a 'smart bomb', the laser-guided bomb has no intelligence – it simply goes where it is told to, leaving the work of finding, identifying and tracking the target to the person operating the laser designator. A laser-guided bomb can be targeted to within a few metres of a given point, so that a single guided bomb can do the job of several hundred unguided ones. Though effective, the new smart bombs were priccy; during the Vietnam era, pilots used to call releasing a laser-guided bomb 'dropping a Cadillac' because each bomb cost as much as a luxury car.

A laser designator is of limited use if it can only reach a few miles. There was a need for a laser using a wavelength which would not be absorbed by the atmosphere, such as ultraviolet.

Ewing's laser was a variation on a gas laser which used exotic materials, rare-gas halides. These are unusual compounds of very unreactive gases which have a lasing action at the shorter wavelengths, into the useful ultraviolet zone. 'Excimer' is a shortened form of 'excited dimer', a dimer being a molecule with two of the same sort of atom in it.

Ewing points out that the gas halides he used were actually not dimers at all but a different type of molecule called exciplexes. However, he thought that 'excimer laser' sounded good whereas 'exciplex laser' was too much of a mouthful, so the incorrect name stuck.

The earlier gas lasers cause a certain amount of 'collateral damage' and carbonized tissue because they heat tissue so much. Excimer lasers are much more useful because the UV light is absorbed directly by proteins, vaporizing them at low temperature without damaging surrounding tissue. This opened the way for LASIK eye surgery, which has been used to correct short-sightedness in millions of patients.

LASIK stands for Laser-Assisted In Situ Keratomileusis. An excimer laser is used to reshape the cornea. Dr Sato in Japan proposed the principle of reshaping the eye to correct short-sightedness in the 1930s but it was not until Dr Svyatoslav Fyodorov carried out extensive work in Russia in the 1960s and 1970s that a viable surgical procedure was established.

Fyodorov started with the use of a scalpel, which led to a significant risk of scarring and impaired eyesight. Infrared lasers were better, but the result was still less than ideal, with a small but significant number of patients ending up with eyesight worse than before. The Excimer laser gave a greater degree of precision and less tissue damage, leading to quick and reliable surgery with a fast recovery time. This has led to laser eyesight correction clinics being set up all over the world.

The first lasers were considered by many to be 'a solution looking for a problem'; like the maser they might have ended up with few practical applications. But the heavy investment in this technology, along with the ingenuity of the scientists and engineers involved, has turned out a steady stream of useful devices. The Wellsian heat ray, vaporizing victims left and right, has yet to emerge. In spite of the high hopes that the maser would overcome the limitations of earlier devices and prove a practical beam weapon, the military benefits have been outweighed by civilian ones.

The laser, which has hundreds of uses and is only used on human flesh for benign purposes, is now well established. Whether this outcome is viewed as a matter of good luck, or the deliberate policy of making use of the fruits of military research, is another matter.

# CHAPTER 5

# Ultimate Power

*Atomic energy might be as good as our present day explosives, but it is unlikely to produce anything very much more dangerous.*

Winston Churchill, writing in 1939

The atomic bomb is the most awesome achievement of military technology in the twentieth century: a weapon so powerful that it threatened to bring history to an end. From the earliest days, nuclear pioneers were interested in more than just bombs. They saw that atomic energy offered the promise of an endless supply of clean, cheap energy.

In the idealistic early days of the 1950s and 1960s there was much talk of nuclear power as the answer to everything. In December 1953 President Eisenhower gave his 'Atoms for peace' speech to the UN, proposing the setting up of what would become the Atomic Energy Agency to encourage the peaceful use of nuclear power.

'The most important responsibility of this atomic energy agency would be to devise methods whereby this fissionable material would be allocated to serve the peaceful pursuits of mankind,' said Eisenhower. 'Experts would be mobilized to apply atomic energy to the needs of agriculture, medicine and other peaceful activities. A special purpose would be to provide abundant electrical energy in the power-starved areas of the world.'

In the US there was a co-ordinated campaign to persuade the public that the mighty atom could solve the world's problems, feeding the hungry and bringing the advantages of civilization to developing nations. There would be nuclear-powered ships, trains, aircraft, spacecraft and even atomic cars. Fossil fuels would be a thing of the past. Power stations would no longer belch out clouds of smoke but would burn emission-free nuclear fuel.

Just a handful of fuel pellets would be enough to provide a family with electricity for a year, the costs so low that it would be a matter of paying a fixed charge for connection. Lewis Strauss, Chairman of the US Atomic Energy Commission, looked forward to a rosy nuclear future from 1954: 'It is not too much to expect that our children will enjoy in their homes electrical energy too cheap to meter,' he said at a science writers' convention.

However, subsequent history showed that harnessing this particular technology for peaceful uses was much harder than expected. Nuclear power is still a topic of heated debate. It provides 16 per cent of electricity generated worldwide, though it has yet to reach the power-starved areas of the world and there has not yet been any sign that it will become too cheap to meter. Arguments continue over the true cost of decommissioning nuclear reactors and how to dispose of waste. Developing nations with nuclear power programs are suspected of building weapons, and some still question whether there are valid civilian uses for this most emphatically military of inventions.

## Origins: the Power of $E = mc^2$

Although the theoretical groundwork for atomic power had been laid out years before, it was only the pressure of wartime that turned it into a reality.

Einstein showed in 1908 that a small amount of matter could be converted into a huge amount of energy. His famous equation $e = mc^2$ indicates just how much energy can be released from matter. Put into English, this equation says that the energy contained in an object is equal to its mass multiplied by the speed of light (c), then multiplied by the speed of light again. 'c' is a huge number – light travels at 186,000 miles per second – and the energy in one kilogram of matter is equal to the energy released by burning over 10 million tons of coal. Or to put it another way, it is equal to 10 million tons of explosive.

The question was how the energy could be released. British physicist Lord Rutherford discovered that this type of energy lay at the heart of radioactive phenomena, but the known radioactive elements only released a slow trickle of energy.

In 1938 a German physicist, Otto Hahn, was experimenting with bombarding uranium with neutrons. Uranium was then one of the heaviest elements known; Hahn was hoping that the uranium might absorb neutrons and be transmuted into even heavier elements. He was surprised to find that it had the opposite effect. Instead of ending up with a heavier element, the neutron bombardment split the uranium into two lighter elements. This splitting is known as nuclear fission.

Further study confirmed that a single neutron colliding with a uranium nucleus produced barium and krypton, and three spare neutrons. One of Hahn's colleagues, Lise Meitner, realized that the amounts of mass involved did not add up. Some mass was missing and had been converted into energy, a significant amount of energy.

Neutron sources could only produce a few neutrons at a time, but the advantage of the fission reaction is that for every neutron fired into the uranium, three more neutrons are produced. If each of these three gets absorbed by a uranium nucleus in turn, they will each set off three more fission reactions, and so on, tripling each time. A single neutron could start a chain reaction which would trigger nuclear fission throughout a large mass of uranium releasing energy from every atom.

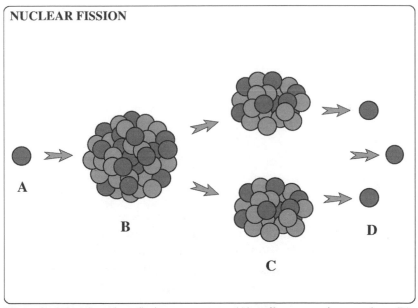

**NUCLEAR FISSION**

A

B

C

D

The principle of nuclear fission: a neutron (A) strikes a uranium nucleus (B), which splits into two lighter nuclei (C) emitting energy and three neutrons (D) in the process.

Other heavy elements could also undergo fission, but of the others it seemed that only plutonium, an element which does not occur naturally, could set off a chain reaction.

How much energy would be released? Only a fraction of one per cent of the mass of the uranium was converted into energy, but according to Einstein's equation it would still be millions of times more than could be produced by any chemical reaction.

The news travelled rapidly around the scientific community, and by the summer of 1939 the press were talking about atomic power. Although the theory was sound, it would take a huge leap from the theoretical model to application, and few physicists were convinced that it would be possible to set off a chain reaction in practice.

Three things stood in the way of a practical nuclear weapon:

1. Natural uranium is a mixture of two different isotopes, U235 and U238. U238, which makes up over 99 per cent of naturally occurring metal, will not undergo nuclear fission. To produce a chain reaction, scientists would have to separate out a quantity of pure U235. The two isotopes are chemically identical and differ only in that U238 is slightly denser. Separating out isotopes on a large scale, rather than microscopic amounts for laboratory use, had never been done before.

2. U235 is highly radioactive, producing its own flux of neutrons. If you bring together a sufficiently large amount of U235 into a ball, this flux of neutrons will set off a chain reaction on its own. This amount of U235 is known as the critical mass. A smaller ball of U235 will do nothing, but as soon as it reaches critical mass, the chain reaction will occur. If the critical mass were very small, you would get an explosion as soon as you isolated a small amount of U235. If it were very large, you would need many tons of it to get any effect at all. Whether atomic power was feasible depended entirely on how great the critical mass was.

3. It was not clear how long the chain reaction would go on for. As soon as it started, the large amount of energy released from the chain reaction would blow apart the mass of uranium in which it was occurring, bringing the reaction to an abrupt halt. A viable atomic bomb would have to somehow maintain a critical mass of uranium in one piece for long enough for the chain reaction to spread through it.

Plutonium was a far more suitable material for a bomb than uranium. But plutonium does not occur in nature. It has to be produced in a controlled nuclear reaction from uranium – but again, there was no technology to do this. These were all serious issues, and it was far from clear that they could be resolved. By mid-1939, the Germans had determined that a bomb was possible in theory and had started on the process of finding out how to separate uranium isotopes.[1]

Britain, France and Russia were working along similar lines. The French effort ended with the German invasion; the Russian bomb was also delayed when the war started, and resources were moved to other projects which were seen as more urgent and more likely to produce results in the short term. In Britain, a committee was set up in 1940 to investigate the possibility of nuclear weapons, aided by the work of Otto Frisch and Rudolf Peierls who had fled from the Nazi regime. The committee, known as the MAUD committee, concluded that a bomb could be constructed within two years, but that Britain did not have the resources for such a project.

The US government did not even have an official nuclear program. Friends encouraged Einstein to write to President Roosevelt personally. Although he was a pacifist himself, Einstein was worried by the possibility of a Nazi atomic program. He wrote:

> In the course of the last four months it has been made probable . . . that it may be possible to set up a nuclear chain reaction in a large mass of uranium, by which vast amounts of power and large quantities of radium-like elements would be generated. This new phenomenon could also lead to the construction of bombs, and it is conceivable – though much less certain – that extremely powerful bombs of a new type may thus be constructed. A single bomb of this type, carried by boat and exploded in a port, might very well destroy the entire port together with some of the surrounding territory. However, such bombs might well prove too heavy for transportation by air.

Einstein's reputation spoke for him, and the President appointed a Uranium Committee. In spite of this, it was not until two years later, when Japan attacked the US, that the atomic bomb program was moved on to a fast track. The report of the MAUD committee was instrumental in persuading the US to go ahead with the bomb. Under the Quebec Agreement of 1943, a team of top British nuclear scientists was sent to Los Alamos to work with the US on the Manhattan Project, the construction of the first atomic bomb.

\*        \*        \*

Various means for separating U235 from its unreactive partner were devised. Several approaches were tried and in the end the work was carried out using a number of methods. All of them relied on the fact that although they are chemically identical, U235 is slightly lighter than U238, and can be sifted out in a high-tech version of separating the wheat from the chaff. All the methods require the enrichment process to be repeated many times, getting a higher and higher concentration of U235 with each pass.

The first was magnetic separation: tiny particles of uranium tetrachloride were propelled along a semicircular racetrack. The lighter U235 particles would pass closer to a magnet where they were collected. This process was not effective and eventually only produced about a gram of enriched uranium.

Centrifuging was also tried, using a similar process of whirling the uranium around so that the denser U238 would be thrown outwards and could be discarded, but this also proved inefficient.[2]

Finally, there was gaseous diffusion. The uranium was converted into uranium hexafluoride gas and passed through a porous membrane; the slightly lighter U235 passed through more quickly.

The question of critical mass was resolved by the theoreticians of the MAUD committee, who determined that the mass needed for an explosion would be a matter of about 10 kg, and not grams or tons. This was in the right range for a bomb.

Further work, both theoretical and experimental, suggested that the chain reaction would go on for long enough to release a useful amount of energy, and that a nuclear weapon was feasible. However, it remained to prove whether it could work in practice.

The mass of fissile material would be bound together using explosives: a series of charges using split-second timing would force several small pieces of uranium together into a single lump for the fraction of a second that the chain reaction would take.

By July 1945 everything was ready for the first test. This was code-named Trinity and took place at a remote desert site called Alamogordo. It succeeded spectacularly, creating a gigantic fireball and an ascending pillar of smoke which formed itself into a mushroom cloud: the symbol of the new nuclear age. The explosion was equivalent to 18,000 tons of TNT, thousands of times more powerful than any existing bomb.

The team moved quickly to create operational weapons. The Manhattan Project succeeded in developing two different types of atomic bomb. One was 'Little Boy', which used 60 kg of highly enriched uranium as fissile material; the other was the more effective 'Fat Man' bomb which used just 6 kg of plutonium.

The war was almost at an end. Nazi Germany had been defeated, and although the Japanese had been driven back across the Pacific, it seemed as though it would take an invasion to force the Japanese to surrender. Such an invasion would have cost thousands of American lives. A massive aerial bombing campaign was being carried out against Japanese cities with the aim of breaking the nation's morale, but this did not seem likely to succeed on its own.

The decision was made to use the atomic bomb to force the Japanese to surrender. Of the two types of bomb, Little Boy was simpler and more reliable, and was selected to be used first. On 6 August, the first atomic bomb was dropped on the city of Hiroshima.

The bomb caused total devastation over a wide area. According to a 1946 estimate, 45,000 people were killed on the first day, with another 21,000 dying of their injuries over the following four months. The flash of heat was so intense that people within a few hundred metres of the explosion were vaporized, leaving nothing but shadows etched on to walls and pavements.

Japan wavered, but did not surrender. Three days later a second atomic strike was ordered. The primary target, the city of Kokura, was obscured by cloud, so the B-29 bomber went on to its secondary target, Nagasaki. The Fat Man plutonium bomb was used this time, 40,000 were killed and another city all but annihilated.

Although the US had a very limited supply of atomic weapons – a third bomb was some weeks away – the Japanese were faced with the steady annihilation of their country. The Emperor ordered surrender.

Fat Man had an explosive power almost 50 per cent greater than Little Boy, but both designs had proven terribly effective. They might not have caused as much destruction as the massed incendiary bombardment of Tokyo, killing 140,000 in the course of two raids, but the atomic weapons showed how much destruction could now be caused with a single bomb. It was obvious that nuclear weapons would change the face of war; whether nuclear power had peaceful uses as well remained to be seen.

## Harnessing the Dragon

A runaway chain reaction means an atomic explosion, but it does not have to be allowed to run on out of control. The nuclear reaction can be tamed. Each nuclear fission absorbs one neutron and produces three; if two of those neutrons can be absorbed before they react, then the fission will proceed at a steady, controlled rate, with a constant flux of neutrons. The uranium will heat up gradually, and this heat can be tapped just like the heat from a coal fire or any other power source. Typically it will be used to heat liquid metal, which in turn heats water turning it into steam to drive turbines.

The first nuclear electricity was generated in the US in 1951, lighting four bulbs. But it was the Russians who achieved the first nuclear power plant at Obinisk in 1954, generating five megawatts of power. The US and Britain were close behind though, with their own reactors starting up in the two following years. The first French reactor went online in 1965, and their policy meant that nuclear energy was to be far more important in France than in the other nations.

Nuclear power has always had a close relationship with nuclear weapons. Calder Hall, Britain's first nuclear power plant, was opened in 1956, ostensibly to bring in the nuclear age and produce cheap electricity. However, it was not until 1961 that the government admitted that the plant was also crucial for producing plutonium for Britain's atomic weapons program, and that its cost of some £35 million was disproportionate for a plant that only produced 40 megawatts for the grid, when coal-fired plants can produce a thousand megawatts or more.

The uranium for a nuclear reactor does not need to be as highly refined as uranium for a bomb. A mixture of inert U238 in with the U235 acts to take up some of the stray neutrons. This type of material cannot be made to explode in the way that weapons-grade uranium can – it just gets hotter and hotter until it melts down and burns. Once vaporized, the density is too low for the reaction to continue – but the radioactive vapour is lethal.

Avoiding this kind of meltdown is one of the main concerns of the nuclear power industry. In a nuclear reactor, the core is made up of interspersed rods of uranium and control rods made up of a material such as boron which soak up neutrons. Raising or lowering the control rods controls the overall level of neutron flux. Reactors are normally designed to be fail-safe, so that in the event of an accident the control rods will automatically fall into place and shut down the nuclear reaction.

However, no system which has human involvement is ever completely safe. The Russian reactor at Chernobyl had safety overrides which allowed the operators to withdraw the control rods when testing safety procedures. In 1986 disaster resulted when the rods were withdrawn and the core overheated, releasing a cloud of radioactive gas. The exact effects on the local population are disputed, but the UN Scientific Committee on the Effects of Atomic Radiation estimated that it caused 1,800 cases of thyroid cancer in children. Others calculated that the accident caused at least 9,000 extra deaths from cancer in the local population, and the effects were felt hundreds of miles away across Europe. Sheep in Wales were found to have measurable levels of radioactivity from grazing in areas in which rain had fallen carrying dust from Chernobyl.

The accident at Three Mile Island in 1979 was not comparable in terms of severity and there were no immediate fatalities, but it showed that even in the US absolute safety cannot be guaranteed.

Another troubling aspect of nuclear power lies in its military roots. Atomic power stations can use uranium that is enriched to a much lesser degree than weapons grade, so a nation with a nuclear reactor cannot use the fuel to make bombs. However, a nuclear reactor produces plutonium as a by-product, and plutonium is ideal for bomb making. If the spent fuel is reprocessed, the plutonium can be extracted.

This creates political difficulties, as a nation's enemies will always assume that their desire to build nuclear reactors is simply a scheme to make nuclear weapons. In some cases such a suspicion may be justified. The problem is that it is hard to separate nuclear power from weapons technology, and an entirely peaceful nuclear programme with no possible military application has yet to be devised.

## Fat Man to Walkman

The Manhattan Project had less predictable spin-offs as well, including, curiously enough, the Sony Walkman. The technology for separating out different uranium isotopes was adapted for other purposes, including extracting the so-called 'rare earth' metals. These are obscure elements like yttrium and samarium, which had not been previously examined because they were so difficult to isolate.

The rare earths proved to be far better than existing materials for making magnets, an essential part of electric motors. There was a big military demand for powerful miniature electric motors for satellites, missiles and aircraft components, and development was pushed ahead.

A key breakthrough was made by Dr Karl Strnat, a civilian researcher at the Wright-Patterson Air Force Base Materials Laboratory, who opened up the whole field of rare earth magnets. In the 1970s he collaborated on the development of a second generation of rare earth magnets including a samarium-cobalt composition called 2–17 (short for $Sm_2(FeCoCr)_{17}$). While in the West the interest in 2–17 was mainly military, in Japan it was commercialized by TDK and Shin Etsu.

When Sony wanted to make a new portable cassette player, it required highly efficient magnets for the motor and headphone speakers if battery usage was to be acceptable. The solution was to use 2–17 magnets in both the tape drive motor and the headphones.

When a better type of rare earth magnet material (neodynium-iron-boron or NdFeB) became available in 1983, the market was well established and the US raced ahead. NdFeB was taken up particularly by makers of computer hard disks and helped boost the growth in this area; it can also be found it in iPod headphones and other such applications.

## Radioactivity

There is no avoiding the fact that nuclear material is radioactive. As has been mentioned, uranium naturally undergoes a slow decay, in which it produces neutrons and releases energy. Radiation comes in three different types: alpha, beta and gamma. All are hazardous to health and can cause cancer.

Alpha radiation has very little penetrating power and can barely go through a sheet of paper. Beta radiation, which consists of electrons moving at high speed, is somewhat more dangerous. Gamma radiation, the most dangerous of the three, is a form of electromagnetic radiation of extremely short wavelength – like X-rays, only more so. Gamma radiation will go through anything and can only be stopped by thick shielding – rock, lead, concrete or similar substances.

The neutrons emitted in nuclear fission are also dangerous; and they have the effect of making other materials radioactive. A containment vessel used for storing uranium will gradually become radioactive itself over time, as will everything else in immediate contact with it. These secondary sources of radiation are known as 'activation products', and although they are much less radioactive than nuclear fuel, they can present a problem because of their sheer mass.

Decommissioning a nuclear power station can mean disposing of thousands of tons of mildly radioactive concrete and steel. It cannot simply be buried, as it would contaminate the local groundwater. It must be safely contained not just for a few decades but for thousands of years. There is as yet no satisfactory solution for how to do this; no human structure has survived for the kind of timescale involved.

The most serious issue is disposing of high-level nuclear waste like spent fuel rods. Spent nuclear fuel can be reprocessed and reused, but the economics are not very attractive. If it cannot be reprocessed, the fuel rods need to be stored safely for a very long time. The quantity of material is daunting; there were estimated to be over 200,000 tons of high-level waste in 2000, growing at over 10,000 tons a year. This will remain hazardous for at least the next 20,000 years.

Nuclear waste disposal poses huge challenges for engineering and also for the law. In The US, a waste disposal site can only be agreed by a state-wide referendum, and nobody wants to have a disposal site in their own back yard. Possible candidate sites have included reservations belonging to Native Americans which already include chemical waste dumps, and where impoverished local communities are willing to exchange the right to build waste sites in exchange for benefits. However, critics have their doubts; even the Yucca Mountain site which is favoured by the US industry experienced an earthquake registering 4.4 on the Richter scale in 2002.

There is also the problem of language. Human languages change over the course of centuries. Scientists have had to ponder the problem of how a nuclear waste dump could be signed to future civilizations with no knowledge of the English language, using some sort of universal danger signal. Nuclear waste disposal is likely to involve many such issues because of the unprecedented timescales involved.

## The Man Who Would Move Mountains

Nuclear fission was only the start. Within seven years scientists had built an even more powerful weapon. Tested on the island of Elugelab, which was a mile across, the fireball from the explosion completely destroyed the island. The scale of the explosion was terrifying even to those who had witnessed earlier atomic tests.

The hydrogen bomb has a fission bomb at its core, and when this is detonated the heat and pressure bring about a second type of nuclear reaction. At extremely high temperatures and pressure – such as those found at the heart of the sun – nuclear fusion occurs. This involves small atoms such as hydrogen fusing together to form heavier elements, in particular helium, and releasing vast amounts of energy. Fusing brings the power of the Sun to Earth.

The man widely known as the 'father of the American hydrogen bomb' is Edward Teller. Born in Hungary, Teller was a gifted scientist and organizer involved in the Manhattan Project. He was also fiercely anti-Communist, stating that 'Stalin's Communism is not much better than the Nazi dictatorship of Hitler.'

After the war it was Teller who insisted that the US needed something more powerful than the atom bomb. Robert Oppenheimer, who was much less hawkish in his outlook, opposed him. Questions were raised about Oppenheimer's loyalty, and he eventually lost his security clearance, finishing his involvement with nuclear weapons. Teller meanwhile went on to create the American hydrogen bomb, soon after which the Russians developed their own 'H-bomb' independently. (The mastermind behind the Soviet hydrogen bomb was the physicist Andrei Sakharov, later known as a human rights campaigner and winner of the Nobel Peace Prize).

Fission weapons – atomic bombs   are limited by the amount of uranium or plutonium that can be squeezed together while the nuclear reaction occurs. In practical terms this means that an atom bomb is limited to a power measured in kilotons of explosive. Fusion weapons – also called thermonuclear or hydrogen bombs – have no such limitation. You can make a hydrogen bomb as big as required, and bombs with a yield of 10 megatons were built, hundreds of times more powerful than the Hiroshima and Nagasaki bombs. A single such weapon would be equivalent to setting off 10 million tons of TNT. An explosion at ground level would leave a crater 3 miles across, and would destroy every building within a 12-mile radius; casualties would run to millions, and the after-effects would be truly unimaginable.

Ten megatons is by no means the limit, and even larger weapons were considered by the nuclear weapons establishment. These might be too big to transport by air, but this need not matter. The 20,000 megaton 'Doomsday bomb' would have the same effect wherever it was detonated, as the explosive effects would be less significant than the fallout which would cover the entire planet with a lethal radioactive cloud, making it the ultimate dirty bomb. This was perhaps the most extreme form of the policy of Mutually Assured Destruction, and thankfully it was never built.

It has not yet been possible to use nuclear fusion to generate power on a practical scale. The reaction requires fantastically high temperatures, and at present these can only be generated easily by a fission explosion in an uncontrolled way. Controlled nuclear fusion will require powerful lasers or other energy source to heat the target material so that it fuses. This is certainly possible, and nuclear fission has been carried out in the laboratory for decades. However, the problem is in setting up a reaction which will last for more than a few microseconds, which generates more energy than is used to start it, and which can be contained. These remain major technical challenges after more than forty years of research.

The latest development is an international project called ITER, involving China, the UK, France, Spain, Japan, Korea, Russia and the United States. This is intended to prove that energy generation from fusion is viable, and will cost something over $5 billion. As of 2004, discussions are still continuing over where the project is to be sited. The building phase will take eight to ten years, and results will start to arrive some time after that.

Fusion power offers relatively clean nuclear power from hydrogen, without the need for uranium or plutonium, and could power the world indefinitely without the need for fossil fuels. But commercial fusion power is still decades away at the very least.

ITER may suffer the funding crises of many multinational projects (the US has already pulled out and then rejoined), or may fail to resolve the technical difficulties that have frustrated previous research. Its future is highly uncertain.

*     *     *

There are other ways of using fusion power, and Teller was a great proponent of hydrogen bombs for other purposes. He took the view that a hydrogen bomb is simply a bigger and better form of explosive that can be used for civil engineering projects on a grand scale. Hydrogen bombs could literally move mountains.

Teller talked about 'the great art of what I call geographical engineering – to reshape the land to your pleasure, and indeed to break up the rocks and make them yield up their riches'.

From 1961 to 1973 the US carried out Project Plowshare, setting off some thirty-five nuclear explosions to test the uses of atomic power for civil engineering. The Russians carried out an even bigger series of experiments with over a hundred explosions.

One idea was that underground detonations could create huge cavities for storing natural gas. Another was that harbours could be created where no natural harbour existed. Point Hope in Alaska was selected as the site for a new harbour; six thermonuclear devices with a total power of 2.4 megatons (more than a hundred Hiroshimas) would gouge out a keyhole-shaped cavity.

Although the idea had some public support and was backed by local newspapers, opposition soon appeared. Biologists, conservationists and Native Americans living in the area did not believe the claims that they would be unaffected by nuclear fallout.

The Atomic Energy Commission was forced to carry out a detailed environmental study, and the results indicated that there might be a radiation problem. Public and political support evaporated and the whole scheme was shelved.

Nuclear stimulation of oil and gas wells was also investigated. Oil and gas can be trapped in 'tight' geological materials; if the rock is fractured, the oil and gas can flow freely. This needs to be done on a gigantic scale to yield useful amounts of oil, and a nuclear explosion was the obvious solution. However, the Project Plowshare tests showed that it was difficult to keep the explosion confined underground, and that radioactive gas would be released into the atmosphere.

The same problem applied to creating underground cavities for storing gas or water. A remote area of Pennsylvania was selected for one such cavity, but again the local people objected. They felt that the locale had been selected because it was marginal and economically under-developed. There was a strong feeling that nobody would be exploding atom bombs under them if it had been a rich and well-populated district. This undermined the message that the AEC was trying to get across, that the explosion would be perfectly safe. Locals argued that if it was safe they could do it in New York. The politicians understandably got cold feet, and again, the project collapsed.

Having failed to find anywhere in the US to carry out his schemes, Teller tried other countries. He even approached Queen Frederika of Greece with a proposal to dig a canal using nuclear means.

'Thank you, Dr Teller,' the Queen is said to have replied, 'but Greece has enough quaint ruins already.'

Other plans included the Kra Canal project in Thailand linking the Gulf of Thailand with the Andaman Sea, using more than a hundred explosions each of over a hundred kilotons. There was a proposal to channel Mediterranean Sea water into the Qattara Depression in Egypt, using the height difference between the two to run a huge hydroelectric power station, but to do it would have required over 180 blasts totalling over 27 megatons. Unsurprisingly, neither project came to anything.

In the USSR, the totalitarian state had a free hand, unhampered by the need for consultation and untroubled by the local population's objections. Local politicians were not in any danger of losing their seats, and the nuclear testing program went on unhindered. Nuclear explosions were tried for seismic sounding, oil and gas well stimulation, creating underground reservoirs, shutting down oil gushers, digging a canal, crushing ore deposits, and even stopping gas escaping from a coal mine.

Most of the tests were carried out in Siberia, and the result was widespread radioactive contamination. Sometimes it spread further than the immediate area; a blast of 125 kilotons which created two new lakes also left a radioactive cloud that was detected as far away as Europe.

The Russians established that geographical engineering was practical, but that it was not possible to limit the contamination. The experiments continued, and an ambitious scheme was proposed to reverse the flow of one of the rivers flowing north from the Aral Sea, in order to increase the amount of water available for irrigation. By this time the environmental consequences were becoming apparent. The leadership of the Soviet Union may have been ruthless, but they did not want to turn their entire country into a radioactive wasteland. The environmental shortcomings meant that further projects were cancelled.

Operations in the US were not on the same scale, but even the Plowshare tests left a legacy of contamination. There was no explosion at Point Hope, but the engineers released a quantity of radioactive material to see how it would spread in the tundra, and there are still measurable amounts of radiation where they worked.

While the US and Russia have now abandoned the idea of 'peaceful nuclear explosions' the Chinese are still in favour. As well as building canals, they believe that nuclear weapons could be used to generate electricity by exploding them in underground caverns and tapping the energy produced.

However, this particular use of nuclear power still seems unlikely to yield useful results in the civilian world.

## Atomic Shipping

By the close of the Second World War, the Germans were building the most advanced submarines in the world. Their most successful design was the Type IXB which displaced almost 1,500 tons and was 76 metres (250 feet) long. On the surface, using diesel engines, it could make 18 knots; submerged, it relied on electric batteries and could only manage 7 knots.

Ten years later, the USS *Nautilus* was launched. Although it was much larger than the size of the Type IXB, it was much faster – more than 20 knots on the surface, and an equal speed submerged. Unlike the earlier submarine which could only stay submerged for hours at a time, the *Nautilus* could run underwater for weeks. The difference was that *Nautilus* was the world's first nuclear-powered submarine. Nuclear power transformed the submarine from a slow, limited craft only suitable for a few specific tasks into a versatile, high-speed vessel that would form a key component of the nuclear arsenal. Nuclear power meant that submarines could stay at sea for months at a time with no need for refuelling.

Other nuclear ships followed. The nuclear-powered cruiser *Long Beach* was launched in 1959, followed a year later by a nuclear-powered aircraft carrier, the USS *Enterprise*, which is still in service. The UK started its own nuclear submarine program, and the Soviets had theirs. By the end of the Cold War, there were over 400 nuclear submarines in operation, although most of these have since been scrapped. The US has ten nuclear-powered aircraft carriers; the US and Russia each have eight nuclear cruisers. While nuclear power made a tremendous difference to warships, in the civilian world it was a different story.

The Atoms For Peace program – the one that was behind Project Plowshare – set out to persuade the world that nuclear power could work in merchant shipping. In 1962, the NS *Savannah* was launched. It was a splendid vessel of 21,000 tons, 545 feet long and described by one maritime historian as 'the prettiest merchant ship ever built'. NS *Savannah* also had a turn of speed – 23 knots – to demonstrate the advantage of nuclear power.

The *Savannah* looked more like a luxury liner than a cargo ship, and she boasted some unusual fittings. As a showpiece, she had thirty staterooms with individual bathrooms, plus a veranda, a swimming pool and a library. She was painted a pristine white to emphasise the fact that she would never be marked by soot.

But *Savannah* was a commercial disaster. The cargo capacity was less than half that of other ships the same size, and the streamlined shape that gave her such speed meant that loading and unloading were labour-intensive and inefficient. Being a hybrid between a cargo and passenger vessel made no sense at all; the passenger berths were empty, but a conversion to all-cargo was too expensive.

She needed a larger crew than oil-fired ships, all of whom were hand picked and given extra training on top of the usual requirements, making operations expensive in terms of manpower. Worst of all, as a one-off ship she required unique spares and maintenance facilities. It was obvious early on that NS *Savannah* was not going to be profitable, but she continued for ten years on a subsidy of almost $3 million a year before being taken out of service.

Three other nuclear cargo ships were launched, including the German MS *Otto Hahn* and the Russian container ship *Sevmorput*, both of which suffered the same fate as the *Savannah*, proving uneconomical to run for similar reasons. In 1983 the *Otto Hahn* was refitted with diesel engines and renamed MS *Trophy*, going on to work as a container ship for a shipping company based in Shanghai. The *Sevmorput* was taken out of cargo service and converted to an icebreaker.

The Japanese had the 8,000-ton *Mutsu* but this had technical problems from an early stage. There were safety problems with the nuclear shielding, and these continued during sea trials. There was also a massive protest against the ship by Japanese fishermen – the country is understandably one where feelings about radioactive contamination run high – and the ship was decommissioned before it ever carried any commercial cargo.

These experiments with nuclear ships occurred in the days before the 1973 oil price shock, when the price of crude oil quadrupled. It has been argued that if there were nuclear-powered cargo ships today they would be far more economical, as fuel prices make up a much bigger proportion of operating costs than they did in the 1960s. However, it seems likely that the operators would also have to factor in the costs of eventually decommissioning the reactor, which might shift the economic balance away from atomic power.

There is one field where nuclear power still gives a crucial advantage – icebreakers. The Murmansk Shipping Company still has seven nuclear-powered icebreakers which clear channels through the icebound northern seas. However, these are relics of the Soviet era and are operated at a loss, so there are no plans to build new vessels to replace the ageing icebreakers. When they retire they are likely to bring to an end the civilian experience with nuclear ships.

## Nuclear Aircraft

The possibility of powering an aircraft with an atomic reactor was considered as early as 1942, during the Manhattan Project. But it was not until 1946 that the subject was given serious consideration, when the USAF set up a project called NEPA – Nuclear Energy for Propulsion of Aircraft. Nuclear power offered two major advantages over conventional fuel: it would be possible to generate a lot of power from a small reactor for a high-performance aircraft, and nuclear energy would give the aircraft almost unlimited endurance.

Instead of being limited to a few hours of flying time, an aircraft would be able to fly for days or weeks at high speed. Conventional bombers spent most of their time on the tarmac, vulnerable to enemy attack, but a force of nuclear bombers would be able to loiter just outside the Soviet Union, ready to strike at any instant.

The initial engineering work went well, and data was gathered about the effects of radiation on the materials used in aircraft. In 1948, a group at MIT produced the Lexington Report which concluded that nuclear aircraft were feasible and would take fifteen years at a cost of a billion dollars.

By 1950 the work had advanced far enough for the next stage, known as ANP (Aircraft Nuclear Propulsion), with the goal of full-scale development of a flying nuclear aircraft. The program was managed jointly by the USAF and the Atomic Energy Commission, and the aim was to have a subsonic aircraft flying by 1957.

The design of the engine was comparatively simple. The nuclear jet is similar to the standard turbojet: cold air is sucked in to the reactor and heated. The hot air expands and is ejected backwards, generating thrust. The airflow provides cooling for the reactor at the same time. The biggest difference between the nuclear version and a conventional jet is that the energy for heating is provided by nuclear power rather than burning fuel.

A variation on the design did not have the air in direct contact with the reactor, but instead circulated liquid metal through pipes through the reactor, and used the pipes to heat the air in the jet. This avoided having the air itself picking up radiation from the reactor.

Although we are used to thinking of nuclear reactors as massive, building-sized pieces of machinery, they can be made much smaller. Even in the 1950s it was possible to build a reactor as small as 5 tons – easily small enough to fit into an aircraft. The problem was the shielding.

Thirty tons of shielding was an absolute minimum for the reactor to be operated with humans anywhere near it, but this was not nearly enough for the long-term safety of the aircrew. Putting adequate shielding all the way around the reactor would have made it much too heavy for an aircraft, so instead the designers relied on 'shadow shielding'. This meant putting a large mass of shielding – another 19 tons of it – between the reactor and the crew, so that they were effectively in its radiation shadow.

Radiation in all other directions would be at a much higher level, but this was not thought to be a serious problem. Aircraft components were re-engineered to ensure that they would not be harmed by the radiation.

In spite of all the shielding, the radiation levels in the crew compartment were not ideal. It was even suggested that the crew should be drawn from older men who had already had children.

The engines and other components would add more weight, but first it was necessary to prove that a nuclear reactor could be flown in an aircraft. Normally the weight of the fuel and the engines can be distributed around the aircraft, but the reactor and shielding for a nuclear aircraft would form a single concentrated weight of 50 tons or so. A flying testbed was needed to prove that this sort of weight could be carried.

There was only one aircraft suitable for carrying a load of this size, the mighty B-36 Peacekeeper. The B-36 was a colossus, with a wingspan of 230 feet. Passing overhead it seemed to blot out the sky, giving it the nickname 'aluminum overcast'. The original B-36 had six propellers, but this left it underpowered, so they were supplemented by the addition of four jet engines ('six turning and four burning'). The B-36 could carry an immense bomb load, some 40 tons – more than the gross weight of a fully-laden B-24 bomber from the Second World War. It would be able to carry the nuclear reactor and, more importantly, the necessary shielding.

A test aircraft was fitted out, known as the NB-36, with a new nose section incorporating the lead and rubber shielding. The lead shielding protected it from hard radiation while the rubber absorbed neutrons. A one-megawatt reactor was installed, although it was not connected to the engines, which were conventional. The NB-36 was intended purely as a testbed to gain experience of flying a reactor.

The NB-36 carried out forty-seven test flights between 1955 and 1957. On each flight a transport plane full of US Marines accompanied it. In the event of a crash, it was the Marines' unenviable job to parachute down and secure the crash site.

The flights were successful, though they showed that more shielding would be necessary. This might have spelled trouble, but the program received a boost from a different quarter. In 1958, it seemed that the Russians were getting ahead. An editorial in *Aviation Week & Space Technology* magazine reported that:

A nuclear-powered bomber is being flight tested in the Soviet Union. Completed about six months ago, this aircraft has been flying in the Moscow area for at least two months. It has been observed both in flight and on the ground by a wide variety of foreign observers from Communist and non-Communist countries.

Appearance of this nuclear powered military prototype comes as a sickening shock to the many dedicated U.S. Air Force and Naval aviation officers, Atomic Energy Commission technicians, and industry engineers who have been working doggedly on our own nuclear aircraft propulsion program despite financial starvations, scientific scoffing and top level indifference, for once again the Soviets have beaten us needlessly to a significant technical punch.

It went on to say that unlike the NB-36 this was not just a test aircraft but 'a prototype of a design to perform a military mission as a continuous airborne alert warning system and missile launching platform'.

Representative Melvin Price of Illinois claimed the Russians were three to five years ahead of the US in this field and that the US program needed to be pushed ahead as a matter or urgency. Sputnik had just been launched, and the US was still in a state of shock that its technological dominance should be at risk. Anything seemed possible, even a Soviet nuclear aircraft.

Although the Russians had carried out paper studies for such an aircraft, they came to nothing. The bogus reports quoted by *Aviation Week & Space Technology* and others came from Congressional sources, the industry and the Air Force, all with their own agenda. But they achieved their goal and the scare story ensured that funding for the ANP would continue, even though there were questions about its lack of progress.

The ANP pushed ahead and by 1961 the prototype nuclear jet engines had been successfully operated on the ground. But just as it seemed as though the nuclear-powered aircraft was ready to fly, President Kennedy cancelled the program.

Kennedy could afford to drop the ANP because new long-range ballistic missiles, launched from submarines or from the ground, meant that there was no longer a pressing need for a nuclear bomber force. The slow progress of the ANP made it look unattractive compared to rockets, which were getting more powerful and accurate every year.

In the end, the failure of the ANP was not so much a matter of technology as bad management. The management of the program was split between the Air Force and the AEC, which resulted in indecisive leadership and a lack of direction. Political interference slowed the program down further, and shifting political demands caused major problems. Was the aim to get a nuclear aircraft airborne as soon as possible as a prestige project, to counter the Russian achievements? Was it to build a functional nuclear bomber? Or was it to develop the technology for a whole range of future nuclear aircraft, rather than simply working with stopgap solutions?

Test facilities were constructed at a cost of millions of dollars and then lay idle because they were no longer required by the project. Contractors took advantage of the confusion, playing the Air Force and AEC against each other. A third body, the Joint Committee on Atomic Energy made repeated attempts to take control of the program.

Bureaucracy thrived, and after spending some $1,040 million without getting into the air, the ANP was looking less and less viable even by military standards.

The Air Force still liked the idea of a nuclear bomber, and wanted to give it a much higher priority than the new ballistic missiles. After all, air forces are about flying, not sitting beside concrete silos waiting for a launch order. But the President was not impressed. In terminating the program, he wrote: 'The possibility of achieving a militarily useful aircraft in the foreseeable future is still very remote.'

In spite of the cancellation, there is little doubt that a nuclear aircraft would be feasible. If rockets had not been available, or if the ANP program had been managed better from the start, the skies of the 1960s might have seen the first nuclear-powered bombers, followed by spy planes, early warning radar aircraft and others. Whether the technology would ever have made it into the civilian sphere is a matter for speculation; what is surprising is that some think it still could.

In May 2001, Professor Ian Poll, Director of Cranfield College of Aeronautics and the President of the Royal Aeronautical Society, argued in *Aeronautical Engineering* magazine in favour of a new form of propulsion for civil airliners,

> The projected growth in air traffic worldwide for the next several years is expected to be 5–7% per annum. If that growth rate continues for 25 years, it would involve the trebling of current aviation consumption levels of kerosene. I believe it is time to consider all the alternatives – and one is the use of nuclear power.

He advocated a closed-cycle version of the atomic engine, which would not involve any atmospheric emissions. Without the need to carry a heavy load of fuel, the aircraft would be very efficient to operate and operating costs would be low. It would also have unlimited range and would be free from the constraints that apply to conventional aircraft which have to conserve fuel. Airlines are normally at the mercy of fuel costs, which can make or break their budgets, varying unpredictably; but the costs of nuclear fuel would be known for years in advance. Also, importantly, there would be no greenhouse gas emissions; a nuclear airliner would be a clean airliner.

'The physics are proven,' wrote Poll, 'so it would just be a matter of examining the engineering issues.'

He accepted that there would be some problems with selling the idea to the public, but believed that it was an area worth pursuing.

Four months later came the events of September 11, when two hijacked airliners were deliberately crashed into the World Trade Center, and two others used in attacks on Washington. Thousands died, but if nuclear-powered airliners had been involved the contamination could have spread over wide swathes of New York and Washington with unimaginable consequences.

As long as the threat of hijackings, shooting down by shoulder-launched missiles and suicide attacks remains, nuclear-powered civilian aircraft are not likely to be on the agenda. In addition, one of the major effects of the September 11 attacks was a general slowdown in the aviation industry. Air travel in the US dropped dramatically, and the airline business started posting losses. Orders for new aircraft fell correspondingly. This makes financing the development of new types of aircraft – already a multi-billion-dollar proposition – even more difficult. It is likely to be decades before nuclear power is even considered in this context.

## Nukes in Space

Small nuclear power packs have long been used to power satellites. These are not reactors, but devices in which a source produces a steady stream of radioactivity which can be converted into electricity. This provides power for much longer than any storage battery, and the high costs are not a factor in a multi-million dollar project. But there is no theoretical reason why a nuclear reactor could not be put on board a spacecraft – in fact, many people have argued that it is the only way to achieve true space travel.

The nuclear thermal rocket program emerged out of the nuclear jet program and worked on a similar principle, except that rather than heating air to provide propulsion the rocket would carry a store of hydrogen for the purpose. The hydrogen is not involved in a nuclear reaction, but was chosen because it is light and when heated it achieves a much higher exhaust speed (and so more thrust) than other gases.

This went through a number of incarnations, including one called Timberwind which was intended as a launch vehicle for the SDI program, but this died out as a result of the scaling-down of SDI coupled with public controversy.

Another approach was based on a nuclear power plant generating electricity which is used to accelerate charged particles. Non-nuclear versions of the 'ion drive' have proven to be reliable and efficient as a means of propulsion, but they suffer from requiring huge amounts of power – most ion engines only produce a few grams of thrust. A nuclear ion drive would still produce relatively little thrust and would have to be boosted into orbit using rockets, but once in space it could motor around the solar system for years or decades.

NASA are considering a nuclear ion drive in a project called Prometheus, using technology derived from the US Navy's submarine reactor program. However, the Columbia disaster has cast something of a shadow over this area: if a spaceship with a nuclear reactor breaks up in the atmosphere, the crew may not be the only ones to suffer, and contamination could be spread over a wide area anywhere on earth.

Far and away the most ambitious and dramatic use of nuclear power to propel a spacecraft was Project Orion which ran from 1957 to 1965. This was a direct offshoot of the atomic bomb program, and worked on the very simple principle of using a series of small atomic bombs to propel a spaceship.

The Orion spacecraft would eject the bombs behind it so that it would be pushed forwards by the resulting explosions. The size of the explosions dictated that a large 'pusher plate' would be needed to transmit the impulse and shield it from the direct effects of the blast, favouring a very large vehicle – one of about 4,000 tons was designed.

Orion could have produced the first Space Battleship, capable of taking off under its own explosive power, travelling into orbit and on into space, and bombarding anywhere in the solar system with nuclear firepower.

Project Orion failed for a variety of reasons, but perhaps the most important of them was political, falling between the Department of Defense, NASA and the Atomic Energy Commission. In the end nobody was willing to fund it, and after seven years and $10 million the plug was pulled. Anybody interested in this astounding project is recommended to read George Dyson's book *Project Orion*.

## The Next Generation: Nuclear Isomers

In the last few years there has been military interest in a new type of nuclear reaction involving 'nuclear isomers'. Molecular isomers are quite well known, being molecules that have the same formula as each other but a different structure, such as the mirror-image forms of some organic molecules. Nuclear isomers are more obscure. The most important here are shape isomers: these are special forms of normal elements in which the shape of the nucleus has been changed.

To use the snooker-ball analogy beloved of nuclear physicists, the protons and neutrons which make up an atomic nucleus can be compared to snooker balls lying on the table. In an isomer state, the same particles are stacked up into a pyramid. By applying a small nudge you can make the pyramid collapse, releasing energy as the nucleus returns to its natural state.

In terms of isomers, this nudge consists of zapping the nucleus with X-rays, causing it to release a tremendous amount of energy. In 1999 a team led by Carl Collins of the University of Texas at Dallas showed that triggered decay was possible with a Hafnium isomer.

This led to a series of claims and counterclaims as some other laboratories failed to reproduce his results and the theoretical basis of the triggered decay was challenged. However, research continued, and the Pentagon took an active interest.

Unlike other forms of nuclear reaction, triggered decay is not a way of producing energy, only of storing it. You have to stack up the snooker balls in the first place, and this is a major challenge. At present, only microscopic quantities of isomer can be produced, though researchers are confident that it will be possible to manufacture on a large scale within five years.

The nuclear isomer reaction does not involve fission or fusion; a weapon based on triggered decay is therefore not considered to be a nuclear weapon under international law. The amount of energy is much smaller than with an atomic bomb, but can still be thousands of times more powerful than a chemical explosive. Unlike a conventional explosive, the energy is released in the form of gamma radiation; some will be absorbed by the atmosphere to give a recognizable explosion, but the rest will be lethal over a significant area. (It has been suggested that the bomb might be useful for neutralizing stocks of biological weapons because of the bacteria-killing potential of the radiation.)

To start off with, the Pentagon's aim is fairly modest: a 23-kg bomb with the power of a conventional 907-kg bomb. In principle though, this research could lead to a 2-kg bomb with the power of a 9,072-kg bomb, or greater. Clearly such weapons would have a tremendous effect on the battlefield, arming the footsoldier with the firepower to destroy a small city – with a weapon which is not considered to be nuclear.

Isomer research is not limited to explosives. Because the decay can be controlled, it could be used as a power source. The USAF has expressed interest in an isomer-powered aircraft which would have a duration much longer than aircraft which rely on jet fuel. Interestingly, they are only looking at unmanned aircraft at this stage, perhaps with an awareness of how sensitive the public are to the idea.

Research on triggered isomer decay is being carried out in several other countries as well, including the UK, France, Japan and Russia, where Viktor Mikhailov, the scientific director of the Federal Nuclear Centre, has suggested that isomers are a possible candidate for a new generation of smaller nuclear weapons.

Clearly nuclear isomers have great potential for the future, but it remains to be seen whether they will emerge primarily as a means of storing energy or a weapon.

We have come a long way from the brave days of 'atoms for peace'. It is hard now to believe that anyone was optimistic enough to suggest the 1958 Ford Nucleon concept car. As the name suggests, this was a stylish vehicle with a nuclear reactor in the back. The original publicity material suggested that it would go 5,000 miles before refuelling. Ford never progressed beyond building a one-third scale mock-up of the car, but even this shows the basic limitation of the concept: the majority of the vehicle is taken up with a large reactor (presumably with plenty of shielding) leaving space for only two seats up front.

The idea was picked up later by the 1966 *Batman* television series, and the fictional Batmobile is likely to remain the only nuclear-powered car.

Nuclear power generation is still important, and as the debate on global warming continues it may take on new significance as a means of producing electricity without carbon dioxide emissions. However, it appears that in every other way the idea of 'atoms for peace' has failed.

# CHAPTER 6

# Flights of Fancy

*I think there is a world market for maybe five computers.*
IBM President Thomas Watson, speaking in 1943

*Man won't fly for a thousand years.*
Wilbur Wright to his brother Orville in 1901

*The bomb will never go off, and I speak as an expert in explosives.*
Admiral Leahy on the eve of the first atomic bomb test in 1945

Hindsight is a wonderful thing, making geniuses of us all, allowing us to conveniently forget how things looked at the time. It is easy to laugh at people who got it completely wrong by underestimating new inventions. Seen with the advantage of hindsight, it is obvious to us that computers, human flight and atomic power all had a golden future ahead of them.

But the course of technological progress is not so smooth. The ENIAC project was successful, but a parallel program carried out by the US Navy ended in expensive failure. Whittle's jet engine was eventually successful, but many others had failed before him. Others, like the purveyors of radio death-ray beams, never had a chance.

The prophets of doom are often right, that there are frequently good reasons why technology cannot be made to work – or not made to work well. The technologies that break through and fulfil their promise are the exception rather than the rule. This is why we need to exercise a degree of caution in predicting what we will see in the future based on what the military are doing now.

Nowhere is this more obvious than with flying machines. As we have seen, the first experimental jet flew in 1939, and ten years later the first jet airliner made its maiden flight. So when the military first proved that an individual could fly with the aid of a rocket belt in 1961, one might have expected that by the 1970s the city skies would be filled with commuters rocketing about from home to business. The first hybrid aircraft/helicopter flew in 1955, so the 1970s should also have seen flying buses on the same urban routes for those not able to afford jet packs. And the high-speed rocket plane was the darling of the 1950s, so we might have expected that three decades later it should have matured to provide mach 5 airliners for intercontinental travel.

None of these things happened, in spite of the first promises of the technology which seemed to be viable. One technology, the tilt rotor, may yet have its day, though this is far from certain.

The three different technologies looked at in this chapter (rocket planes, personal flight and tilt rotors) all have their roots in the Second World War. Any of them might have been as successful as the jet aircraft, and could have had as big an impact on society. But in every case, in spite of military development, which in some cases has spanned several decades, none of them has ever come to fruition. The reasons for these failures were different for each of the three, and by looking at what happened to them we may get an idea of how promising technology can go off the rails.

## Faster than a Speeding Bullet: Rocket Planes

During the 1950s and 1960s, rockets were all the rage. Jet airliners were all very well, but the science fiction writers of the day had their eye on something better. What they wanted was an aircraft which would make the run from London to Sydney not in twenty hours, but in two. (Years earlier, the Austrian rocket pioneer Max Valier, had written an article 'Berlin to New York in One Hour'.) Their enthusiasm was grounded in hard science and engineering, and the remarkable feats being achieved by the X-plane program. However, in spite of decades of development, the rocket airliner never materialized.

Although the first steps towards rocket planes were taken in the Second World War, in theory it could have emerged much earlier. As aviation writer Bill Gunston has pointed out, there were heavier-than-air gliders for many years before a suitable power source was found. Hiram Maxim tried an aircraft with a steam engine, and it was only with the invention of small, powerful, internal-combustion engines that propeller power became viable. But rockets had been around for centuries, and if some enterprising inventor had put rocket motors on a glider he might have beaten the Wright brothers to powered flight by decades.

As we have seen, jet engines were being developed in Germany in the 1930s, but this was in parallel with an ambitious rocket program. While the jet would be powerful, new liquid-fuelled rocket motors matched with aircraft would be even more impressive. These aircraft would be incredibly fast and agile, as far ahead of the jet as the jet was ahead of the propeller.

It was clear that any rocket would have a very limited burn time, probably only a few minutes. The rocket plane was therefore envisaged as a small interceptor rather than a fighter. Work progressed quickly and the first rocket plane, built by Messerschmitt, flew in 1940, a year after the first jet.

The airframe was a stubby design with sharply swept wings and no tailplane. Designated the Messerschmitt Me-163 it quickly broke records, being the first aircraft to break 1,000 k.p.h. (625 m.p.h.) in 1941. An operational version, the Me-163B Komet was ordered immediately. The Komet was the first rocket aircraft to see action, and the results may have surprised even the designers.

At the heart of the Komet was a rocket motor designed by Hellmuth Walter. It mixed an oxidizing agent (a hydrogen peroxide mixture known as T-stoff) and a fuel (hydrazine hydrate, methyl alcohol, and water, called C-stoff). As soon as these two were combined the effect was explosive. Though small, the motor was tremendously powerful, generating an impressive 1,500kg of thrust for an aircraft that only weighed 1,900 kg. This gave it twice the thrust-to-weight ratio of the Me-262 jet fighter, itself considered awesome for the time.

By 1944 the first operational Komet unit went into action, intercepting the massed allied bomber formations that were pounding Germany. Other German fighters had difficulty getting past the protective screen of Allied fighters, and then had to deal with the overlapping fields of fire from the machine-gun turrets of several bombers. But the Komet could not be stopped.

The aircraft's performance was phenomenal. In three minutes it could reach an altitude of almost 30,000 feet, in preparation for an attack run which involved swooping through the Allied bomber formation, blazing away with a pair of heavy 30mm cannon. With a maximum speed of 560 m.p.h., it was much too fast for Allied fighters to catch, too fast even for Allied gunners to track with their machine-guns.

However, the Komet was also too fast to engage the enemy. Going through enemy formations at a relative speed of 200 m.p.h. or more, the pilot needed good reflexes to have any chances of hitting a target. The cannon had a low rate of fire, and because the Komet only had fuel for eight minutes of powered flight, a pilot was unlikely to get more than a single pass at the bombers.

A total of 370 Komets were built, and they destroyed a total of nine Allied bombers between them — in itself very disappointing for the Luftwaffe. But this result is only half of the story. The other half is about just how dangerous the Komet was to the pilot.

One problem was with the hydraulics, which were developed quickly without adequate testing. They had a tendency to lock up, leaving the plane going in a straight line. If this happened during the climb, the aircraft climbed high into thin air, where the pilot would pass out in the unpressurized cockpit. If the hydraulics failed during the dive, the Komet would accelerate to very high speed; if it broke the sound barrier, the shockwave would cause the aircraft to disintegrate. Otherwise, the speeding Komet would plough into the ground like a thunderbolt.

There were also some problems with the plumbing of the exhaust, which was vulnerable to cracking due to the forces of take-off. The exhaust was steam, harmless enough except that it was at very high temperature. A small leak into the cockpit clouded the windows and made it impossible to see; a major leak would flood the cockpit with steam at several hundred degrees, cooking the pilot like a lobster.

Fuel leaks could be equally dangerous. C-stoff was unpleasant, but T-stoff, concentrated hydrogen peroxide, was highly corrosive. A damaged pipe could mean that the pilot was dissolved where he sat.

However, these were minor considerations compared to the major safety issue of using a liquid-fuelled rocket engine in the first place. The mixing of the fuel and oxidizer results in a controlled explosion, driving the rocket. But if they mixed unpredictably, there was an uncontrolled and frequently catastrophic explosion.

This could happen at any time, but the most frequent were accidents on landing. The Komet had a skid rather than wheels, leading to some very bumpy landings; many pilots suffered back injuries. The danger was that if there was any remaining fuel in the tanks, the shock of landing would cause it to slosh around and mix, with lethally explosive results.

Of those that were lost, 5 per cent were brought down by Allied fire in the air; 15 per cent were lost due to problems with the controls and hydraulics; the other 80 per cent were victims of explosions. No wonder pilots nicknamed it 'The Devil's Sled'.

Much has been written about the technical sophistication of the Komet by aviation enthusiasts who are sometimes carried away by its performance figures. But the key point about the Komet is not that it was so very advanced, it is that it was so very useless. Rarely can any air force have fielded an aircraft that killed so many of its own pilots and so few of the enemy. If the Third Reich had fielded more, the Luftwaffe could virtually have wiped itself out without the need for combat.

In spite of the obvious flaws in the Komet, towards the end of the war a desperate Nazi government was pouring resources into an even more dangerous aircraft. The Bachem Ba 349 Natter was to be a simpler, cheaper, rocket-powered aircraft, with a wooden airframe. The nose would have carried two dozen unguided air-to-air rockets, and the whole thing was 'semi-reusable': rather than landing, the plane broke apart after it had carried out its attack run. The pilot and the valuable rocket motor assembly parachuted back to earth; the body of the aircraft simply came crashing down.

The Natter was designed to operate without runways, with a rocket powerful enough to launch it vertically. On the first test flight with a manned pilot, it went disastrously wrong: the canopy broke away during the launch, the pilot lost control, and the plane crashed to the ground with a huge explosion.

One Natter prototype was completed and allegedly went into action against a formation of B-24 Liberators, downing two and damaging a third. However, there is no documentary evidence for this claim, and it seems unlikely under the circumstances.

The rocket interceptors were small stuff compared to what might have been built. Hitler was more interested in bombers than fighters, and the Nazis lacked a long-range bomber force of their own. To fill the gap Dr Eugene Sanger proposed a most radical solution. For maximum speed and minimum air resistance, an aircraft should fly as high as possible. If it did not need to have oxygen for the engines to burn, and if it did not need air for propellers or jet engines to work, it could go to any altitude at all, and the higher the better. This thinking produced plans for a rocket-powered stratospheric bomber, one of the most ambitious aircraft ever planned.

In 1939 Sanger had constructed a one-twentieth scale model and tested it in a wind tunnel. Building a full-scale version would be more of a challenge.

Sanger's bomber started on a sled, travelling along a 3-km stretch of track. The sled was driven by giant rockets generating 600 tons of thrust for eleven seconds, which would boost the aircraft to a take-off speed of 1,100 m.p.h. Now powered by its own rocket engine, the bomber would soar to an altitude of almost 500,000 feet (a hundred miles). At this altitude the atmosphere is almost non-existent and the bomber would be travelling at phenomenal speed – Sanger estimated about 13,000 m.p.h. With no atmosphere to provide lift, the bomber would descend in an arc until it struck the atmosphere where it would bounce up again, like a stone skipping across the surface of a pond. Technically the bomber could be described as a rocket-glider in the two different phases of its flight.

The design was wildly impractical. With a range of perhaps 15,000 miles, the stratospheric bomber would have ended up on the other side of the world – turning around was not practical. And it would have carried a very small bomb load, probably just one 454-kg bomb – less than a medium bomber and a fraction of the 2,722 kg load of a single B-17 Flying Fortress. The chances of actually finding a target given the speeds involved and relatively primitive guidance can only be guessed at, but aiming at anything smaller than a city would have been wildly ambitious.

Further, Sanger had little idea of how to deal with the kind of problems that his aircraft would have faced. Chief among these was atmospheric heating, which would have been too great for any materials available in the 1940s. Implementation may have been impossible, but Sanger's ideas were proved to be influential after the war.

## Post-war Rocket Planes

Five Komets were transported to the US after the war, along with one of the designers, Dr Alexander Lippisch. The Komets were used for glide testing only, towed into the air and released by carrier aircraft. Unlike the Nazis, the US government could not afford to be careless of the lives of its test pilots.

A series of US test aircraft followed, known as the X-planes. Their mission was to explore the technical and engineering issues of breaking the sound barrier in a manned aircraft. They caught the public imagination and in 1947 Captain Chuck Yeager became a national hero as the first man to break the sound barrier in the rocket-powered Bell X-1.

Flying the rocket planes was a hazardous business and called for the 'Right Stuff'. In 1951, an explosion on an X-1 destroyed both the plane and the Boeing B-50 which was carrying it. A second X-1D was lost when fuel being jettisoned caught fire, and a third was lost to another explosion.

Another tragedy nearly claimed Chuck Yeager in 1953 when the X-1 he was piloting tumbled out of control. Yeager was thrown about in the cockpit so violently that he thought he had cracked the canopy with his head – there was no canopy release, and no ejector seat. Yeager's only chance of survival was to bring the aircraft back under control; amazingly, he did it. This confirmed his status as an American hero, but did nothing for confidence in the safety of rocket planes.

If these tests had shown that it was possible to break the sound barrier and survive, they also showed that a rocket-powered aircraft were extremely dangerous.

In spite of the obvious risks, there was a vogue for rocket power among the air forces of the major powers. This was because they were all having to face up to the new challenge of the post-war era. During the Second World War, intercepting bombers was a matter of attrition: if 10 per cent of the bomber force could be shot down on every raid, the enemy would soon run out of planes. But in the nuclear age, every single bomber needed to be intercepted before it could reach its target. Just one of the new H-bombs would do more damage than a thousand-bomber conventional raid, devastating an entire city.

The British contribution to this field came from the Saunders-Roe aircraft company. In response to an RAF requirement of 1951, they put forward plans for a hybrid interceptor which had both a jet engine and an additional rocket engine. Using the jet alone it could fly at mach .95; when the rocket cut in it accelerated to over mach 2. The rocket had short endurance, estimated at seven minutes, but this would be enough to boost the plane to high altitude and make the interception. A prototype, the SR.53 flew in 1957 and the more advanced SR.177 was due to fly the next year.

The Saunders-Roe plane was finished by the infamous defence White Paper of 1957. This declared that henceforth the interception mission would be carried out by long-range missiles, and that the Lightning then being built was to be the last manned British fighter. Much has been written about the damage done to British aviation by this decision, and the rocket interceptor was just one of many casualties.

In the US, an armed version of the X-1 was proposed but never built. The Republic XF-91 Thunderceptor was another hybrid craft with a jet engine and a rocket motor. Although it had a better endurance than other rockets, with a limit of half an hour, it was still not adequate to cover the distances demanded for defending the vast continental US.

The most ambitious scheme in the US was a strategic aircraft called the DynaSoar (short for Dynamic Soarer). This was a multi-role craft bearing a market similarity to the Sanger project, launched by a powerful rocket and then gliding unpowered at phenomenal speed for the rest of the flight. Maximum speed would range from mach 5 to mach 25 depending on which version was deployed. Development started in 1958, but it was doomed because of a lack of strong support.

The plan was for DynaSoar to evolve from a research vehicle to a reconnaissance plane much faster than the Blackbird (discussed in the next chapter) and then on to being the ultimate bomber, carrying nuclear weapons around the world at more than mach 20. Unlike a ballistic missile, the DynaSoar could be re-targeted or recalled after it had been launched, and by gliding it would be able to approach targets at low altitude giving little warning on radar.

If everything had gone to plan, the US would have had an orbital bomber by 1974. Later plans called for further developments that would have turned DynaSoar into the first weapon of the space age, an orbital vehicle that could ferry payloads into space or bring down enemy satellites. Ultimately this might even have produced civilian spin-offs for space launches or other purposes – sub-orbital airliners or space tourism.

Unfortunately the program lacked a powerful sponsor when Kennedy came to power. Defense Secretary Robert McNamara cancelled Dyna-Soar in 1963 because he saw no requirement for an expensive new rocket plane.

Research into rocket propulsion continued, reaching its zenith with the North American X-15 in the 1960s. This was a joint program between the USAF, US Navy and NASA to explore the limits of aircraft performance at high speed and altitude. The X-15 still holds the record for being the fastest aircraft (as opposed to a spacecraft) ever built. It was launched from under the wing of a B-52 bomber at an altitude of 40,000 feet.

The X-15 embodied the latest in technology. Its skeleton was titanium, light and strong, and its skin was a special nickel alloy to withstand the extreme temperatures caused by air friction. The rocket motor had to be small and powerful, and, crucially, it had to be safe. Pumping more than a hundred litres of fuel a second into the reaction chamber meant that even a fraction of a second's unburned fuel would cause a lethal explosion if it did not combust properly. The rocket was powered by ammonia with a liquid oxidizer.

The X-15 was basically a manned missile; the top speed was mach 6.7 (7,300 k.p.h. or 4,500 m.p.h.) and it soared to an altitude of 108 km (67 miles, 354,000 feet), the very margins of space.

In 200 flights there was one fatal accident, when an electrical fault caused uncontrollable oscillations and the aircraft broke up at high speed. For such an extreme experimental craft this might be considered a good safety record, but it still underlines the hazards associated with rocket planes.

The X-15 was cancelled in 1968. It was an invaluable research tool, and helped to lay much of the foundation work for high-speed flight. The original ideas about hypersonic airflow had to be worked out again after X-15 showed many of them to be mistaken. In particular, the craft's stability at high altitude was not as expected and changes were needed to bring it under control.

One of the factors that spelled the end for the X-15 was the rise of NACA which later became NASA. The agency was committed to space travel with rockets, and it was not in favour of rocket funding which did not contribute towards the space effort. Almost as soon as the X-15 took off it was eclipsed by the Russians' achievement of putting a man into space. NASA wanted money for the manned space program, and flight within the atmosphere, however glamorous, was not going to convince the world that the US was keeping up with the Soviets.

It seems that the future of rockets lay outside the atmosphere, and that the X-planes were a dead end. Reusable rockets like the Space

Shuttle, which takes off like a rocket and glides back to earth like an aircraft, owe a lot to the X-planes, but the original idea of rocket aircraft has been left behind.

The prospect of London to Sydney in just a few hours continues to be dangled in front of the traveller, as we will see in a later chapter – though if it ever happens it will be a rather different technology that is used. Rocket planes were never truly tamed, but it remains to be seen whether the right lessons were learned from their evolution.

## 'You'll Believe a Man Can Fly' – Jet Packs

Strap on a rocket belt and take to the skies . . . the concept dates back to Buck Rogers, a hero from the Golden Age of Science Fiction. Featuring in a pulp magazine called *Amazing Stories*, Buck was a First World War fighter pilot transported to the twenty-fifth century to battle evil. The rocket belt was his preferred mode of transport, as essential a piece of kit as the ray pistol. Similar devices became a staple of pulp science fiction including the movie serial *King of the Rocket Men*.

The idea of personal flight was irresistible. In 1933 a German inventor used a backpack with rockets on it to propel him as he roller skated, though staying upright proved to be more of a problem than he expected. A few years later German pioneers experimented with the idea of a pack which would allow the wearer to leap over barbed wire, minefields or other obstacles.

The pack consisted of two units, one strapped to the user's chest, the other to his back. Their combined thrust was sufficient to lift the user into the air, but they had to be fired simultaneously. In tests, users were said to have been able to leap a distance of 50 metres, but the project does not seem to have ever reached an operational stage, and there is little evidence to back the claims. Whatever boasts were made for Nazi technology, this one was not a giant leap forward for mankind.

The idea went to the US with Wernher von Braun's team, and the US Army showed some interest in 1949. Theoretical work suggested that the small rockets of the era did not have enough power to keep a man in the air for any length of time, and the idea was shelved until more power was available.

The idea was never far from the minds of those who worked with rockets. In 1953 Wendell F. Moore was working on small hydrogen peroxide rockets for Bell Aerosystems. He tells the story that late one evening he sat doodling at his kitchen table and drew a picture of a man between two rockets, with fuel tanks on his back. The doodle grew to include valves, lines, controls and gauges. It was entirely fanciful, but Moore started wondering exactly why it could not be done in practice. He concluded that there was no reason it should not work, and put forward a proposal to the US Army. They liked it enough to advance $150,000 for a 'Small Rocket Lift Device' capable of carrying an individual soldier.

The technology involved was simple enough. All it needed was a container of hydrogen peroxide, and pressurized gas to drive it. The pressurized gas forced the hydrogen peroxide over a silver catalyst; the catalyst broke down the peroxide into steam and oxygen in an explosive reaction.

The components for the first rocket belt were scavenged from other projects. It closely resembled Moore's original doodle, consisting of little more than three tanks attached to the back of a corset worn by the pilot. The middle tank was the nitrogen propellant, the other two contained the peroxide. The only controls were a throttle governing the rate of fuel flow, and a directional controller.

As with the rocket-powered roller skater, stability turned out to be a major headache. Pitch and roll (the tilting movements side to side and back and forward) were controlled by the movements of the pilot's body and had to be learned. Moore injured his knee during an early test and stopped flying. The first free outdoors flight using the prototype rocket belt took place in April 1961, for a total distance of 35 metres – a little less than the Wright Brothers at Kitty Hawk, and half the distance of Goddard's first rocket in 1928.

However, while Goddard could experiment with bigger and bigger rockets with small payloads, the rocket belt had to work to other limitations. The whole contraption had to be small enough to be portable by one man. This limited the fuel supply, and even under ideal conditions the maximum length of time that the rocket belt could operate was just twenty-one seconds – enough for a very short trip indeed. After design improvements, this was increased, but only to thirty seconds.

The rocket belt was extremely popular and the demonstrations were wonderful. It featured memorably in the James Bond film *Thunderball*, providing 007 with a convenient escape from a villain's lair in the pre-title sequence. Flying over the rooftops, Bond lands besides his getaway car and says of the rocket belt that 'No well-dressed man should be without one.'

But he still needed a getaway car as well, since the limited flight time means that the rocket belt could not get him more than half a mile away from danger. As with the laser in *Goldfinger*, the rocket belt was added by screenwriters and did not appear in the book. Bond was just keeping up with the technological trend.[1]

The spectacular demonstration was, as it turned out, all the rocket belt was ever good for. It featured in Michael Jackson concerts and in the opening ceremony of the 1984 Olympic Games. However good it looks, the rocket belt has not yet found any practical use, even for the military. But the army loved the idea, and the quest for personal flight moved on to new technology. Instead of a rocket, the flying machine would be powered by a jet engine.

The jet project was funded by ARPA, under a $3 million contract, also to Bell Aerospace. Again, the principle was simple enough: instead of a rocket, the system would have a backpack unit with a single, vertically mounted jet engine. Thrust was channelled through two nozzles projecting downwards behind the operator's shoulders. Steering was more sophisticated than the rocket belt, and flight times would be measured in minutes rather than seconds.

The Bell Aerospace jet belt took to the air in 1969, and after a series of increasingly ambitious flights it was travelling at 30 m.p.h. for four minutes at a time. Bell believed that it would eventually be capable of twice that speed for up to half an hour.

The flight demonstrations of the jet belt were even more impressive than the rocket belt. Unfortunately, even by this time the limitations of the concept were becoming apparent.

The jet belt weighed more than 75 kg, making it extremely cumbersome. The weight made landings tricky, and the last thing you want to do is fall over when you have a jet engine strapped to your back. If the intake sucked in dirt or stones the results could be catastrophic.

By 1970 Bell decided to abandon the project, and sold the rights to the jet belt to Williams Research Corporation. The Army were still keen on getting their personal flying machine, and Williams were funded to develop a new generation. Instead of having a backpack unit, the Wasp was more like a flying pulpit, or perhaps a flying dustbin.

The pilot stood astride the engine and no longer had to rely on his legs as landing gear.

Having a jet engine running at 63,000 revs per minute between their legs caused some concern among pilots. If the engine ever threw a rotor blade, some very sensitive parts of the pilot's anatomy were likely to be severely injured; additional Kevlar shielding was added to protect against this possibility. However, the Wasp did not match the performance of the jet pack. After many years with little to show, the Army cancelled the program in 1982.

After the failure of three successive programs, the idea might be expected to have died. But Williams went on to develop a more advanced version of the Wasp called the X-Jet. This featured an advanced and highly efficient engine modified from a cruise missile, giving it a top speed of 60 m.p.h. and an endurance of half an hour. By 1989 the X-Jet was fully operational – but the Army turned it down anyway.

Their reasons for rejecting the X-Jet were not released, but it is possible to make an educated guess. Williams put the cost of production at 'more than a jeep, less than a helicopter', the exact price depending on the size of the production run. This highlights the issue that the X-Jet was in competition with helicopters – which are bigger, faster and carry a larger payload over a greater distance and for a longer period. Given that it was likely to be a specialist item, the cost per unit of the X-Jet was not likely to be competitive with helicopters, even if the Army accepted the original cost estimates. (It is unusual for the finished item to be as cheap as envisaged.)

The one area where the X-Jet might have had an advantage over helicopters was in the niche of tactical reconnaissance. Smaller, quieter and more agile than a helicopter, the X-Jet might have been a success – except that the place had already been taken. Small unmanned aircraft carrying cameras and other sensors were already proving a big hit, and they had the additional advantage of not exposing the operator to hostile fire.

And still the idea did not die. In 1997 a civilian company called Millennium Jet took up the challenge, aided by funding from the Defence Advanced Research Projects Agency – to the tune of $5 million this time – and NASA.

Millennium Jet's design was called the SoloTrek XFV. Rather than a jet engine, this featured two enclosed fan blades like small helicopter rotors behind the pilot. The whole thing was built as an exoskeleton which completely surrounds the pilot so he did not need to carry the weight. The design goals were a speed of 75 m.p.h. for two hours. Cost was always going to be a factor though: Millennium suggested that it would be 'similar to a high-end sports car'.

It was planned that United States Special Operations Command, which is in charge of all US Special Forces units, would be field testing the XFV in late 2003. However, it would have taken an incurable optimist to believe that the SoloTrek XFV was likely to be a success.

'The thing was way too fragile and the soldier too exposed for it to ever make it as a combat zone vehicle,' commented one ex-employee. 'I believe that a sparrow or large bug could have caused the fans to scatter.'

His comments were borne out when a loose tether apparently damaged the test vehicle beyond repair. DARPA dropped their funding; in an effort to stay afloat, Millennium attempted to auction the prototype on the Internet. The prototype was put on sale 'for display purposes only', with the stipulation that the purchaser should not fly it. However, the company omitted to mention that it was actually in a 'non-operable condition'. When the state of the machine was revealed, any chance of a sale was lost. XFV failed to raised the needed $7 million, bringing an end to the engineering side of the SoloTrek program.

What remains of the company is Trek Entertainment Inc, which offers licensing opportunities for the entertainment industry – in other words, the rights to use the image of the XFV in films and other media. Their homepage now shows a picture of the XFV being flown by an 'action superhero'.

Seventy-five years after Buck Rogers took to the skies, personal flight has come full circle and is back in the hands of superheroes and video-game characters.[2] However, the technology of personal flight is just one of the issues involved; making it work in the civil world will be another matter entirely.

Restrictions on noise level mean that your neighbour cannot fly a helicopter from his back garden, and the same would apply to a jet pack. Civil aviation safety rules prohibit aircraft from flying at low altitudes, where there are hazards from overhead power lines, trees and other obstacles. This would apply equally to personal flyers.

The question of air traffic control would also be a serious issue, with the chances of an accident increasing rapidly with the number of flyers involved. Then there are issues about bad weather and darkness, and how flyers would be licensed or certified for different conditions.

One could almost believe that conspiracy exists to prevent such a device from ever being developed. However, given the technical issues involved, it seems that no conspiracy is necessary.

## Tilt Rotors: 'Any Year Now'

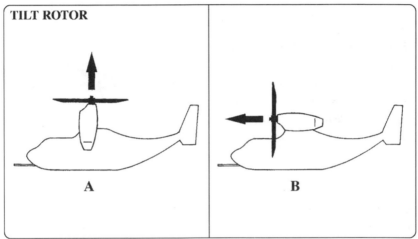

Tilt rotor aircraft in 'takeoff' mode (A) with rotors horizontal. In 'flight' mode, rotors tilt forward by 90° (B).

According to Boeing, the US government and industry have invested $5 billion developing a technology which has no equivalent anywhere else in the world. It is a proven technology, made only in the US, with decades of work behind it. This is no short-term project but, as Boeing say, 'A forty-year investment'.

However, it is not an investment that has shown any return so far. After all this time there is still not a single production aircraft using tilt-rotor technology. Boeing is confident that the technology will bear fruit in the very near future, transforming both military transport and civil aviation. Boeing has its reasons for taking this line, whereas anyone considering investing in the program might like to look at its history.

Helicopters can only rarely compete with fixed-wing aircraft. For the few specific jobs where hovering or vertical take-off and landing are critical, the helicopter wins almost every time. But for any other task they lose out.

Helicopters have to run very fast just to stay still. The lift from a helicopter is provided by the rotor blade blowing air downwards. Even if the helicopter is going nowhere, it takes a lot of power just to keep it in the air. This contrasts with fixed wing aircraft which get their lift by airflow under the wing and this does not require much power – gliders can fly for hours on end with the help of a few thermal updraughts from the ground.

This makes helicopters inefficient in terms of fuel consumption, limiting them to comparatively short range. In addition, the aerodynamics of helicopters mean that they are limited to relatively slow speeds. The Westland Lynx is an advanced design and represents one of the best helicopters available. It has a top speed of 200 m.p.h. and a range of 800 miles. However, compare this to a fixed-wing aircraft from forty years previously. The later Spitfire had a similar weight to the Lynx, but it was more than twice as fast and had a much greater range.

But however great their performance advantages, fixed wings could not perform the sort of vertical take-off that could be so vital for battlefield operations. For supplying troops in the field, or the quick evacuation of casualties, for landing Special Forces in hostile territory, the helicopter was still indispensable. If only it could be improved so that it could have the range and speed of a fixed wing but still be able to take off from a patch of grass.

Engineers have been trying to find ways around helicopters' weaknesses since they started flying. A number of solutions were proposed for 'convertiplanes', hybrid aircraft that shared the characteristics of the two types of aircraft.

One answer was the tilt-rotor, which combined the best properties of helicopter and fixed wing by having its engines on mounts which could be rotated through 90°. During take-off, the engines would be vertical, so their propellers acted like the rotor blades of a helicopter, and the craft could lift off vertically. Once in the air, the engines would rotate forwards to act like propellers, and the tilt-rotor would fly like a fixed-wing aircraft. On reaching its destination, the tilt rotor could land like an aircraft, or the engines could be swivelled to the vertical for a helicopter-like landing.

The idea of the tilt-rotor originated in the Second World War. The Luftwaffe wanted an aircraft which could keep up with the speed of the

Blitzkrieg, and which would not have to rely on airfields, and they came up with the Focke-Achgelis FA-269. This had a large engine in its fuselage and a complicated transmission system to drive two huge propellers on the wings. In normal flight the propellers would face backwards, but they could be rotated through 85° for take-off and landing.

Models were built and tested in wind tunnels which seemed to validate the theory, but in June 1942 the Focke Achgelis factory was destroyed by bombing. Faced with starting from scratch, and with a number of engineering problems still to solve (the control system was particularly troublesome), the 269 was cancelled. However, the aero-dynamics of the tilt rotor made perfect sense. The promise of the best of both worlds was still there, and as helicopter design progressed after the war, building a tilt-rotor did not look too difficult.

In August 1950, the USAF and Army announced a tilt-rotor design competition. This was won by Bell, which was a leading helicopter manufacturer, and which was awarded a contract to build the XV-3.[3] This was to be a prototype able to take over the helicopter's military roles of observation, reconnaissance and medical evacuation.

Manufactured and initially flight-tested at Bell's facility in Fort Worth, Texas, the world's first tilt-rotor took two years to build. Tests in a wind tunnel confirmed that a pair of normal helicopter rotors operating at a reduced speed could propel the craft, and only minor design changes were recommended after the initial testing.

What the wind tunnel tests did not look at was the effect of the large, slow-turning rotors on the bending motions of various flexible components on the aircraft, known as aeroelastic effects. These effects were to plague the XV-3 throughout its early life. There was no computer modelling in the 1950s, and the tools were not available to analyse what was going on.

The XV-3 was rolled out in 1955, and first flew on August 11 of that year, but flight-testing took longer than expected. The interaction of the various components of the aircraft – the rotors, the wings, and the rotating engine mount – meant that at certain positions the craft would become unstable. Every time one of these instabilities was encountered, modifications had to be made to the craft to smooth it out and testing started again.

On 25 October, during a test flight, one of the two prototypes suffered severe rotor instability and crashed. Luckily the pilot survived, although he was seriously injured. Bell was forced to reconsider the entire rotor system, and ultimately decided that its basic design was unsatisfactory. Major changes were made, and the new version did not start flight-testing until 1958. Further instabilities were encountered, and when the tilt-rotor finally made its first conversion in flight – going from helicopter mode to fixed-wing while in the air – it was three years behind schedule.

In 1959 the prototype was handed over to the military for testing. They found that the maximum forward speed was 175 m.p.h., but the craft was not stable beyond 130 m.p.h. – not enough to give it an edge over a helicopter. The weight of the prototype had increased with each modification made, and this ate into the capacity for carrying fuel and payload. It had reached a point where only the pilot and half a tank of fuel could be carried, without any cargo. While the XV-3 showed that the tilt-rotor idea was feasible, the technology was not mature and there was still no real understanding of aeroelastic coupling that made it unstable.

However, further testing of the XV-3 helped Bell engineers to understand the dynamics of the air craft, and they continued working through the 1960s. As more advanced computers became available it was possible to model tilt-rotors without having to build them.

Bell were not the only ones experimenting with tilt-rotors. In 1962, the Army ordered five prototypes of a new design from a team including LTV, Hiller and Ryan. The aircraft, the XC-142, was more ambitious than the XV-3, having four engines instead of two, and with the capacity to carry a platoon of thirty-two fully equipped troops, or twenty-four stretcher cases in the medical evacuation role.

The new design was much more successful. In 1965 the XC-142 converted from vertical to horizontal flight and went on to fly at over 300 m.p.h. A civilian version, the Downtowner, was proposed as a solution to airport congestion problems. Acting as an airport taxi, it would have carried forty passengers at over 250 m.p.h., and could have landed in city centres without the need for a runway.

All did not go well during flight-testing and there were a number of 'hard landings' causing damage. In 1967, one flight test was carried out which simulated a pilot rescue under combat conditions; the aircraft crashed, killing three crew members.

Instabilities were also found with the XC-142, especially while flying at very low altitude – the most dangerous time for the aircraft to become unstable. Pilots found that there were excessive levels of noise and vibration in the cockpit; combined with the high workload, they further increased the risk of accident.

The military were not inclined to believe that the mechanical problems could be resolved, and so the craft never progressed beyond the prototype stage. The Downtowner remained a dream, appearing occasionally as an ornament to pictures of futuristic building projects.

Meanwhile the Russians were also working on their own version of the hybrid aircraft, although their solution abandoned the idea of a tilting engine. Instead, they had a conventional propeller aircraft with the addition of large helicopter rotor blades mounted on the wingtips. The helicopter blades were used for take-off, and the propellers for normal flight.

The Russian aircraft was the Kamov Ka-22 Vintokryl, known to NATO by the code name Hoop. It soon broke the world records for speed and climb for its class. But it suffered from the same safety problems as the tilt-rotors: two Hoops were lost in fatal accidents. The investigators were unable to discover the exact cause of the crashes, and the program was discontinued.

The Russian experience strengthened the conviction that the hybrid concept could be made to work, and money flowed into tilt-rotors. The Composite Aircraft Program ran through the 1960s, with the aircraft in question being a composite of helicopter and fixed-wing. After this came a more ambitious scheme, to finally solve the tilt-rotor's problems for good.

It was 1972 and Bell were involved once more. This time their offering was the XV-15; again, design and construction took some time, and it was 1977 before the air craft finally made it into the air, and a further two years before it converted to flight. There were further delays with the flight-testing, and testing and demonstrations continued into the 1980s.

The XV-15 appeared to have finally solved the stability problems that had plagued earlier tilt rotors. Flight was smooth and quiet, and it was easier to control in hover mode than a helicopter. The pilot's workload was low, and the conversion between modes was very straightforward. However, funding was modest until a new war persuaded the military to go ahead with a full-scale tilt-rotor program. Curiously, the war did not even involve the US.

In 1982, a British Task Force successfully recaptured the Falkland

Islands from the Argentines. It was the first seaborne assault since the Second World War, and technology had changed radically. For the first time, the Argentines successfully attacked ships at a range of 30 miles or more using Exocet missiles. The US military doctrine of the time was that assault ships would remain 15 miles offshore during a landing, sending marines in by helicopter and landing craft.

The Falklands experience showed that this type of assault would be suicidal in the face of land-based anti-ship missiles. The Argentines had a very limited supply of Exocets; if the defenders had been armed with batteries of anti-ship missiles, the landings would have been disastrous. In future, assault craft would have to remain further back from the threat, and troops would have to fly in from greater distance at higher speed. Over-the-horizon assault was the new tactical imperative. Helicopters did not have the speed and range for the job; tilt-rotors did.

In 1983 Bell and Boeing received a US Navy contract to develop a larger version called the JVX (for Joint Service, Vertical Lift, Experimental).

The other lesson learned in the Falklands was the need for airborne early warning to detect incoming fighters. While large helicopters can carry radar, they can only reach a ceiling of a few thousand feet; the new Bell-Boeing JVX would be capable of loitering at 30,000 feet for extended periods.

So the development process started again; unfortunately, this time it was a joint program which started with the huge disadvantage of having to be all things to all men. The Army, Navy, Marines and Air force were all involved with the JVX, each with their own set of requirements. Worse, all of them were suffering from budget pressures. Given a choice, the Army was less interested in the JVX than in its own LHX (light helicopter, experimental) program; the Air Force announced that if the Army pulled out, the increased cost meant that the program would no longer be worthwhile for them.

In 1987 the different interests resolved themselves at last. The Marine Corps needed their long-range assault capability, and in spite of severe budget cuts and interference from the Department of Defense (which favoured helicopters), the program continued. The Army and Air Force were out; the Marines and the newly formed Special Forces Command became the sole customers.

The JVX became the V-22 Osprey, and it made steady progress until

1989 when a new disaster struck: Secretary of Defense Dick Cheney cancelled it. A series of arguments and counter-arguments took place, with politicians keen to preserve what was seen as a major source of employment in their states. Congressional supporters even threatened to take Cheney to court over the matter, but it was politics that saved the Osprey. Texas, a key state for Bush's election campaign, was where the aircraft would be built, and the President did not want to alienate voters. Cheney was persuaded to restore funding.

It looked as though a tilt-rotor aircraft was about to enter service for the first time. Then, in June 1991, one of the prototypes crashed during its maiden flight demonstration. There were no deaths, and the crash was blamed on wiring problems.

In 1992 another prototype suffered a mechanical failure which disabled one engine and it crashed into the water at a USMC base, killing seven crew members. The problems were traced and corrected.

In April 2000 there was an even worse crash, with nineteen crew and passengers killed. The cause was found to be pilot error, as the Osprey was descending too quickly – more than twice as fast as the recommended maximum. This caused a condition where lift was lost and the aircraft was out of control.[4]

In December 2000 a fourth Osprey crashed, this time killing four. Another problem was found, a leaky hydraulic line combined with a software defect. The software was supposed to compensate for the hydraulic loss, but instead it reacted the wrong way and aggravated the problem.

Although the complications which led to these crashes appeared to have been fixed, an operational evaluation carried out in 2000 found other problems.

The findings of the evaluation included 'marginal mission reliability, excessive maintenance manpower and logistic support requirements and inadequate availability, interoperability, human factors documentation, and diagnostics capabilities'. The Marine Corps requires 75 per cent availability; the Osprey was only fully prepared 20 per cent of the time.

More changes had to be made, and a further two years of flight-testing were ordered. Whether it will survive and actually make it into service remains open to question. In 2003 one of the test Ospreys made an 'unscheduled landing'; the cause was traced to a manufacturing defect in one of the hydraulic lines that gives power to the control surfaces. 'Uncommanded yaw oscillations' – in other words, the aircraft wobbling unpredictably – also appeared during flight-testing. Again, the problem was resolved but it does not necessarily increase confidence.

There are still doubts about the Osprey's actual capabilities. It is supposed to be able to transport twenty-four fully equipped troops, but some reports have suggested that the actual capacity will be more like eighteen, which makes some of the comparisons with helicopters invalid. There are also questions about its survivability; critics have suggested that technical factors will make it more vulnerable to ground fire than a helicopter.

Spiralling costs, a common enough feature of the procurement process, have been evident in the program. In 1986, the V-22 was to have cost $24 million per aircraft. By 1989 it had reached $35 million. Since then the price has increased further, and the 'flyaway' cost per aircraft, which excludes the billions spent in development, is now up to almost $70 million per aircraft. This will be the cost if the full production run goes ahead; the latest small batch cost a staggering $105 million each. Further delays will increase the costs more. Increased costs will lead to the total order being reduced – and this will push up the unit cost in turn.

The financial argument for the Osprey against helicopters has changed. The delays have also given helicopter technology the chance to move forward. New rotor designs have dramatically improved fuel economy. Helicopters remain a tried and tested technology, whereas tilt-rotors are new and very much unproven.

However, there is still an effort to produce a civilian version of the tilt-rotor. In partnership with Italian helicopter manufacturer Agusta, Bell is producing the BA609 civil tilt-rotor. This is intended for coastguard search-and-rescue missions, corporate transport, flying ambulance and transport to and from offshore oilrigs.

The BA609 has shadowed the Osprey, with the program being mothballed for a period while problems with the military tilt-rotor were being resolved. The first prototype made its long-anticipated maiden flight in 2003. A test program which will result in 'type certification' – clearance so that the aircraft is deemed safe for civil operations – is expected to continue until 2007.

In 1998, Bell were planning to deliver the first production aircraft in 2001. The BA609 is already five years behind schedule. As with the Osprey, the cost will depend on the number manufactured. In 1997, the manufacturers were estimating a cost of $8–10 million. This is much higher than an equivalent helicopter, so the tilt-rotor's advantages of speed and range would have been decisive. By 2003 the estimate was $10–12 million for a basic configuration; fully equipped aircraft would be more expensive.

There are already orders for sixty-five BA609s – a mixture of corporate, offshore and emergency services – but the market is far from sure. Even if the economics work out, unless customers can be persuaded that the tilt-rotor concept is now a safe method of travel, Bell Agusta will have trouble finding takers. If the military do not take to the Osprey, it is unlikely that the civilian market will be convinced.

'A forty-year investment,' says the Boeing website. In fact, it is now fifty years since the first US tilt-rotor, and sixty years since the wind-tunnel tests at Focke-Achgelis. Whether we are now any closer to seeing the tilt-rotor enter service is another question. Boeing might be happy to see the investment stretch to more decades, as long as funding continues to flow from the US government.

The Marine Corps enjoys a greater level of political support than the other armed forces from politicians who used to be in the service. 'Once a Marine, always a Marine,' is a proud boast, and it helps ensure that whatever happens to the Army, Navy and Air Force, the Corps always gets the funding it needs for the programs it wants. Unfortunately, this applies whether the project is a good one or not.

The US government may be happy to continue the funding as long as the money flows to factories in key electoral areas. The *News and Observer* of Raleigh, North Carolina, found subcontracts for the Osprey spread across 45 states and 276 congressional districts, giving nearly two-thirds of Congress a direct interest in it. Forty billion dollars of government money buys a lot of goodwill.

Perhaps the most ominous development is the suggestion that there were moves within the upper echelons of the Marine Corps to cover up problems with the Osprey. Officers were encouraged to falsify maintenance records to make the Osprey look better; an investigation identified the guilty parties in this case, but there are concerns that this is just the tip of the iceberg, and that there is a determination to get the Osprey into service whatever it takes. The Marine Corps does not want what they see as bureaucrats at the Pentagon to deny them their new aircraft.

Doubts persist over how safe the Osprey really is. The families of those killed in the April 2000 crash do not believe that pilot error was the real cause and have attempted to sue Bell. If they are right, then the Osprey may be less airworthy than anyone realizes; and if they win their case, the design will have to be amended again.

Meanwhile, the US Marine Corps still relies on helicopters for its over-the-horizon assault capability. And commuters will continue to dream of a Downtowner flying to the heart of the city at 250 m.p.h., while they wait for yet another delayed train.

PART 2

# Looking Forward

# CHAPTER 7

# UFOs and Secret Technology

*President: Regardless of what the tabloids have said, there were never any spacecraft recovered by the government. Take my word for it, there is no Area 51 and no recovered space ship.*
Chief of Staff Nimziki suddenly clears his throat.
*Nimziki: Uh, excuse me, Mr President, but that's not entirely accurate.*

From the screenplay of *Independence Day*
by Dean Devlin & Roland Emmerich

## The Black World

Few things are quite so mysterious as the secret world of 'black' research. These are the Pentagon's classified programs. Some of them, like the F/A-22 Raptor fighter, are relatively open – the aircraft is thoroughly publicized, although some of the technology used in its construction is still secret. Others, like the stealth bomber, are murkier, with only the vaguest details being given away of the plane's actual characteristics. Others are darker still, and blackest of all are the unacknowledged programs whose very existence will not be confirmed or denied. Such programs are covered by a level of security known as 'Special Access'. This is higher than 'Top Secret', and compartmentalized so that individuals may be aware of one program but not another.

Spending on black programs runs to billions of dollars every year. Security is tight, and speculation is rife. What does the Air Force keep in its secret hangars, in remote bases far away from prying eyes? This is fertile ground for speculation, paranoia and outright fantasy. Some people claim that the most important secret military technology is light years ahead of the rest of the world, having come from another planet.

The *X-Files* may have got there first, but the scene above from *Independence Day* was the sign that a story that had been in UFO circles for decades had reached the mainstream. According to this legend, the US government had recovered debris from a crashed alien flying saucer in Roswell, New Mexico, in 1947. The technology was being analysed so that it could be duplicated to build a fleet of US military spacecraft.

The legend was fairly detailed, specifying that the secret activity was taking place at an installation in the Nevada Desert called Area 51, where witnesses had seen strange craft flying around, craft that did not correspond to any known human construction. There were even accounts from people who had worked at the top secret base themselves, telling of an antigravity technology that was far ahead of modern science. The whole thing was so secret, in fact, that even the President was kept in the dark, with the real power lying in the hands of an inner cabal within the military-industrial complex.

These stories were completely denied by the government. There was no crash at Roswell, and no Area 51, so there was no question of craft powered by weird alien science. But the evidence kept piling up.

One thing is admitted, and that is that the US Air Force spends a huge amount of money on secret research. The exact figures are not published, but can be deduced from the gaps in the published accounts. According to aviation journalist Bill Sweetman, during the 1990s so-called 'black' programs accounted for over $7 billion a year in research and development and a further $7 billion a year in production. The last production aircraft to come out of the dark side was the B-2A stealth bomber, rolled out in 1992. Nobody outside the US defence establishment knows what all the billions have been spent on since then.

In spite of the lack of visible products, the US government is approving even greater sums. The 2004 black budget was estimated at over $23 billion out of a $300 billion defence budget.

What we do know is that there is secret technology out there, some of it already flying at desert bases. It may transform air travel in the future; but it is not necessarily what the flying saucer buffs have in mind.

## The Roswell Incident

One component of the myth focuses on events in 1947. On 14 June, a rancher near the small New Mexico town of Roswell in New Mexico found some strange debris on his land. It consisted of sticks, metallic paper and tape with what appeared to be hieroglyphic writing in an unknown alphabet. The rancher, Mac Brazel, reported his find to the local sheriff, who passed the report to the intelligence officer at the nearby army airfield.

The military came round to pick up the debris and were understandably puzzled. The term 'flying saucer' would be coined a few days later, but this was the golden age of science fiction and there were plenty of reports of sightings of disk-shaped craft. For reasons which have never been entirely clear, the commanding officer at Roswell Army Airfield put out a statement that they had retrieved parts of a 'flying disk'. The statement was quickly taken up by the press, and photographs of the mysterious recovered items made it to the newspapers. (It has been suggested that Roswell was not the most exciting posting, and that the 'flying disk' idea was the result of boredom and over-active imaginations.)

When the story was published, more senior officers were outraged by what they saw as tomfoolery. A second statement was put out saying that the find was nothing but a common weather balloon. But the hue and cry had already been raised, and questions were asked about how the pieces on display could have been part of such a balloon. There was no explanation of the tape with the cryptic writing on it, so the military must be covering something up.

And, as it happens, they were covering something up, the truth of which did not come out until many years later. It was known that sound could travel for vast distances under the ocean in 'sound channels' caused by undersea currents. Scientists theorized that a similar effect could occur in the jet streams of the upper atmosphere, so that a colossal explosion – an atomic bomb – might literally echo around the world. A project was instigated to test this idea as a means of detecting Soviet atomic tests, under the name Project Mogul.

The way to pick up these echoes would be to launch an acoustic sensor up into the jet stream to listen for them. There was no single balloon large enough to carry the instruments for this, so a cluster of smaller balloons was used; in order to track it from the ground, radar reflectors were attached to the apparatus. However, it turned out that the radar reflectors were too flimsy, and had to be reinforced. The reinforcement was hastily improvised from metallic gift-wrapping tape with flower patterns embossed on it.

A string of two dozen weather balloons with radar reflectors was launched from Alamogordo in New Mexico on 4 June 1947. Contact with the balloons was lost when they were less than 20 miles from the Brazel ranch.

Given these facts as we now know them, the coincidence between Project Mogul and what was found at Roswell is hard to deny. Hard, but far from impossible: once you accept that the government puts out cover stories, everything they say becomes immediately suspect. If you believe that aliens and flying saucers are involved, then the Mogul story is just another attempt at misdirection.[1]

The belief in the flying saucer crash did not just survive, it flourished and grew. In the 1970s there was a renewed interest in the case, and more witnesses came forward. They reported seeing not just the debris – and much more substantial remains than the 3 kg or so that Mac Brazel had recovered – but also alien bodies. Some even suggested that one or more of the alien pilots were still alive when they were captured by the military.

The story gained momentum, and in 1995 Ray Santilli, an English TV producer, came up with some footage of what had been offered to him as a genuine 1947 film of the autopsy of an alien from the Roswell crash. The Fox TV network bought the film and turned it into a major documentary. Sceptics were unconvinced and pointed out details (such as a safety sign in the background) that were not consistent with 1947, but the believers loved it.

Alien autopsies became an established part of the Roswell story and this mythology was taken up by the writers of *Independence Day*. In their version the secret base at Area 51 houses not just some wreckage but an intact alien craft the size of a jet fighter as well as the preserved bodies of several aliens.

This has also fed into the myths surrounding 'alien abductions'. There are very detailed and elaborate descriptions of how of the US government has undertaken secret deals with extra-terrestrials in which the aliens are allowed to abduct humans in exchange for giving the US advanced technology. According to this version, the military are involved in covering up the whole abduction phenomenon, or may even be involved in it.

In this fertile soil for conspiracy theory the smallest seed of fact can grow into a whole paranoid world.

## The Beginnings of Black Aircraft:
## Muroc and Paradise Ranch

When the US started development of their first jet fighter in 1942, it was clear that the new aircraft would have to be test-flown somewhere as far away from possible enemy spies as possible. The Bell plant in Buffalo, New Jersey, used for other aircraft was too public.

Fortunately America possesses many wide-open spaces, and thousands of square miles of them belong to the military. The Southwestern Desert is a patchwork of reservations, national parks and military areas. At the opening of the war, a dry lake bed called Rogers Dry Lake was being used as a gunnery and bombing range by the US Navy. The range included a mock-up of a Japanese cruiser in the middle of the lakebed as a target. The nearest settlement was a place called Muroc and within months it was the site of two new airbases, the first of their kind.

The jet aircraft, XP-59 Airacomet, were dismantled and shipped out to the remote location, a journey taking several days. Their first flights took place away from prying eyes, and it was not until 1944 that the existence of the aircraft was revealed. The remote bases had proved their ability to keep a secret.

In 1954 when the Cold War was in full swing secrecy became even more important. American secrets could be concealed in the vast military areas, but the Russians had a whole country for their own. The Iron Curtain was an impenetrable barrier – there was no way of telling how many ballistic missiles the Russians had, how many aircraft, or how many aircraft carriers or submarines were under construction. Estimates of the Soviet economy were based on guesswork, as there was no reliable information on how many mines or factories were in operation, or even how many million hectares of wheat were being grown.

Soviet airspace was closed to Western spy planes. A few attempts were made involving dashes at high speed across parts of the USSR close to the border, using bombers especially stripped down for the purpose, but these were hazardous and could achieve little. What was needed was an aircraft that could fly across the whole expanse of the Soviet Union, photographing everything, without being intercepted.

Lockheed took up the challenge to design a spy plane for the CIA capable of operating at high speed above the reach of Russian interceptors or surface-to-air missiles. The existence of this aircraft, whose sole purpose was to violate international agreements on airspace, was to be kept top secret. Originally known by the code name Aquatone, it later became the Angel, and eventually received the innocuous designation of Utility aircraft, model 2 or U-2.

The U-2 required an unusually long runway to reach take-off speed, and would have to be kept somewhere remote from any human occupation while it slowly climbed to its cruising altitude of almost 60,000 feet, three times the altitude of the 1950s airliners. A team from Lockheed scoured the south-west for a suitable location.

They found the ideal site, another dry lakebed which formed a natural airstrip, 'as smooth as a billiard table without anything being done to it', at a place called Groom Lake. This was another wartime gunnery range, but apart from some empty shell cases it needed little clearing up.

The Groom Lake site reached over a hundred degrees in summer with fierce dust storms, and was one of the most rugged, isolated and inhospitable locations imaginable. To encourage staff to move there, Lockheed manager Kelly Johnson christened it 'Paradise Ranch'. The location and nature of the site were highly classified, so people only found out what they had let themselves in for after they arrived.

The base was generally just known as 'the Ranch' or 'Groom Lake'. The radio call sign of the base gave it another nickname: 'Dreamland'. By 1960 the whole area had been renamed in accordance with a new system for designating military test sites in Nevada and it became 'Area 51'. However, the site has remained classified since its inception, and as far as official statements are concerned, it does not exist.

There were a few scant references to the U-2, which was officially a scientific research aircraft used by NACA (the National Advisory Committee for Aeronautics, the precursor to NASA) for gathering information about weather conditions at high altitude. However, sightings of UFOs by airline pilots and air traffic controllers increased in the area around the Nevada test site. The early U-2s were silver, and especially at dawn and dusk they became highly visible as a bright light against the dark sky. They showed up on radar, and were clearly flying much higher than any known aircraft. Their altitude made them difficult to see clearly, and it was difficult to see more than the silvery glint, so it was not obvious that they were aircraft rather than saucers.

At the time, there was some interest by the military and intelligence communities in the flying saucer phenomenon and whether it might present a threat. A study of flying saucer reports was being carried out under the name Blue Book, and the investigators were able to check with CIA staff. According to one estimate, as many as half the otherwise unexplained UFO sightings of the period were caused by the U-2.

Later U-2s were painted a dark grey, and measures were taken to reduce the plane's radar signature. Although intended to make it harder to detect by Russian air defences, these also reduced the number of UFO sightings.

U-2 overflights of Russia started in 1956 and were highly successful: the Russians could see the plane, but they could not reach it. U-2s flew with impunity over Russian military bases, photographing rows of aircraft and hangers, getting pictures of runways and naval shipyards and everything else of interest. Missiles were fired at them, but exploded harmlessly without getting close. Russian MiG interceptors circled helplessly thousands of feet below, unable to reach the altitude of the U-2. If they tried, their jet engines 'flamed out' in the thin air, so they would fall back earthwards.

The Russians were not about to admit publicly that the Americans could violate their airspace with such ease – to do so would have been a confession that they were technically inferior to the hated capitalists. They kept quiet about the overflights, while maintaining their efforts to intercept one. Bringing down a U-2 would be a major propaganda achievement.

Given the huge success and continuing secrecy surrounding the U-2, the CIA could not allow it to be known that UFO sightings were actually their new aircraft. The American public were fed a certain amount of deliberate disinformation which suggested that the sightings were caused by mirage effects, ice crystals in the atmosphere, meteors and other natural phenomena. Many believers in flying saucers did not believe any of this, and there was a growing conviction that the CIA was covering up flying saucer sightings. This was quite true, except that what was being concealed did not originate from beyond the stars, but from Lockheed's manufacturing facility in Burbank, California.

## The Fastest Thing on Earth

The U-2 was only the first of the black aircraft to come out of Groom Lake. The CIA knew that it was only a matter of time before the Soviets caught up and built a plane or a missile capable of intercepting the U-2, and they were already working on something better.

The U-2s replacement would fly even higher, but rather than altitude it was speed which would make it invulnerable. Again, Lockheed won the contract and set out to build a revolutionary new aircraft. Initially known as 'Oxcart' as an ironic reference to its speed, it was later called Archangel and was eventually given the military designation SR-71 (for Strike/Reconnaissance). The plane was popularly called the Blackbird because of its dark paint finish, the Strike/Reconnaissance tag being given because in addition to spying, the Blackbird would be able to carry a one-megaton nuclear device. This was intended for 'second strike'; after a nuclear exchange the Blackbird would be able to fly over Russia, identify targets which had not been destroyed and finish them off with a one-megaton bomb.

The Blackbird would have to overcome major technical hurdles to achieve its goals. Even in the thin air of the upper atmosphere, air friction would heat the wing edges to high temperature – over 250°C. Normal materials could not tolerate this kind of temperature, and the airframe would have to be made of titanium alloy.

The program required titanium of a quality never used before, and many discoveries were made about engineering with the metal. In one case a worker wrote on a piece of titanium with a normal pencil, and found a week later that the graphite had caused etching in the surface. Considerable experience was gained in new methods of working titanium, but although there have been some niche applications, especially in aerospace, expense has limited its use. Wearers of replacement hip joints may owe something to the SR-71 though.

Special fuel was another requirement; at high speeds it would be exposed to temperatures which would cause ordinary fuel to boil and explode.

The airflow around the engines was critical, especially the problem of being able to operate at low speed during take-off and high speed at altitude. A novel engine design allowed a series of vents to act as intakes at low speed, but at higher speed the flow was reversed and the vents provided thrust.

Disaster struck the US manned spy plane program in 1960 when the Russians finally succeeded in shooting down a U-2. The US government initially denied that the aircraft was one of theirs, but were trumped when the Russians triumphantly paraded the captured pilot, Gary Powers. It was a major embarrassment for the US, and manned overflights were banned thereafter.

However, the Blackbird went ahead. There were many areas apart from the Soviet Union where the US needed spy planes. Vietnam, North Korea, Cuba, Libya and the Middle East were all areas of interest that could only be reached by an aircraft with exceptional speed. The Blackbird remained secret until 1964 – this time it was politics that broke down the wall of secrecy.

Right-winger Barry Goldwater was challenging incumbent Lyndon B. Johnson for the Presidency. One of Goldwater's charges against the President was that he had allowed America to lose its technological edge and that he was soft on defence. To counter this claim, Johnson announced the existence of the new aircraft:

> Tested in sustained flight at more than two thousand miles an hour and at altitudes in excess of seventy thousand feet . . . far exceeds any other aircraft in the world today. The development of this aircraft has been made possible by major advanced in aircraft technology . . . the existence of this program is being disclosed today to permit the orderly exploitation of this advanced technology in our military and commercial program.

This was not entirely true; the statement seemed more geared towards ensuring Johnson's election victory. The technological achievements of the Blackbird were considerable, but none of them translate easily into the civilian world. High-speed engines, exotic metal alloys and high-temperature fuels have little application outside the world of supersonic aircraft.[2] America gave up the idea of a supersonic civilian airliner at an early stage, judging it to be an uneconomic and quixotic idea that could be left to the Europeans.

The Blackbird's speed, somewhere in excess of mach 3, has never been beaten by a jet aircraft. For a time there was a contest with the Russians for the world air speed record. Every time the Russians took the record back with a modified MiG-25 Foxbat, a Blackbird pilot would be given clearance to put his foot down a bit more and beat the record again.

The world air speed record for a jet aircraft stands at 2,193 m.p.h., set by a Blackbird in 1976. That works out at a mile in just over one and a half seconds. In 1974 the Blackbird captured the record for flight between New York and London, taking under two hours, compared to three hours for Concorde or seven hours for a normal airliner.

This highlights the difference between the Blackbird and other military jets. While other aircraft can fly at mach 2 for very brief periods on full afterburner, they use up their entire fuel supply in minutes. The Blackbird can fly at higher speeds for hours, and the afterburner provides most of the thrust. The fuel economy was not that good though, with up to six in-flight refuellings being required on each mission.

Being in the white world also gave Blackbird pilots some opportunities for harmless amusement. On one occasion, Los Angeles Airport air traffic control was contacted by a military pilot requesting permission to go to 60,000 feet, higher than any airliner could reach.

'And just how are you going to get up to 60,000?' inquired the disbelieving air traffic controller.

'Not up,' replied the pilot, 'I'm going DOWN to 60,000.'

In its day the Blackbird was invulnerable, something that was proven time and again as barrages of SAMs were fired in a futile effort to stop the spy flights. But no overflights of the Soviet Union were attempted, or at least none that have been officially acknowledged. Even at almost 100,000 feet there was always a chance that the Blackbird might be in danger from the later Russian missiles. This has done nothing to dent the iconic image of the Blackbird as the fastest plane on earth.

In 1990 the Blackbird was officially retired from service. Even then the USAF could not resist the temptation to break the Los Angeles-Washington air-speed record before the aircraft was mothballed.

There is an interesting comparison with the nearest Russian equivalent, the Mig-25 Foxbat. Originally developed as an interceptor to tackle the cancelled B-70 bomber, the Foxbat evolved into an effective reconnaissance plane. For several years they flew with impunity over Israel. One was actually tracked at over mach 3, which caused some alarm – the Foxbat appeared to be faster than the Blackbird. It later emerged that this speed was achieved during an emergency manoeuvre and effectively destroyed the Foxbat's engines. The sustained speed was more like mach 2.8, slightly less than the Blackbird.

When a defecting Russian pilot landed his Mig-25 in Japan, Western analysts were given a chance to take the new Russian aircraft apart, and were astonished to find only a small proportion of the aircraft was titanium, the airframe being made of steel.

While the Foxbat was not quite as fast as the Blackbird, it still holds a number of aviation records such as climbing to over 120,000 feet. And while the Blackbird was retired with only thirty-two built, hundreds of Foxbat variants still serve in several different countries, as well as Foxhound interceptors developed from the Foxbat. The need for hard currency is very much a factor in Russia; while you can only dream of flying in a Blackbird, you can take off in a Foxbat and soar to 'the edge of space' (as it says in the advertising) for a cool $12,000. Thrilling rich tourists probably is not what the MiG designers had in mind, but it's another example of the accidental spin-offs of military technology.

Many aviation watchers assumed that a new secret plane was being brought in to replace the Blackbird. There was talk about 'Aurora', an aircraft said to fly at mach 6, or mach 8, or even faster. Mysterious contrails, with a shape described as 'doughnuts on a rope' were sighted in California. Some aviation experts believe that these contrails were the signature of a novel type of jet engine capable of operating at very high speed. Loud sonic booms, 'skyquakes', not associated with any known aircraft, were thought to be related to the mysterious Aurora.

But as the years passed, it seems increasingly unlikely that there is a super-Blackbird out there. The Blackbird was briefly brought back into service in 1995 and then retired again in 1997.

The reconnaissance needs of the CIA and the USAF can be met more easily using increasingly advanced spy satellites and a range of 'stand-off' sensors capable of peering from a distance without the need to fly immediately over the target. Unmanned spy planes like Global Hawk and Predator can carry out tactical reconnaissance without risking human pilots.

Although Aurora the spy plane may be a myth, there may be other high-speed aircraft out there which might account for some of the observations of strange aircraft in the area, including the mysterious contrails and skyquakes (discussed in a later chapter). But it was not just aircraft sightings that fed the Area 51 myth.

## Lazar and the Saucers

The Area 51 myth has taken on a life of its own. Flying saucer watchers hike out to the mountains to get a look at the air activity over Groom Lake. Freedom Ridge was the preferred viewing point until 1993 when the USAF took over the area to try to stop people from overlooking their activities. But the watchers still go, in spite of the signs warning 'DEADLY FORCE AUTHORIZED' against trespassers.

Security is provided by a civilian agency, and although there are no fences, there are cameras and other sensors. The guards, known as 'Cammo dudes' for their camouflage uniforms, drive $4 \times 4$ jeeps and round up trespassers or chase them off. Deadly force has not been used yet, but there are many colourful stories.

The watchers come back with stories of golden orbs, blue streaks, zigzagging lights and other phenomena. Evidence of flying saucers is lacking; a television crew sent to bring back footage of the weird craft at Area 51 showed the flashing lights of an unidentified flying object over Groom Lake, an effect somewhat spoiled by the fact that it could easily be identified it as a Boeing 737 – one of the regular flights from Las Vegas carrying the work force to and from Groom Lake.

In 1989 the story received its biggest boost ever. A man called Bob Lazar gave interviews on local television in Las Vegas describing how he had worked at Area 51 as a physicist on a secret project – working on the propulsion system of extra-terrestrial flying discs.

According to Lazar's account he only worked there for a few months. However, in addition to finding out how the craft worked (it generated a form of antigravity using a new element with an atomic number of 115), he was also given access to material about the visiting aliens who come from a star system called Zeta Reticuli. According to Lazar these aliens had been involved in genetic manipulation of humanity for at least 10,000 years and had planted several major religious leaders, including Jesus Christ.

Many took Lazar's story at face value and he was widely considered to be one the most credible witnesses to the government-alien conspiracy theory. Everything he said fitted with existing theories. His physics and engineering sounded convincing, and he propounded a detailed ex-planation of the antigravity drive system.

Getting any supporting evidence for his claims was difficult, however. Naturally there is no official record of Lazar's work. He also claims to have received degrees from MIT and Caltech, but there is no record of these either; Lazar points to this as evidence of how thorough the security apparatus is when it needs to discredit someone.

Lazar is a flamboyant character whose other interests include jet-powered cars, firearms and explosives. He used to run a photography business in Las Vegas, specializing in real estate which was then a thriving occupation. He once pleaded guilty to pandering (recruiting women for prostitution) and admitted that he had been involved in modernising the computer systems at a legal brothel in Nevada.

Although it might be easy to forgive Lazar his lifestyle and accept that his qualifications as a theoretical physicist are valid, what ultimately undermined his case was bad science. His physics may sound good to people who grew up with *Star Trek* dialogue about matter-antimatter reactors and dilithium crystals, but genuine scientists have found it laughable. Progress in the last few years has not helped Lazar either. In the 1980s, the idea of a stable element, 115, was much more tenable than it was in the 1990s when research into such superheavy elements progressed. Element 115, christened Ununpentium, has now been created in a Russian laboratory, but as predicted by nuclear physicists it only lasts for a fraction of a second before decaying.

Lazar's story has anomalies. Although he was the most junior member of the team, he had insights into alien technology that allowed him to leap ahead of his colleagues who had been studying it for years. How exactly did he make that leap? More strangely, why was he given information about the aliens themselves and their relationship with humanity which had nothing to do with his work?

Although Lazar has lost all but his most hard-core fans, he has played an important role in promoting the idea that Area 51 is the home of weird science. Long after Lazar dropped out of the public eye, the idea remained that They Are Hiding Something.

And of course They are.

## Now You See It . . . Stealth

The designers of the U-2 and Blackbird knew that radar was the main threat. The distance at which an aircraft will show up on radar depends on several factors including its size, shape and material. Size is obvious –

a large object reflects back more radio waves than a smaller one. Composition is equally easy to understand by analogy with vision; a radar is like a searchlight shining a narrow beam of light out into the dark. If the object it is shining on is black, it will be much harder to see the reflection than if it is white. But why should the radar return depend on the shape?

Visible light has a wavelength that is measured in millionths of a metre, but radio waves can have a wavelength of tens or hundreds of metres. This is crucial, because it means that the objects being detected by radar are on the same scale as the radar waves. Shine a light on a surface, and the light will be scattered back in all directions, so the surface can be seen from any angle. But if you shine a radar beam on the same surface, it will reflect back much more strongly in some directions than others. From some angles, hardly any of the radar beam will be reflected back to the source. From these angles, the radar will not be able to 'see' the surface.

Ever since the invention of radar, aircraft designers knew that some aircraft were more visible on radar than others. Certain features such as square corners and junctions are very good radar reflectors; smooth curves much less so. The designers of the Blackbird attempted to reduce the radar visibility of the aircraft as far as possible, by avoiding features which were known to increase the radar signature. It was a very hit-and-miss affair, as there was no set of rules to determine exactly how to work. But they succeeded fairly well, and the Blackbird has a very small radar signature for its size. The idea is that when combined with its high speed, its low radar visibility means that by the time it is close enough to detect the plane on radar, it should already be too late for air defences to engage it.

In 1962 a Russian mathematician called Pyotr Ufimstev published an obscure paper in a journal on optics, called 'Method of Edge Waves in the Physical Theory of Diffraction'. This paper set out a new method of calculating exactly how electromagnetic waves (including light and radio waves) would be reflected from different shapes. The paper was seized on by researchers at Lockheed's Skunk Works who realized the implications.[3] Ufimstev's paper would give them the tools they needed to calculate exactly how to shape an aircraft so that it did not reflect any radar energy back to the source. Such an aircraft would be effectively invisible to radar. The new science of radar stealth was born.

Unfortunately, the process of calculating radar reflections required huge amounts of computing power, but the computers of the day were simply not powerful enough to carry out the calculations needed for three-dimensional objects. They could only work in two dimensions. Undaunted, the Lockheed engineers set about the task of designing an aircraft whose external surface consisted entirely of flat, two-dimensional shapes.

The result was an extremely unstable, ungainly craft called Hopeless Diamond. It would not have won any prizes for its aerodynamic qualities, but it proved to have a radar visibility that was thousands of times less than any existing aircraft.

The distance at which an aircraft can be detected depends on the fourth power of its radar visibility; in other words, if it has 10,000 times less radar signature, it can be detected at one-tenth the distance. Stealth aircraft are not completely invisible to radar, but their shaping along with a coating of radar-absorbent material (the radar equivalent of black paint) means that they can only be detected at close range. Air defence radar are strung out over many miles to cover an approach; stealth meant that by plotting a course that stayed away from individual radar, aircraft could slip through the holes in the defences.

The implications were mind-boggling. For decades, the Cold War had relied on the theory of deterrence and Mutually Assured Destruction. If one side tried to attack the other, there would always be a few minutes warning before the missiles hit, long enough for the other side to launch a retaliatory strike. Stealth meant that there would be no warning of an attack until after the stealth bombers had dropped their bombs.

Stealth aircraft would also be effective in a tactical role. During the Vietnam War, a strike mission had to be accompanied by extra aircraft to jam enemy radar, 'defence suppression' aircraft which would deal with surface-to-air missile sites, and fighter cover. Airborne radar was needed to keep track of all these aircraft and any enemy fighters in the area. A stealth tactical aircraft would be able to operate on its own without any of this additional support.

The political implications were considerable. Stealth would make existing aircraft obsolete. More importantly from the point of view of the defence industry, the B-1 bomber project had only just started. This was an expensive and controversial bomber program; stealth meant that it would be outdated before it ever got off the ground.

In the end politics and money beat technology. The B-1 was allowed to continue until a stealth bomber could be built. Meanwhile a stealth fighter based on Hopeless Diamond was already on the way.

Although the world was aware of the existence of the stealth fighter, its appearance was a closely-guarded secret. As soon as it was seen, the distinctive covering of flat panels would be obvious as the key to its radar stealth. Extensive security precautions were therefore undertaken.

As part of a disinformation campaign, the squadron operating stealth fighters was also issued with a number of A-7 strike aircraft. These were equipped with empty napalm cylinders on the wings which were painted black and fitted with flashing lights. A prominent warning sign on the cylinders warned: 'DANGER – REACTOR COOLANT OUTLET'.

Whenever the A-7s used another airbase, ground crew were ordered to lie down on the ground and look away. Armed military police would move any onlookers as far away as possible. Sure enough, the rumour got out that the new top-secret stealth technology was some sort of nuclear black box device. The actual secret of stealth was not revealed until the Nighthawk stealth fighter was unveiled in 1990, nine years after it had first flown.

The stealth fighter was followed by the B-2A Spirit stealth bomber. Later aircraft such as the next-generation F/A-22 Raptor have been built with a high degree of 'stealth'. Details of stealth technology are still highly secret, but other nations have shown little interest in following suit. The costs are simply too high, and the compromises between the demands of stealth and the demand of aerodynamics are too great, for it to be practical for any except the world's richest nation.

Stealth has found little application outside of the military arena. One unusual offshoot has been a proposal for radar-absorbing material on the blades of wind turbines, as a way of preventing them from interfering with military radar. Although one might have expected that an air defence system costing millions might be able to overcome the effects of a stationary turbine, the Ministry of Defence have objected to a number of possible wind farm sites because of possible interference with their radar. A stealth coating on the wind turbine blades should remove the problem.

Meanwhile work continues at Area 51. There is no doubt that there are advanced stealth concepts, unmanned aircraft and other projects going on. But no new black aircraft have been unveiled. There have been some appetizers like Tacit Blue, a prototype stealth battlefield surveillance aircraft that was unveiled in 1996, but that was a contemporary of the early stealth fighters and the program had finished by 1985.

The question is what have the black aircraft builders been working on since 1990? An aerospace consultant from Jane's came up with an answer in 2001 which harked right back to the days of Bob Lazar: antigravity flying saucers.

## Podkletnov, Nazi Science and US Flying Saucers

Although Nick Cook's claims bear some resemblance to those of Bob Lazar, the two could hardly be more different. Cook was editor at the prestigious industry magazine *Jane's Defence Weekly* for fourteen years and is an award-winning journalist, with a high reputation in the aviation industry. So his book *The Hunt for Zero Point – One Man's Journey to Discover the Biggest Secret Since the Invention of the Atomic Bomb* brought high expectations with it.

Written like a thriller, *Zero Point* opens with a 1956 news article arriving mysteriously on Cook's desk. The article explains how American scientists are developing an antigravity engine; Cook decides to find out what happened to this technology.

He travels to Poland, where a local man shows him the site of German experiments during the Second World War. Cook concludes that the Nazis were working on antigravity and that after the war an SS officer called Kammler made a deal with the US and giving them the technology, like the V-2 scientists in Operation Paperclip.

Cook also talks to a Russian scientist, Podkletnov, who has published controversial papers claiming discovery of an antigravity effect using rotating superconductors. Although not widely accepted in scientific circles, this work was considered to be significant enough for NASA to investigate it as part of their Breakthrough Propulsion Physics Program into revolutionary new drive systems for spacecraft. Cook traces the origins of Podkletnov's work back to the same source as the Nazi experiments in Poland.

Cook also finds a 1947 memo from General Twining saying that flying saucer-type craft were 'within the present US knowledge'.

It all adds up, in Cook's view, to a solid case that the US possesses secret technology that could transform the world. The claims are amazing; the big difference between this and earlier versions is that in this case the astounding new technology has been captured from the Nazis rather than from aliens.[4]

Does this mean we should be buying shares in antigravity research and looking forward to weightless aircraft in years to come?

Cook mentions the Roswell flying saucer crash and Bob Lazar, and many others. He admits that there is plenty of exaggeration, outright lying and disinformation out there, but his guiding principle is that there is no smoke without fire and there must be something to it all.

The evidence does not look good under close examination. The fact that it is never established how Witowski, his Polish informant, knows that the supposed secret Nazi site was anything to do with antigravity, is never disclosed. Other informants are left unnamed and uncheckable.

Cook's source for the Nazi involvement in flying saucer technology comes from Viktor Schauberger, an Austrian forester who features in various versions of the Nazi UFO story but has no track record in actual science.

The Twining memorandum may look like a smoking gun, but on closer examination it turns out to be something else. Twining lists six traits for the flying discs sighted in UFO encounters: metallic or light reflecting, no trail, circular or elliptical, formation flying, no associated sound, level speed above 300 knots. He also says that they could achieve a range of 7,000 miles.

Could US technology of the time have built something with these features? They could, and they did – all of these traits except for speed were shared by balloons in the secret US balloon program (and in some cases the balloons were tracked at over 200 knots). And, as it turns out, the man in charge of the balloon program was the same General Nathan Twining. It is not surprising that he should be referring obliquely to the balloon program.

For all his hunches, Cook lacks substantial evidence. A story in 2002 that Boeing was working on an antigravity drive flared briefly but disappeared when the company publicly denied it.

Meanwhile Podkletnov's work has led nowhere. NASA did not succeed in replicating his experiments in antigravity, and in January 2003 their Breakthrough Propulsion Physics program was scrapped.

In the end, we are left with nothing. Sometimes there is smoke without fire – especially when there is a smokescreen. There may be something behind the smokescreen though. This is a technology known as plasma aerodynamics, in which an electric field is produced around an aircraft. This has beneficial effects both in reducing drag and in mitigating the shockwave created at supersonic speed; interestingly, it also increases radar stealth by absorbing incoming radar energy. Both the US and Russia have significant plasma aerodynamic programs, but the subject is rarely discussed in the white world. As one industry insider put it to me, 'I know of no unclassified uses of this technology.'

## The Art of Concealment

'Sometimes the truth must be guarded by a bodyguard of lies,' said Winston Churchill. Maintaining the secrecy of black programs requires similar measures.

Security requirements are estimated to contribute up to 20 per cent of the overall expenditure on black projects. It is not just a matter of the physical security of installations and computers, and the need to ensure that all staff are reliable. There is also the question of how to handle the inquisitive media.

But there is more than one way to block investigation. At the beginning of *Zero Point*, a witness describes the crash of a secret stealth fighter. The crash site is quickly surrounded by masked commandoes brandishing assault rifles, who drive off the spectators with threats and reduce the witness to hysteria. She is awestruck by the scale of the cover-up afterwards.

'I read they sieved the dirt for a thousand yards from the impact point,' she says. 'They were damned thorough.'

This would mean that some 27 million square feet were sieved; even if it was only sieved to a depth of one foot that would be a million tons of material.

On the other hand, just circulating the story that the area had been sieved would put off most souvenir collectors.

An account of the stealth fighter crash not mentioned by Cook says that after removing as much of the wreckage as possible the recovery team sowed the ground with parts of a crashed F-101 Voodoo. Whether this is true is irrelevant: the story is enough to cast doubt on any fragments found at the crash site. Any traces of secret technology slip away into the shadows of deception and misdirection.

The public are entranced by stories about aliens and antigravity. Ever since the silver UFOs seen by airline pilots in the 1960s, the USAF has taken care to ensure that the real story remains hidden. If you do see something strange, the USAF would far prefer you to report it as a flying saucer (alien or otherwise) than give away the secrets of classified aircraft.

# Real Black Programs

The black world continues to thrive in the Post-Cold War world, partly because of the continuing war against terror. The black budget for 2004 was estimated to be the highest since the 1988 peak of the Reagan administration, according to the Center for Strategic and Budgetary Assessments. The estimated $11 billion earmarked for production might be expected to buy whole squadrons of aircraft – in 2003 the pre-production Raptors cost a phenomenal $200 million each, so the black budget would buy fifty of them, or hundreds of more reasonably-priced planes. Even if the black aircraft are giant, multi-billion-dollar craft there ought to be a lot of them about by now.

With no new classified plane revealed since the stealth bomber, the nature of black procurement remains a mystery. Although anti-gravity programs may be going on, it is more likely that the main expenditure is in other directions.

There may be some spy plane research, although the known US reconnaissance assets leave few gaps. The U-2 spy planes are still in operation, although they have been greatly upgraded and carry much more sophisticated sensors which allow them to look down from a distance without having to fly over an objective. The Blackbird was finally retired in 1998, as the operating costs were high – said to be more than $100,000 for each flight – and it faced competition from faster, higher spies in orbit. The U-2's long flight endurance means that it is still useful, although increasingly it is in competition with the Global Hawk (discussed in a later chapter).

The US can dominate the airspace virtually anywhere in the world with its existing assets. It is highly unlikely that a black program would trespass on the territory of an unclassified project. The F/A-22 Raptor, the next-generation air supremacy fighter, is the USAF's flagship program. It has come under repeated attack by Congress as a 'Cold War relic'. The numbers planned have been slashed repeatedly from an initial purchase of almost 800 aircraft, and the projected number is now down to less than 300. Because of budget limitations, any increase in the development costs will be funded directly out of the production costs.

If it was discovered that there was a better option being developed under a black program, the Raptor would be threatened – it is hard to see the USAF allowing anything to endanger its most cherished program. The lessons of the B-2/B-1 are likely to have been learned and black aircraft will not undermine white ones.

But if the money is not going on 'platforms' – the aircraft themselves – where is it going? Based on discussion with those within the industry, there are several candidate areas:

**Advanced targeting systems.**   The ability to merge data from several different sources – radar, thermal imaging, millimetre wave (see Chapter 13) sometimes called 'hyperspectral' imaging because it uses a wide range of the electromagnetic spectrum – would lead to a significant improvement in the ability to locate and identify targets.

For example, a tank parked in a clump of trees with its engine running will be invisible in the visible spectrum due to the cover. But its heat emissions will show up in infrared, and it will be visible to foliage-penetrating radar. Looking at three separate images and comparing them would be difficult in real time. But a system which could combine and analyse the three together would be able to say that there is an object under the trees which has a radar signature and is generating heat, making it unlikely to be a dummy or a natural object – rocks show up on radar and animals are warm, but the combination is a stronger indication.

By combining this sort of data with other sensors, such as millimetre wave or terahertz, acoustic and chemical sensors, a far more complete picture can be built up.

**Satellite launches.**   As we have seen, the military are a major user of satellite launch services. Unfortunately, the limitations of the civilian facility does not always suit them. Since satellites sometimes need to be launched in response to external events – such as wars – and since it is often desirable for such launches not to be made public, it is difficult for them to fit in with civilian agencies. And while a commercial user may opt to have their package put into space by another country – Russia, for example, is a commercial leader in this field – this is not an option available to those interested in covert reconnaissance operations.

There have long been suggestions that the USAF possesses some capability of its own to launch packages into low earth orbit. This capability would be extremely expensive and would not necessarily produce visible results, but it would make a significant difference in terms of coverage by spy satellites and communications systems.

It might take the form of a 'piggyback' launch. This would be a rocket carried by a fast aircraft like a large Blackbird. The aircraft would effectively be a reusable first stage, carrying the rocket to high altitude at mach 3 or more before releasing it.

The possibilities for future launch technology are also discussed in a later chapter.

**Deep attack.**   The war against terror has given a renewed emphasis on ways of attacking deep underground bunkers. Bombs leave a shallow crater on the surface but hiding underground has been good protection since the first Zeppelin raids.

To improve capability against hardened targets, the USAF developed bombs with a thick steel casing and a time-delay fuse, so that they penetrated through earth and concrete before detonating. During the 1991 Gulf War these proved inadequate against some of the stronger targets, in particular a series of underground barracks protected by about forty feet of earth and two feet of reinforced concrete. An emergency program was carried out, making a bomb using an old artillery barrel as a casing. This was the 2,087-kg BLU-113 which was effective against Iraqi bunkers, but a better solution was needed in the long term.

In the late 1990s a new dimension was added with the perceived threat from 'weapons of mass destruction', chemical and biological weapons which might be produced or stored in underground facilities. There was a demand for a new type of weapon capable of neutralizing WMDs without spreading them, known as an 'agent defeat warhead'. This would have to produce very high temperatures without a massive explosion; alternatively a 'bleach bomb' could use chemicals to clean up WMD sites.

The USAF is known to have looked at a large number of different options for bunker attacks. Systems under development include various high-velocity missiles and special bombs which use a series of explosions or projectiles to tunnel into an underground complex before exploding. These have already produced a civilian spin-off in the form of a new type of drilling machinery which fires expanding projectiles to break up rock, and is claimed to be faster, quieter and cheaper than other methods.

There are also more theoretical studies for 'subterranean platforms' – unmanned vehicles which burrow like moles to reach an underground target, and can either carry sensors to spy on the target or a warhead to destroy it on command.

We will also see in later chapters how nanotechnology and hypersonics are being used to contribute in this area.

**Directed-energy weapons.**   In the long run, directed-energy weapons could make many existing weapons obsolescent.

The airborne laser is in some ways a primitive system. It is a Boeing 747 freighter carrying a large chemical laser, intended to shoot down incoming ballistic missiles. The chemical laser only has about thirty shots because of the need for large amounts of chemical fuel, and it is a massive and cumbersome system. But if it works, it will have huge implications because as well as shooting down ballistic missiles, such a laser could also destroy anti-aircraft missiles long before they were close enough to threaten the aircraft.

A laser-armed aircraft would rule the skies. It could not be shot down with surface-to-air missiles, intercepting fighters would be equally helpless, with their air-to-air missiles knocked out of the sky before they could hit, and the laser aircraft could shoot down the would-be interceptors with ease. The laser can have 360° coverage, and works at the speed of light. Dogfights would be a thing of the past; no manoeuvre could ever be fast enough to dodge a laser, and there is no need to line up on an opponent's tail.

Laser weapons research is now oriented towards building a powerful solid-state laser. This would be able to draw power directly from the aircraft's engines, giving it an effectively unlimited supply of ammunition.

The precision and control of laser weapons would also make them suitable for attacking ground targets. At present, attack aircraft do not usually approach close to targets, preferring to stand off at a distance of several miles and a height of 15,000 feet or more to avoid the threat of defensive missiles. But a laser-armed strike aircraft would not suffer from these limitations, and would be able to come in close to attack vehicles, buildings and even troops with surgical precision. The Advanced Tactical Laser is intended to fulfil this role. This is another chemical laser which will probably be mounted in a Hercules or similar low-flying tactical aircraft. The precision fire is being promoted as an option for use in urban areas, where it will be able to take out pinpoint targets without collateral damage. The ATL has been suggested as a 'non-lethal' weapon which is so precise that it can immobilize vehicles by melting their tyres without harming the occupants, and destroy weapons without harming their crew.

The ATL is a white program, but it is possible that this is the visible aspect of a much larger program which could be a generation or more ahead of existing laser technology. If it performs as advertised – always a major consideration with untried weapons – it could represent the next big leap in military technology.

**Invisible aircraft.**   Whilst the existing stealth technology is adequate for most threats, there are still some areas where it falls short. In particular, aircraft are still easy to detect by the most obvious method of all, what the RAF refers to as the 'Mark One eyeball'. A truly stealthy aircraft would not just have a low radar signature and reduced heat emissions and noise, it would also be hard to see.

Stealth fighters and bombers are painted a very dark grey. They fly at night; during daytime, they are simply too vulnerable to enemy attack, lacking any defensive armament and not being fast or agile. At night they can slip through defences and launch devastating attacks without warning, but in daytime they are just more slow attack planes.

During the Second World War, US Navy aircraft flying anti-submarine patrols were being spotted 10 miles away or more, giving U-boats the chance to dive before they could be attacked. Even white aircraft show up as a dark shape against the bright sky. The solution was the 1943 Project Yehudi, in which an aircraft was fitted with lights shining along the leading edges, a technique known as counter-illumination. These could be adjusted to match the background light level, with the effect of reducing the range at which the aircraft could be spotted from 12 miles to 2.

Yehudi was never put into production, but the same idea was tested during the Vietnam War, using a modified F-4 Phantom codenamed Compass Ghost. A more advanced version appeared in the 1990s involving an F-15 fitted with fluorescent panels. Aviation expert Bill Sweetman interviewed a technician on the project at Tonopah Test Range.

'According to the technician,' says Sweetman, 'the fighter virtually disappeared as it lifted off the runway.'

Other reports from Area 51 describe a jet which can be heard clearly but appears only as an indistinct blur. This may be linked to advances in the area of special camouflage coatings which change colour and brightness when a current is passed through them, similar to an LCD panel.

This technology could be applied to future aircraft, manned and unmanned, giving them greater ability to launch surprise attacks, appearing out of nowhere and then disappearing from sight.

Which of these areas is really getting all the money remains to be seen, and the possibility of civilian spin-offs in the near future is remote. When billions of dollars are spent on developing a capability that will put the US ahead of the rest of the world, it is unlikely that the technology is going to be shared around freely. The secrets of Area 51 and elsewhere seem likely to remain secret for the time being. All we can say with any degree of confidence is that the alien involvement is just a story, that there is no sign yet or weird technology, and airliners powered by antigravity systems do not appear to be in prospect.

# CHAPTER 8

# The Next Computing Revolution

*Frogstar Robot: 'You ain't seen nothing yet. I can take out this floor too, no trouble.'*
SOUND EFFECTS: More demolition. The Frogstar robot falls through the floor with a deafening cry which dies away as it falls through several lower floors as well.
*Marvin: What a depressingly stupid machine.*

Douglas Adams, *The Hitchhiker's Guide*
*to the Galaxy* (Fit the Seventh)

Computers are an integral part of the war machine, and ever since the days of Colossus and ENIAC, the military have sought more powerful computers. Since most computers come from the commercial world, the military are some way behind. The long period of testing and certification means that commercial technology, like new processor chips, can take years to be introduced into military service. If there are special requirements involved, such as 'ruggedizing' a system for field use, this may add further years. The infantry officer's laptop is likely to be a generation behind the slimline version toted by his cousin in the business world.

Fighter aircraft have state-of-the-art avionics when they are built, but updating them is a lengthy and expensive business, and it is not practical to put in a new computer system in the entire fleet every few years. The F-15C, currently the USAF's front-line fighter, entered service in 1979; by the 1990s there were cars with more computer power. When it was upgraded, the main computer processing system only had one megabyte of memory. This compares with the thirty-two megabytes on the Playstation 2 with which armchair pilots fly F-15 flight simulators. The new F/A-22 will have the equivalent of a supercomputer on board – but by the time it eventually gets into service this will seem pedestrian.

If it takes longer to get a product to the front line, in peacetime anyway, the military compensate by being very advanced in their research. But the technology flow from the military to the civilian world is very slow these days. In areas like artificial intelligence, networks and human interfacing the military can be well beyond the cutting edge of the commercial world. By looking at some of their less classified projects we may be able to see the future of computing in these areas.

## The Need for Artificial Intelligence

Computers are ridiculously stupid. They can calculate faster than any human being, but they lack the common sense of a two year old. Computers can barely manage the simplest conversation, and for all their megabytes and gigahertz they can only carry out instructions given in the plainest language. The problem is deeper than a lack of intellectual or linguistic skills; machines are incapable of carrying out the kind of mental task that is basic to every animal from insects upwards.

This lack of intelligence has led to the field known as Artificial Intelligence (AI) – designing machines which can carry out the equivalent of mental processes rather than just shuffling numbers. Progress in AI has been sporadic, with whole decades passing without significant progress in some areas. Research has been concentrated in academia rather than industry. Although there have been commercial AI systems of various types, there is no sign that IBM, Intel or Microsoft will be moving into the AI business rather than more conventional computing. Industry will sink billions into developing ever-faster silicon chips because of the proven market, but funding for serious AI research has to come from somewhere else.

Much of modern AI research comes from projects started by DARPA and the Office of Naval Research in the 1960s. According to one estimate, as much as 80 per cent of US AI research is currently funded directly or indirectly by DARPA. Unlike industry, the military do not need to see an immediate payback on their investment, and while commercial products may meet the requirements for military computers, there are some areas that only they can afford to finance.

Programmers discovered at an early stage that operations which are basic to human life – say, being able to recognize an apple – cannot be easily described in strict formal terms. A computer can carry out any set of instructions given to it, but the instructions need to be in a logical, mechanical form. The day-to-day world in which we live does not work like that.

A computer could recognize an apple if it could be defined as something with a specific size, shape and colour; by defining every possible appearance of an apple from any point of view, we could program the machine to determine whether any given object was an apple. Unfortunately for the computer, apples come in many shapes, sizes and colours, and their traits may overlap with other objects. A human probably wouldn't mistake a tomato for an apple even if it was the right size and colour, but we cannot find how to put into words what it is that constitutes 'appleness'.

Before it can even start on the apple problem, the computer has to know how to decide what constitutes the object and what the background. This is a seemingly impossible task, but it is one which humans and animals carry out every second of their waking lives. It is only under very taxing conditions – picking out a flower against floral wallpaper, for example – that we even notice we are doing anything at all difficult.

Similar problems have afflicted researchers into robotics. Walking and picking things up are not mentally demanding exercises for a human, but it is a different matter for robots. Instead of dealing with the clean and straightforward world of manipulating numbers, they have to work with countless unknown variables. We literally take this kind of thing in our stride, walking effortlessly from road to pavement without having to pause to calculate the size of the kerb and how high we need to lift our feet. Millions of years of evolution have equipped us with sophisticated natural software to deal with this situation, but robot makers have to start from scratch. The best robot walkers can still only cope with fairly even surfaces.

A similar challenge faced programmers trying to design software to use and understand human language. It is easy enough to build a computer, which can accept input in the form of typing or speech, and output as well, but the process of using language has proven quite opaque. Humans have built-in linguistic skills that allow us to pick up languages naturally, but nobody knows how this works. There are no formal rules to turn English into something a computer could use. We can use language with ease, but again we cannot explain exactly how we do it.

The quest for 'intelligent machines' which would be more than idiot-savant calculating machines began shortly after the Second World War. Alan Turing (mentioned earlier as part of the Bletchley Park team) is often considered to be the leading pioneer in this area. This led to much philosophical discussion on what we mean by intelligence and how we can recognize it. These questions have never been satisfactorily resolved.

Turing himself came up with the one of the best solutions, side-stepping the question by means of the 'Turing test'. In this test, an observer communicates with a subject by means of a teletype machine. The subject might be another human being, or a computer; any machine which can successfully carry out a conversation with the observer and persuade them that it is human could be considered to be intelligent. It's not a question of whether the machine really is intelligent or not, just whether it can convincingly appear intelligent.

There have been many efforts down the years to make machines which can pass the Turing test. Every year the Loebner Prize Competition is held, offering $100,000 and a gold medal to the first computer whose responses could pass for human. Although there have been some interesting competitors the prize does not seem in any danger of being won just yet.

Mainly, these attempts have made scientists aware of the richness and subtlety of human communication. This does not mean that no machine will ever pass the test; just that using language naturally, like many other human mental processes, is a complex matter.

Rather than having one machine to do everything, AI researchers have branched out into several different specialized approaches. The most notable of these are listed below:

**Expert systems,** also known as knowledge-based systems, are computer programs that incorporate a mass of information about one particular field. They can work successfully because unlike the broader field of human knowledge, small specialities can be described in terms of exact definitions.

A very simple expert system on geometrical shapes, for example, would contain all the human knowledge on what constitutes each shape. Given the information that a shape has four equal sides and four 90° angles, the system could deduce that it was a square, as no other shape in the system fits these criteria.

Expert systems have to be programmed with human knowledge, and while this is useful in some domains such as medicine, where an expert system can store more information than any one doctor, there are other areas where an ability to learn is important.

**Neural networks**   mimic the way that the human brain works. Biological brains are made of cells called neurons, each of which is connected to many other neurons forming a dense network. As the brain learns, information is stored by strengthening the connections between some neurons. A neural network is a device for learning by association. As the network is trained, it adjusts the connections until its output matches the correct answer.

For example, a neural network learning how to recognize a square will have strong connections between its concept of 'four straight lines' and 'four right angles' and 'square'. As more and more squares are shown to it, it will remove the connections between, say, 'blue' and 'square' when it finds that not all squares are blue, and that blueness is not connected to being square. But since all squares have four right angles, there will be very strong links between its internal representation of right angles and its representation of a square.

**Genetic algorithms**   are also based on a biological idea, that of evolution. An algorithm is a set of instructions for solving a problem. In a computer, the algorithm takes the form of a program. With genetic algorithms, various different programs are tried as a means of solving a problem and their outputs are compared. The programs that get closest to the solution are then combined together, in a process analogous to the way that animals breed to produce different offspring by sharing their genetic material. The next generation are then tried on the problem again, and the most successful are again retained for breeding the next generation.

Genetic algorithms could also be used for the problem of square recognition. Starting with a large pool of programs, each identifying a different feature, some would be more successful than others. The program for identifying a curve would be rejected, but the one for spotting a straight line would be retained – it is not always reliable in identifying squares, as it will also be triggered by triangles, but it is good breeding stock. Subsequent generations will get increasingly subtle, and different combinations of genes will filter out other confusing shapes like diamonds and parallelograms.

Eventually a winner will emerge after many generations as the 'genes' for spotting four lines of the same length all at right angles to each other are all combined together. Testing the winner against a large number of squares and other shapes shows that it will successfully work every time, and the new algorithm can then be used in practice – or as a gene to be a building block for future projects, such as identifying a cube.

**Intelligent agents,** sometimes called bots (short for robots) or droids, are autonomous programs that can carry out actions without human input. The term is usually applied to software that carries out Internet searches. They are also important in simulations, where agents can take the role of human beings and interact with each other. The idea of software agents was taken up in the *Matrix* film trilogy, in which the agents were literally embodied as FBI-style agents in a virtual reality world. Real agents are not nearly as advanced, but they can still carry out useful work beyond what conventional techniques can achieve.

The amount of development angled towards military use can be seen from the major projects undertaken with AI. A list of major neural network programs carried out in the 1990s included an underwater listening system from General Dynamics, which trained a network to be able to distinguish between different types of ships, boats and sub-marines by the sound of their engines. Another network from TRW was trained to recognize aircraft at a distance. Other programs worked on 'hyperspectral data fusion for target detection' – combining images from infrared, radar and video to identify objects – and 'real-time target detection'.

Target recognition is one of the great shortfalls in computer ability, and by developing neural network-based devices, the military hope to overcome these limitations. However, neural networks did not advance as quickly as their makers hoped, and the aim of having missiles and bombs identify their own targets still has not been achieved.

Early good results from military neural networks could sometimes be misleading. In one program a computer was being trained how to find camouflaged enemy tanks concealed in a landscape. The computer was presented with a series of pictures and told whether or not the picture contained a tank. The neural network would then gradually discover which features indicated the presence of a tank; rather than being told to look for a turret with a gun, caterpillar tracks and rectangular shapes, the computer would find its own set of rules.

The computer learned quickly, and was soon scoring 100 per cent hit rate. As a test, a new set of photographs was introduced, and the success rate dropped to no better than guessing. The reason eventually emerged: when the original pictures were taken, the ones with tanks in were taken on a sunny day, while those without were on a dull day. All the machine had learned was the difference between light pictures and dark ones.

There has even been some reaction against the idea of missiles finding their own targets. The USAF's Low Cost Autonomous Attack System (LOCAAS) is a small missile with a range of tens of miles, with a sophisticated radar system for detecting and identifying targets. It is intended to be small and cheap enough so that it can be used in large numbers, and the original concept called for swarms of LOCAAS to be launched at formations of enemy armour. Unlike existing missiles, the target would not need to be specified in advance; each missile would seek out its own target and destroy it. The program was started in 1998, but five years later the request was made for a 'man-in-the-loop', so that a human operator would be able to confirm the target and give approval for the attack.

The Dominator, under development by Boeing and the Air Force Research Laboratory, is a small unmanned air vehicle intended to operate in packs of up to twenty-four vehicles, working co-operatively to identify and destroy enemy vehicles. Dominator will patrol a given area for up to forty-eight hours, and initially it was suggested that it might be able to operate without supervision, but later it became clear that a human supervisor was considered essential.

This decision may have been motivated by 'friendly fire' incidents where US and Allied soldiers were killed by their own side. Giving the responsibility for life-or-death decisions to a machine may still look like too much of a gamble until the technology is completely proven.

Intelligent agent technology is also particularly valuable to the military. Agent-based systems are used for military simulations where it would be impractical to involve large numbers of human participants. This makes it possible to have large war-gaming exercises where officers can command subordinates with a realistic level of initiative (or lack of it) and where opponents behave in an unpredictable but intelligent fashion.

This technology has also found a use in projects with agent-based systems for the 'war against terror' where they are used to predict how crowds would behave during a terrorist incident, or whether a chemical attack would be able to bring a city to a standstill. These simulations are good for predicting the actions of large numbers of people based on simple factors, such as their knowledge of where the nearest exit is and how great their fear is, but they fail when complex behaviour and intelligence is at work.

Sandia National Laboratory, best known for nuclear weapons development, has a lot of experience in computing and is working on a simulation of how a million people might react to an attack, modelling physiological and emotional responses. Vice President of Sandia Jim Tegnelia describes their work as being distinct from what has gone before: 'Modelling cognitive processes . . . this is really exciting stuff. It's a brand new area with it's own nickname – aug-cog for augmented cognition.'

Whether aug-cog can overcome the limitations of previous agent-based systems remains to be seen.

In other fields AI has fallen foul of basic limitations. Expert systems can be very effective within their own area, such as diagnosing one particular type of medical condition, but attempts to create more general expertise have not been successful. Marvin Minsky, one of the pioneers of AI in the 1970s, found that each expert system had a very limited realm.

'For each different kind of problem,' says Minsky, 'the construction of expert systems had to start all over again, because they didn't accumulate common-sense knowledge.'

As a result, we have a proliferation of special-purpose AI systems, but no general-purpose system that can match human performance. When it comes to recognizing apples or speaking English, humans still reign supreme and the intelligent computer has yet to emerge. However, given the investment involved it's a fair bet that machine intelligence (like the original computers) will not emerge from the commercial world, but from the section of academia that draws its funding from the military.

## The Fighting Network

As we have seen, the Internet developed from a DARPA requirement for its computer installations to communicate with each other in the 1960s. The future of the Internet may lie with a system being developed today for the US Army.

Tanks are out and networks are in for the new vision of 'network-centric' warfare. The heavy armour is being replaced with a networked collection of vehicles called the Future Combat System. Instead of heavy metal, data is going to be at the cutting edge of future warfare. The driving factors are size and weight: the 70-ton M1 Abrams is simply too heavy to be transported to the theatre of operations; its replacement will therefore have to be small and light. The current spec calls for it to be around 16 tons so that it can easily be flown around the world, and if necessary, parachuted into action.

Tanks are heavy because they carry a lot of thick armour plate, enough to stop high-velocity anti-tank rounds and shaped charges. The frontal armour of the Abrams can stop pretty much anything that the battlefield is likely to throw at it, but achieving the same level of protection with less armour is practically impossible. The Abrams also carries a powerful gun with a large store of ammunition.

Instead of trying to pack all the ability of a heavy tank into one small vehicle, the US Army is going to build a network to do the job. The network will have sensors, weapons, vehicles and people – not necessarily all in the same place. The network will provide a total situational awareness to its commanders, and this is what will take it into a league of its own.

The enemy will not be faced with a line of vehicles; instead, all they may see are miniature robot helicopters and remote sensors. As soon as the enemy are located, vehicles miles from the front will direct smart missiles at them from behind cover; updated in flight by the network and equipped with sophisticated sensors of their own, the missiles will fly to the target area, pick out the highest value targets and destroy them.

Enemy attempts to engage the FCS will be frustrated by counter-measures to blind, dazzle or confuse radar, infrared or other targeting methods. FCS vehicles will merge into the scenery using active colour-changing camouflage. If the enemy fire a missile or shell, it will be tracked and intercepted in flight by anti-missiles or tactical laser systems, destroyed or disrupted before it can hit the target.

Every move the enemy make will be seen, predicted and countered before it can be completed. The network will allow the FCS force to perceive, think and act much faster than the enemy, like an agile boxer who can hit their opponent at will while easily avoiding the counter-punches.

This is the dream; realizing it will take several leaps forward in technology. The foundation that it all rests upon is the network; without it the FCS cannot see, attack or defend itself. A mobile network capable of carrying all the necessary data, voice and imagery traffic is far ahead of what exists today, but it is under development under a program called MOSAIC – Multifunctional On-the-move Secure Adaptive Integrated Communications.

MOSAIC will not just have to carry a massive amount of traffic, but it will have to be capable of setting itself up and reconfiguring itself on the move. Vehicles joining the network will have to be patched in immediately, and the network will have to heal itself if any gaps appear.

In 2004 the program is expected to demonstrate a self-organizing cluster of at least fifteen nodes, with a data transmission rate between 56 Kbps and 15 Mbps – somewhere between the speed of a modem and an office network. The rate depends on the distance between nodes, which may be up to 100 km apart. The system will have to be secure and highly robust in the face of electromagnetic interference or jamming, while at the same time it must have a low chance of being detected by the enemy. This probably means highly directional antennae operating across several different radio bandwidths; and stealthy reconnaissance units may rely on even more secure means such as laser communication.

Bandwidth management is a major challenge. The FCS will poten-tially have gigantic masses of information flowing around it from multiple video feeds and other bandwidth-hungry inputs. It will only be able to handle these with a combination of powerful compression tools and intelligence built into the sensors so that they extract key information rather than transmitting raw data.

MOSAIC is still very much unproven; nobody could accuse the makers of a lack of ambition. If it can be built, then the battlefield will be transformed and the network-centric war machine will be a reality, outgunning anyone without the same network capability.

## The Human Interface

As mentioned, one of the great goals of AI is to improve communication between humans and computers, for example by speaking in plain English. This would free operators from the mouse and the keyboard, so that they can carry out other tasks. This is especially important for people like pilots and tank crews, who already have more than enough on their hands without having to type in commands in the stress of battle.

DARPA are seeking to overcome this by means of the Brain Machine Interfaces Program. One of the best funded of their bio-engineering projects, the aim of this is to construct means for humans to interact directly with machines. The expression DARPA uses is quite opaque: 'New technologies for augmenting human performance through the ability to non-invasively access codes in the brain in real time and integrate them into peripheral device or system operations'. Put in simple terms, this means plugging your brain into a computer.

There are several different areas under investigation in this program, as ultimately it is not just a matter of extracting information from the brain but also of putting information in. Consider, for example, the idea of a robot arm used for remote work such as the handling of hazardous radioactive materials or bomb disposal. At the moment, control is awkward and challenging; those fairground machines that invite you to 'pick up the prize' with a miniature crane make good money from people who think remote manipulation is easy.

The DARPA program will not be able to read your mind – at least not at first – but it will have to identify a pattern of neural activity that corresponds to a particular action, such as 'reach out'. The technology to do this already exists in principle: sensitive electrodes attached to the scalp can pick up 'mu rhythm' brain waves which are associated with limb movements. Just thinking about moving your arm generates the appropriate mu rhythm.

Unfortunately, the mu rhythm will not be the only thing going on in your brain at the time. At any given moment it is a seething mass of electrical impulses, not just from your conscious thoughts but also from all the activity in the autonomous nervous system. It is not simply a matter of training your brain to be quiet and discarding irrelevant thought; your reaction to background sounds like a bell in the distance is enough to set off the electric pulses.

One way around the problem is to get a sensor close enough to the scene of the action. Having an electrode on the scalp is like trying to listen to one person shouting above a crowd. By moving closer to what you are listening to you have a better chance of making out what they are saying. This was the approach taken by a team at Emory University in Atlanta, Georgia led by neurosurgeon Roy Bakay. They drilled a hole in a patient's skull so they could implant a tiny sensor the size of a grain of rice. The sensor was placed very precisely next to the part of the brain responsible for a specific limb movement.

The implant was coated with nerve tissue from the patient's leg, encouraging the brain's own neurons to grow around the implant, allowing the implant to sense the activity of the relevant part of the brain only, without any background noise.

There is an easier and simpler shortcut to this approach. Rather than having an electrode on the brain and trying to pick out a particular signal in the chaos, it is much simpler to have an electrical sensor on the muscles of the hand. One of the most successful approaches in this area has been electro-oculography, which measures the tiny electrical impulses associated with voluntary movements of the eyes. By attaching electrodes to the forehead and under the eyes, it is possible to accurately track the horizontal and vertical components of a person's gaze.

It takes a lot of training before it can be controlled with any sort of reliability. The group who are at present in the best position to take advantage of the technology are patients who have been paralysed by back injuries or other conditions, and if nothing else they do have time and patience – and plenty of motivation. A system developed by Syracuse University is enabling such patients to learn how to control a computer using brain power, giving them access to communications and other facilities. In principle any system, from an electric wheelchair to domestic appliances like TV, video and other devices, could be controlled this way. Given further advances in automation and domestic robotics, even severely disabled people could gain a high degree of independence. This may become more important as the demographic shifts and the population ages.

Commercial systems are already available that utilize this technology, such as the 'hands-free mouse controller' from BioControl Systems, designed to give computer access to physically challenged users.

Given the power of the Internet, this is a useful ability to have, but it is still at a level of laboriously moving a pointer one step at a time. What DARPA want is not simply the ability to give a two-dimensional off/on command – 'start moving the cursor up, stop moving the cursor; start moving left, stop moving' – but an entirely natural type of movement. The information is all there in the brainwaves, but the technology for pattern recognition does not exist yet so that it can interpret 'move the cursor quickly in that direction until it reaches the square.' To use the jargon, the sensor needs to be 'force dynamic'.

The second part of the challenge is for the system to provide feedback to the user. While it is possible to operate a robot arm using just visual cues, when handling something naturally we rely on our sense of touch and kinesthesia, the sense of position and muscular tension that lets us know exactly what our limbs are doing. Working without these cues is like being completely numbed and some tasks become virtually impossible. It is easy enough to tie your shoelaces in the dark, but just try doing it when your fingers are numbed by cold and you will discover how much you rely on tactile feedback.

Providing feedback will be essential for more sophisticated forms of control. At present DARPA are only considering robot limbs performing simple actions like picking up and manipulating objects, but in the longer run they have much bigger ambitions. Ultimately they would like to be able to provide sensory feedback for all five senses, to make the maximum amount of information available to a soldier in the most natural form possible.

In theory this could lead to a tank driver getting all his sensory information from his vehicle. The driver would 'feel' the surface beneath the tracks, so he could tell how muddy or slippery the surface was; he would 'see' everything detected by radar and infrared sensors. The presence of enemy radar could be represented by virtual 'sound', indicating the direction, strength and type – from the terrifying howl of a weapon system achieving lock-on to the gentle murmur of surveillance radar. Even taste and smell could be invoked as useful aides: a smell of petrol reminds the driver that fuel is low, a taste of liquorice warns of a system failure.

The main non-military uses of these technologies to date have been in the area of helping the less physically able. As the population ages, and as computers become more useful in everyday life, this will become increasingly important. In theory it could gain wider currency, but in practice DARPA have yet to show that the technology is viable.

There are also questions over public acceptance. While it may be tolerable for soldiers to be fused to their equipment, it is not clear whether the public will be quite so casual about breaking down the barriers between man and machine. In fact, it is far from certain that they will wholeheartedly embrace the idea of cyber-warriors consisting of man and machine in equal parts.

## Brain Stimulation

The human-machine interface is not just about communicating with computers; the computers may also start to work with us in more profound ways. It will not simply be a matter of passing information between human and machine, but the human brain may be directly 'programmed'.

During the Second World War, millions of 'pep pills' were issued to troops who had to maintain their concentration and stay awake for long periods. Although the dangers of amphetamines are well known, the practice has continued until the present day. The influence of these pills has been blamed for accidents and friendly fire incidents, such as one in Afghanistan in 2001 when four Canadian soldiers were killed by a US aircraft. The pilots said that they had been ordered to take 'go pills' (dexamphetamine) to stay alert, and that this had impaired their judgement.

The use of go pills seems to be quite standard; stealth bomber pilots reported that during long missions they were routinely ordered to take the pills before crucial parts of the missions such as refuelling and weapons release to ensure that they were at a peak of attentiveness.

In an effort to find alternatives to chemical stimulants, DARPA is investigating the possibility of 'Extended Performance War Fighter', a soldier who can stay active for several days at a time with the aid of direct brain stimulation. Research is currently concentrating on an area of the brain over the left ear, and contracts have already been granted. The results of the first 'Assisted Performance' tests should be delivered by 2005.

The studies will build on existing knowledge of rTMS – 'repeated Transcranial Magnetic Stimulation' in which a powerful magnetic field is applied to the brain via a special helmet. rTMS has already been used with some success to alleviate depression, and it shows promise for treating many other conditions. It may even be able to help patients recover from brain damage by stimulating new brain circuits. Researchers are hopeful that people suffering from strokes or other damage may be able to regain function more quickly and more completely than before.

DARPA's interest is in boosting existing functions rather than replacing damaged ones. Vital sign monitors will be able to keep track of the soldier's state of awareness. When he starts to flag, a device built into his helmet will be able to deliver a magnetic jolt which will provide the reviving effect of a shot of espresso without any chemical downside.

The possibilities for the kind of technology are obvious, as the huge market for stimulants shows. It is not just long-distance lorry drivers, students and shift workers, and others who need to concentrate on their jobs either, although this is a primary market. Anyone who has had a hard day at work and wants a pick-me-up before they go out partying is going to be interested.

This is only the start, as other brain functions are also up for grabs. Deep brain magnetic stimulation would open up the whole brain to influence. A team at the National Institute of Mental Health has developed a new type of antenna to stimulate the deeper regions of the brain more than a few centimetres from the surface.

Cancelling out fatigue is already feasible, in the laboratory at least. But deep stimulation offers more possibilities for morale boosting and otherwise creating the perfect soldier. Low morale could be countered by stimulating the pleasure centres; fear or cowardice could be neutralized. In the civilian world the possibilities for electrical Prozac or an addiction to pleasure-stimulation have not yet begun to be seriously considered.

As well as being an 'upper', brain stimulation is of military interest as a 'downer'. In the excitement of combat, hands shake and firing becomes much less accurate. A soldier's heartbeat may rise to 300 beats a minute, compared to a rest rate of 72 or the usual maximum of 200 for athletes. Physical training and mental conditioning can reduce the physiological effects of the adrenaline rush, but few soldiers are capable of good marksmanship under battle conditions, however competent they are on the firing range. Brain stimulation may be a way of damping down the excitement and giving the soldier cooler nerves and steadier hands – if this is what the Army wants.

During the Falklands war, British pilots were reportedly issued with sleeping tablets to ensure that they would rest properly before going into action.

Patients with brain lesions sometimes undergo a personality change, becoming angry and aggressive. This has led to the discovery of an aggression centre in the brain. If this is stimulated, the most peaceful individual can be driven to acts of violence. Studies carried out in the Second World War suggested that something like 80 per cent of infantrymen never fire an aimed shot in combat. Fear overcomes aggression and they prefer to keep their heads down, firing in the general direction of the opposition rather than picking targets. Stimulating the aggression centre slightly, so the soldier feels a genuine desire to do violence to the enemy, is likely to produce a larger proportion of directed fire.

In future, the soldier's helmet may not just be there to protect his head. If DARPA's work is successful, the soldier will be able to go into battle with a clear head and no trace of fatigue; he will be calm and untroubled by fear, but with a steely determination to do some serious damage to the enemy.

As Wellington is reputed to have once said about his men, 'I don't know what effect they will have on the enemy, but by God they terrify me.'

There is undoubtedly a market for products capable of influencing brain activity. A Korean company is already offering Peeg, a 'personal encephalogram' which is supposed to stimulate brain waves into different patterns. The device has settings for improving memory, relaxation, sleep, alertness and jet lag recovery. Peeg may be greeted with scepticism, but later versions may be more convincing. If the technology works, then anyone not using it will place themselves at a disadvantage. Sportsmen already go through a series of psychological exercises to key themselves up for the big match, but brain stimulation could provide the same effect more easily. Salesmen needing the 'killer instinct' to close a deal could programme themselves into the right frame of mind.

On the domestic front, it might be more positive: candlelight and soft music could be supplemented by dialling up a romantic mood for two. A bad day at the office or at home could be quickly dispelled by a session under the machine, and domestic harmony could be restored easily. The limits and ethics of such mood control will rapidly become a major issue. One partner wants the other to switch their mood to 'uncontrollable lust', the other suggests the mood-change equivalent of 'cold shower'. Science fiction has paved the way, but it is likely to be military technology that brings the reality.

## Quantum Supercomputers

While there are major gains to be made in the areas of artificial intelligence and human-computer interfaces, another approach may transform computing power itself. Current hardware research is mainly aimed towards providing more of the same, but faster and cheaper. Moore's Law mentioned previously states that computing power for a given price doubles every eighteen months. This has given us a slow but steady progression of computing power over the last three decades. But the next generation may leapfrog this and catapult us into a new realm altogether by replacing the digital computer with a quantum device.

Quantum physics is a set of rules that apply to very small particles. The quantum world is unlike the physical world that we are used to: instead of being distinct objects with definite outlines, subatomic particles are fuzzy and are not located in one place but exist in a cloud of possible locations. In particular, it is possible for a particle to be in what is known as a 'superposition of states'. In a normal digital computer a binary digit is set to either one or zero; superposition of states means that a quantum particle can be one and zero at the same time.

In theory this means that a quantum computer could be built with a processor handling quantum bits (qubits) which would work in an unusual way. One qubit exists in a dual state, one and zero; put two qubits together and you have four states (in binary, 00, 01, 10 and 11), add a third and you have eight states (000, 001, 010, 011, 100, 101, 110 and 111). A calculation carried out using qubits will calculate every possible state involved, so using three qubits you can take into account all eight combinations. Three qubits is not very powerful, but move up to thirty qubits and there are over a billion combinations. In a single step, the thirty-qubits computer would carry out a process that would take a conventional computer a billion steps.

Quantum computers become incredibly powerful as more qubits are added. A sixty-qubit computer would make a billion billion calculations with one step, and could carry out calculations instantly that existing computers could not complete in millions of years.

The military are interested in quantum computers for the same purpose that the original Colossus was built for: breaking codes. Enigma generated a cipher with 10 million billion combinations, a cipher which could be cracked with additional help from a computer. Modern security uses a much more powerful code which has a key comprised of 128 binary bits rather than Enigma's ten gear wheels; using a normal computer, it should take many years to break the security on a single message by checking all the possible key combinations. (The NSA and other security organizations have banks of supercomputers which are believed to be able to break through these codes rather quicker, but they are working at the edge of the possible.) However, with a quantum computer this kind of code could be cracked in a fraction of a second.

Increasing the size of the key, which is how security has been improved in the past, will not help. Every bit added to the length of the key doubles the number of combinations and so doubles the amount of computing time taken for a normal computer to break the code. However, even a thousand-bit key will be child's play for a thousand-qubit quantum computer.

Given a usable quantum computer, security agencies could read not just every encrypted message sent, but every intercepted message they have ever recorded. If this was not enough, a quantum computer would also be able to break through all existing computer security that relies on keys: every online banking system, every transaction carried out by computer.

Quantum computers have other uses as well. One of the classic problems of computing is the 'travelling salesman' problem, in which the salesman has to visit a number of customers in different locations and wants to find the shortest possible route that covers all of them. Problems of this type are difficult because the number of possible permutations (visit A, then B then C; or ACB, CBA, CAB etc.) become astronomical as the number of customers increases, and calculating all of them becomes impossible. A quantum computer could do this with ease.

Many computing tasks which are currently impossible, such as long-range weather prediction, would become practical with quantum computing. Detailed modelling of the flow of air or water over cars, aircraft or boats on a microscopic scale would become possible, leading to a whole new generation of designs. The same benefits would flow to areas like materials design and other fields of engineering. There are also possibilities for using quantum computers in artificial intelligence research, with the suggestion that it will take a quantum machine to pass Turing's test.

Quantum computing is still very much in its infancy, and there are technical arguments about whether a quantum computer using more than a few qubits will ever be feasible. This is an area in which DARPA are known to be highly active, and where the military have most to gain. At present, terrorist groups can use the Internet to send encrypted messages to their cells around the world with little fear of interception, and in that sense it is the perfect untraceable communications medium for them. If DARPA do succeed in building a quantum computer, it remains to be seen whether they share their successes with others or, as with Colossus, keep it to themselves. Letting the terrorists know that their communications are no longer secure would be a tactical mistake; but keeping quantum computing from the rest of the world might be a bigger mistake in the long run.

## Picking Winners

It is impossible to tell whether any of these initiatives will be successful, just as it was impossible to tell whether the builders of ENIAC would eventually succeed in making a working digital computer or whether Colossus would be able to contribute to cracking the Enigma cipher. Until we have a working product, we cannot tell whether the problems are difficult or actually insurmountable.

Artificial intelligence was fashionable in the business world in the 1990s but has largely disappeared, apart from a few niche applications. Having been initially oversold, it failed to live up to its early promise and there is little interest in this area. A working AI system would change the situation completely.

For example, a natural language interface – a computer system that could speak and understand English – would instantly be a competitor for the hundreds of call centres manned by thousands of human operators. Early systems would be unlikely to replace human operators entirely, but they would take a significant slice of the market and this would increase sharply as further investment improved the technology.

Expert systems have already shown some usefulness in niche areas like medicine and law. A full-blown system that would be able to pull out all the information relevant to a particular case could revolutionize both professions. It could potentially threaten both of them: if a computer system gives a client full access to everything (and can translate it into normal English), the role of the professional is severely reduced. The system does not need to be capable of making a complete diagnosis or putting together a legal case; if it just does some of the basic legwork then the requirement for a doctor or lawyer is halved.

Robotics has never really made it out of the factory, because robots need a structured environment with a well-defined task. Apart from occasional toys and devices like the Trilobite robotic vacuum cleaner, robots have not invaded our homes. An AI with the ability to recognize and manipulate objects would revolutionize factory robotics, with a vastly increased ability to take over from humans. Given the ability to walk as well as a human would open up very wide horizons in the commercial and domestic markets.

Everything from robot window-cleaners and street sweepers to automated farming equipment and construction machinery would become possible. Suddenly a robot capable of carrying out domestic chores would become viable.

The MOSAIC network would do well in the commercial market. A secure network that can be set up anywhere within minutes, linking mobile users many kilometres away would be the answer to a network engineer's prayer. It is the dream solution for any business; when you buy a new PC, printer or phone all you need to do is bring it to the office and it is immediately set up and connected to the company network.

The data capacity of MOSAIC would be enough for most businesses; bigger ones will use a series of networked networks and the 100-kilometre range is more than adequate for the urban areas that comprise the vast majority of companies. It would also allow workers to take their computer between office and home working without even disconnecting.

It would be useful for other requirements too – for example, providing a permanent mobile videophone connection to family and friends within a wide radius. Telecoms licensing issues are more likely to be a barrier than technology, and the only delays in the next revolution will be caused by our attempts to control it.

DARPA's man-machine interface projects look even more like science fiction. At the very simplest level they could result in a mobile-phone implant allowing you to hear and talk without moving your lips – effectively a form of telepathy as far as users are concerned. You could be in contact with someone as easily as thinking about them; and the exchange of verbal information would just be the start. The impact on society can barely be guessed at; look at texting and photo messaging and raise it to another dimension.

In fact, the impact of any of these would be impossible to estimate. Even when ENIAC was working, the Chairman of IBM did not think there would be a market for more than five computers; instead we have tens of millions. When DARPA connected their three computers in a network, they did not anticipate that the Internet would become a part of everyday life for millions of people. Future steps in computing evolution may well have equally dramatic consequences.

# CHAPTER 9

# Non-lethal Death Rays

*Set phasers on stun.*

Star Trek, passim.

*These weapons have the potential to incapacitate everyone in a building, allowing us to disarm them and sort out the hostiles from the non-combatants . . . The effect of the directed energy weapon system on incapacitated individuals is potentially visually disturbing and could cause negative publicity if the intent is improperly conveyed.*

US Marine Corps Manual

Sometimes the military problem is not so much a matter of killing people as not killing them. Armed force is as much the last resort in national politics as in international politics, the final blunt instrument for dealing with popular unrest. But a blunt instrument is not capable of the fine graduations in force that may be needed to deal with a situation. Some incidents from recent history illustrate the problem:

- In 1986 there were mass protests in the Philippines against President Marcos, who had just announced himself to be the winner of presidential elections. The elections were widely denounced as fraudulent; opposition leader Cory Aquino called for a mass campaign of civil disobedience, and people took to the streets in large numbers.

  On 22 February, Marcos ordered a column of tanks to dislodge the protesters' main encampment in Manila. The tanks were faced with a dense crowd blocking the road. Faced with the choice of killing unarmed protesters or backing off, the Army withdrew. Helicopter gunships were sent in, but they too refused to open fire. General Ver, commander of the armed forces, ordered the air force to bomb the protesters, but again the pilots refused to attack their own people.

Within days the Marcos regime was swept aside. Even though some of the armed forces remained loyal to Marcos, their refusal to open fire on civilians meant that he could not stop the wave of protests.

- In the summer of 1991, communist hardliners in the Soviet Union decided to act against the reforms put in place by President Gorbachev, which they believed were endangering the country. Gorbachev was taken prisoner at his dacha in the Crimea, tanks occupied strategic points on the streets of Moscow, and the plotters announced that they were in charge. The coup seemed to have succeeded, until Boris Yeltsin, the mayor of Moscow, led a counter-coup from the parliament building in Moscow.

    The Russian air force was ordered to bomb the building, but refused. Ground forces were massed for an assault, but the troops could not be persuaded to open fire on the building or the crowds of demonstrators who had come out in favour of Yeltsin and protected the building with their bodies. The coup collapsed.

It does not always happen this way. In 1993 Yeltsin was President, and was also faced with a revolt from parliament. He too resorted to force, putting it down with 146 dead and over 1,000 injured (according to official figures). Nor are we likely to forget the events in Tiananmen Square, where the Chinese government sent in tanks to suppress a popular protest in 1989. The official death toll was 300; witnesses put the figure much higher. The brutal suppression of the pro-democracy movement threatened to create a backlash which would bring down the government.

Situations where troops are faced with massed civilians are not confined to repressive dictatorships. In Somalia in 1993, US helicopter gunships were sent to support Pakistani peacekeepers. The Pakistanis were part of a UN force attempting to remove barricades in Mogadishu, and were facing an angry crowd. Soldiers had been killed in an earlier riot, and because of the perceived danger the helicopters opened fire with automatic cannon with lethal effect. Some sources put the number of deaths at more than a hundred. The US maintains that there were gunmen in the crowd and that the UN troops were in real danger. Whatever the truth, the population of Mogadishu turned against the peacekeepers. A few weeks later an American helicopter was downed and eighteen American servicemen were killed fighting against a large number of local militia, forcing the US to leave Somalia. (The story of this action is told, from a pro-US point of view, in the film *Blackhawk Down*).

Even the world's most advanced superpower cannot satisfactorily handle encounters where less than lethal force is required. Overwhelming firepower may work, but it is rarely an option for a peaceful democracy, and even dictators like Marcos may regret trying to use it.

The situation would be changed if there were a way for the military to turn aside crowds without using lethal force. Just as the death ray was seen as the answer to the great military problem of shooting down enemy bombers after the First World War, the non-lethal directed energy weapon is being promoted as the solution to some of today's toughest military challenges.

## History: Rubber Bullets and Tear Gas

The alternatives to live ammunition are currently limited to variations on the themes of baton rounds and tear gas.

Baton rounds were intended as an extension of the traditional club, a way of delivering a powerful but non-lethal blow from a distance. Bullets are small and have very high velocity; to be safe baton rounds need to be slower and bigger. They are typically ten times heavier than conventional bullets and fired at a tenth of the speed. Because they are made of lighter materials, they are a hundred times bigger than live rounds.

The first baton rounds were used by the British in 1967. Made of teak, they were nicknamed 'knee knockers' because they were supposed to be bounced off the ground to injure the target's legs. Used against rioters in Hong Kong they were found unsatisfactory because of the risk of injury if they were fired directly at a target. In the heat of a riot, soldiers could easily fail to take the rioters' safety into account.

The teak round was replaced and a new version made from hardened rubber. The name 'rubber bullet' was adopted to suggest something harmless and 'slightly humorous'.[1] Tens of thousands of these bullets were fired during the troubles in Northern Ireland, causing a number of deaths. The design was changed and plastic bullets were adopted from 1973, but the deaths continued, leading to yet another version of the baton round being introduced in 2001. This is smaller and lighter than earlier plastic bullets, but is fired at a higher velocity. It is more accurate, so there is less risk of accidentally hitting the head, which is the main cause of fatalities. Critics point out that the higher velocity and small size make head injuries much more serious, with the possibility of particularly nasty eye injuries.

Purely apart from the effect on the target, baton rounds are inadequate from a military point of view. They are inaccurate, making it difficult to hit the stone-thrower rather than the person next to him, and they are single-shot weapons. Their maximum range is no more than 70 metres, so it is impossible to keep crowds throwing stones or other projectiles out of range.

Tear gas was used from the start of the twentieth century. A substance called DA (short for DiphenylchloroArsine) was used by Germany during the First World War. This caused severe irritation of eyes, nose and throat, and penetrated the gas masks in use at the time. Combined with other chemical agents, the DA caused victims to sneeze and cough, and take off their gas masks, exposing them to more deadly gases.

There are now two types of tear gas in common use, CN and CS.[2] Neither is actually a gas, both being spread as a fine dust or liquid droplets in the air.

CN, first developed in 1923, is the active ingredient in the Mace sprays used by US police forces. CS was first isolated by Carson and Soughton, two American chemists who gave their initials to the discovery, but it was converted into an effective riot control agent by the British Porton Down Chemical Defence Experimental Establishment in 1956.

CS and CN have similar effects, causing pain, burning and irritation of exposed mucous membranes and skin, and will incapacitate an individual in agony for up to ten minutes. After being gassed by riot police in South Korea, the writer PJ O'Rourke described the effects as being like 'trying to breathe fish bones'. However, it is extremely variable – people under the influence of drugs or alcohol may be completely immune to their effects.

CS is more powerful and less toxic but has the drawback that it is much more persistent. A subject sprayed with CN can be taken away in a police van after a few minutes, but anyone wearing clothing contaminated with CS will continue to spread it for some time.

Like anything that rides on the wind, tear gases are indiscriminate and will affect anyone in the area. A change in the wind can blow the gas cloud in an unexpected direction, and the effects are unpredictable. Tear gas may be better than rubber bullets for clearing a wide area and keeping a crowd away, but it is still far from ideal.

Non-lethal weapons are a growth industry in the US military. As the line between peacekeeping and peacemaking gets increasingly blurred, troops are more likely to be facing civilians than enemy soldiers. Winning over the local population is an important part of any military operation, and there are few things more calculated to turn the people against you than shooting at them.

The challenge is to develop non-lethals that are as precise and controllable as possible, so the level of force can be exactly matched with the threat. There is no such thing as a truly non-lethal weapon, but lethality should be controlled as far as possible. Also, the military want something that can be used at long range, so troops are not subjected to a barrage of bricks or petrol bombs.

Kinetic energy weapons like baton rounds have their limitations, as do chemical agents. But directed energy weapons open up a whole new world.

## Microwaving the Population

The idea of using a beam of radio energy as a death ray was described in an earlier chapter. While even powerful radar beams are not enough to kill people, they can make them very uncomfortable, and this principle is used in a new type of non-lethal weapon. In 2001 the US Marine Corps revealed their Vehicle Mounted Active Denial System (VMADS). Mounted on a truck, it fires a beam of millimetrewave radiation, similar to microwaves. Quickly dubbed the 'People Zapper' by the media, the beam works like a microwave oven, heating the target.

The frequency involved is higher than microwaves – 95 GHz, with a wavelength of a few millimetres rather than a few centimetres. The USMC claims that the beam only penetrates a fraction of an inch into the skin. It causes a sensation like touching a hot light bulb, painful but harmless, encouraging the victim to retreat. They say that the worst injury their tests have produced is a small blister.

VMADS is to be used as a type of barrier. An area can be marked out with tape and signs warning people not to enter; anyone straying into the marked zone will be swiftly irradiated by VMADS until they retreat.

Whether VMADS will work remains to be seen. There have been some suggestions that the beam could be blocked by simple expedients like a wet blanket or a sheet of tin foil, but this is by no means certain. In any case, the USMC have taken a fairly robust line on the matter. Their policy of ensuring that there is lethal back-up to the non-lethal barrier means that anyone trying this kind of trick might regret it.

It is also unclear whether the beam is as harmless as has been claimed. Virtually nothing had been released about VMADS in terms of the power levels, beam diameter or other details, so there is no independent verification of its effects. Research into possible eye damage and carcinogenic effects is still ongoing.

VMADS is at present the only known product of the US military's non-lethal radio-frequency (RF) program. But there are hints of a host of other devices, including some said to interfere directly with brain function. However, the most likely route of influence is to be heating. One such device is the Penetrating Direct Energy Weapon (PDEW). Rather than heating a small spot, the PDEW appears to work by heating the entire body and causing an artificial fever, leading to rapid unconsciousness.

PDEW's main advantage is that the beam can go through walls. Storming a building which contains a mixture of enemy troops and terrorists with civilians or hostages is a tactical nightmare. In 2001, Chechen terrorists took over a theatre in Moscow, taking hundreds of hostages and threatening to blow up the building. The Russian authorities pumped an aerosol containing a compound related to Fentanyl which rendered everyone inside unconscious before storming the place. What could have been a triumph turned into a disaster when 120 of the hostages died as a result of exposure to the chemical agent and inadequate medical back-up to reverse the effects.

PDEW, mounted on a vehicle like VMADS, would be able to neutralize a building from a range of a hundred metres of more, incapacitating everyone inside – 'harmlessly', in theory.

As PDEW has no official status (at the time of writing it does not officially exist), there is hardly any information available about it. The health effects, both short and long-term, are unknown. If it works by raising body temperature it may affect people in different ways, especially children and the elderly. A temperature of 26°C may be needed to effectively incapacitate the victims, but 28°C can be fatal, giving a very narrow margin for error, especially considering variations in exposure and other factors.

There is unlikely to be any public discussion of PDEW or other systems, or independent verification of their effects until, like the Fentanyl aerosol, they are used in action, so any potential ill effects may not be discovered until they are used in action.

In the long-term, the secrecy will delay any transfer of the technology to the civilian sector. While its most obvious application is in law enforcement – being able to disable gunmen in a siege or hostage situation could save many lives – there may be other areas where such a device could be used. Pest control could be one such area, bringing the ability to destroy infestations of insects deep within building structures without having to resort to chemicals, but other uses are limited only by the imagination, when and if the technology is made available.

## The Laser that Packs a Punch

Lasers started out as a military technology, but laser weapons proved technically difficult as we saw in Chapter 3. After forty years of work death rays have still not made it to the front line. However, some other prospects have arisen out of laser research.

During the Star Wars program of the 1980s, many tests were carried out firing lasers at pieces of metal simulating ballistic missiles. The object was to find out how powerful a laser was needed to puncture the missile's skin and destroy it. Experiments seemed to show that even very powerful laser pulses were having much less effect than expected. Somehow the laser energy was being dispersed and only doing superficial damage.

A thorough investigation found out the cause of the problem. When the laser beam initially struck the metal, the high temperature instantly vaporized the outer layer. This created a cloud of hot, ionized gas (a plasma) in which the gas molecules shed their electrons. A phenomenon known as brehmsstrahlung can occur in hot gas, in which electrons slow down and emit light; the opposite, inverse brehmsstrahlung is when electrons absorb light. That was what was happening with the tests: the cloud of gas was absorbing the rest of the laser beam so it never reached the target.

On the one hand, this meant that scientists would have to make sure that the conditions for inverse brehmsstrahlung did not occur if they wanted to shoot down a missile. But on the other hand, this effect itself can damage the target.

As the hot gas absorbs laser energy, it expands rapidly – extremely rapidly, in fact, with the force of an explosion. The laser does not damage the target, but the shockwave from the expanding gas might. In 1992 a program was initiated to look at whether an 'impulsive kill' laser (as opposed to a 'heat kill' one) would be effective.

The explosions created would be fairly small, as they would only release the same amount of energy as the laser beam at best. A kilogram of explosive releases over four million joules of energy, but a laser pulse will only contain a few thousand joules. This is not the sort of explosion which is going to destroy tanks and battleships, but a small controllable explosion might be effective for non-lethal purposes. Certainly a non-lethal laser weapon would be useful, as it has an inherent precision and accuracy that is lacking in rubber bullets.

The Pulsed Impulsive Kill Laser (or PIKL) used in the 1992 tests fired pulses of between a hundred and a thousand joules, and was tested against a variety of targets simulating clothing and skin. The test results are still classified, but the laser showed that it scorched the surface of the target and produced an impact force roughly equivalent to a golf ball falling one metre. This would not deter a rioter, but it proved that the principle worked.

There was some cause for concern about its use as a non-lethal weapon though; one leaked report revealed that a series of pulses would 'literally chew through the target material' – not good news if the target material happens to be your body.

By 1998, a California company, Mission Research Corporation, were building a pulsed chemical laser under a contract with the Joint Nonlethal Weapons Directorate. This was 'a tuneable weapon with effects ranging from nonlethal through lethal', described as follows: 'The basic mechanism is to produce a laser breakdown of the surface of the target to produce a high-temperature plasma that is essentially a "flash-bang" with varying kinetic energy.'

This reveals one of the big selling points of the pulsed laser, the fact that it is a 'tuneable' weapon. What the military want is something that can be turned up or down depending on the kind of target it is being used on.

Peaceful but stubborn protestors can stop a convoy of tanks from crossing a bridge simply by lying down in front of it. A tuneable weapon could drive them off with a series of loud but harmless 'flash-bangs'. The same weapon could disperse a crowd throwing stones on a higher setting, delivering a series of painful blows accompanied by brighter flashes and deafening bangs. More serious threats, such as the gunman sheltering in the crowd, would be engaged on the highest, lethal, setting, with the precise surgical accuracy of a laser.

A few years later on the pulsed chemical laser has evolved into the Pulsed Energy Projectile.[3] The program is still highly classified, but a few pieces of information can be gleaned.

The PEP is a long-range weapon, effective at one or two kilometres, and capable of firing several shots in rapid succession. It is vehicle mounted, currently in a 'Hummer' vehicle, but it might also be fitted to a helicopter or even to a fixed-wing aircraft like the AC-130 gunships used by US Special Forces. This would allow troops to deal with riots without ever having to put a man on the ground.

At the same time as the technical development and evaluation of target effects is taking place, the PEPs effects are being described in ever-more favourable terms.

'It's the nearest thing we have right now to phasers on stun,' said Colonel George Fenton, during his time in command of the Joint Nonlethal Weapons Directorate. 'The device directs an invisible in-duced plasma pulse at a target that will create a flash-bang *near* the intended target.' [Author's emphasis]

An article in *Time* magazine said that the PEP 'superheats the surface moisture around a target so rapidly that it literally explodes'.

It seems that there is a reluctance to acknowledge that skin is being vaporized here. More importantly, 'flash-bangs' involve high-pressure shockwaves which are inherently dangerous. A hit in the mouth could result in lung rupture from the blast pressure, and an impact on the chest or abdomen could damage internal organs. A strike anywhere near the ear would be literally deafening.

There is also the question of what the effects might be on eyes. An explosion on the surface of the eyeball would probably result in blinding, and might be lethal. However, no details of the PEP have been released. There is no information on power levels, duration or beam diameter, or anything that could help researchers to assess the possible risks. Similarly, the US military studies describing the effects of such laser pulses on living beings have not been released.

The PEP will probably be on the streets some time before 2007, most likely in a country where the US is engaged in military intervention and needs to control the civilian population. It will undoubtedly be less lethal than spraying crowds with water cannon fire, but whether it is truly 'non-lethal' remains to be seen.

Although the PEP uses a chemical laser, future plans call for a semiconductor laser which would run on electric power and so would have an effectively unlimited ammunition supply. This will require great advances in semiconductor laser technology – but there are signs that the military are moving ahead in this area. As with previous decades when technology like the gas-dynamic laser was at an advanced stage without the rest of the world being aware, it is possible that the Pentagon is testing lasers which are many years ahead of their civilian counterparts.

Again, the results are likely to remain classified for some time to come, but the ultimate impact may be as great as the PEP itself. Powerful lasers not requiring chemicals and running off cheap electricity could have all sorts of industrial and commercial applications from precision drilling and cutting to tunnelling and construction. But such lasers have already been tested for a more imaginative use.

## By Laser to the Stars

An unfortunate effect of the secrecy surrounding PEP technology is that it is not available for other uses, which may ultimately be of greater importance. For example, there is a plan to use a laser to launch a craft into orbit. In 1987 Professor Leik Myrabo invented the concept of the Lightcraft. This uses the same principle as PEP, with a ground-based laser being aimed at the underside of a spinning, metallic craft. Professor Myrabo was working for the SDI at the time, so it is likely that his idea came from the same experiments.

The Lightcraft tests involved model craft weighing about 150 grammes; multiple laser pulses propelled them to a height of over 70 metres, showing that the idea worked in principle. In order to propel a larger craft into orbit, a more powerful pulsed laser would be needed – the kind of laser being used in the PIKL and subsequent tests.

The leaked results from PIKL show that it could generate a more effective push than the lasers used on the Lightcraft. By 1997, the Lightcraft had already beaten the first flight of Robert Goddard's first rocket of 1926. Myrabo calculated that with a multi-megawatt laser it would be possible to push a payload of a kilogram into orbit. This can be a useful size for a satellite in its own right, and a new class of satellites dubbed 'nanosatellites' are being built for both military and academic research purposes.

The Lightcraft would also offer a way of resupplying the International Space Station or other orbiting platforms. Supplies might only arrive 1 kg at a time, but this gives tremendous flexibility. If you need antibiotics urgently it is better to have them sent up immediately than waiting weeks or months until the next space shuttle launch.

However, laser propulsion can do more. Because the launching system remains on the ground, it can be made as large and cumbersome as necessary. It can launch satellites using nothing but cheap electricity, and they can be sent up in rapid succession. A Lightcraft launch base could send up hundreds or thousands of small components which could be assembled in orbit into a larger structure.

Previous plans have called for the construction in orbit of large space stations, or craft for exploring the outer solar system, but these have always been hampered by the high cost of launching into orbit. While the human crew might still have to travel by rocket, the small components making up a larger vehicle could all be sent by Lightcraft, and a tiny fraction of the $10,000-a-kilo price of conventional launches.

The other possibility would be for the laser to act as a launcher for nanosatellites – see Chapter 16.

In spite of this huge potential, the Lightcraft is in abeyance. When I contacted Professor Myrabo about the comparisons between the PEP and the laser launcher he used, he said he was unable to comment on the matter. It is possible that there is a military Lightcraft launcher out there which is being jealously guarded; but it is equally possible that the potential of pulsed-impulse lasers is being stifled because of its applications in weapons. This may be a great loss; technology that provides a virtual 'tractor beam' being able to apply small nudges at a distance could be very useful in space and elsewhere.

## The Electric Zap Gun

The discovery of electricity was closely tied in with the discovery of its effect on the nervous system. The Italian, Luigi Galvani, showed that frog's legs could be made to twitch by applying an electric current – this came to be known as the galvanic response. By the twentieth century, electric fences and cattle prods were being used to control livestock. It did not take a huge leap of the imagination to see that similar devices could be used on people.

Electric 'stun guns' producing a jolt of 20,000 volts or more are quite effective, but have the limitation that you have to be standing next to the target. What is needed is a way of delivering an electric shock at a distance.

The Taser is a popular weapon in use with US police forces for subduing criminals. It fires two darts trailing wires; when the darts strike the target, the Taser pumps a 50,000 volt shock through the wires. This is usually enough to subdue the target so that they can be arrested without a struggle. The makers claim that their latest model has a success rate in excess of 80 per cent in bringing down suspects with a single shot.

Human rights groups have campaigned against the Taser and other electric shock weapons. It is alleged that they are regularly abused, as a way of punishing or coercing suspects. In 1991, Rodney King was shocked twice with a Taser by LA police officers who claimed that they were having difficulty subduing him – they also struck him with batons at least thirty times.

Although controversial, the Taser is an effective weapon as far as it goes. But it does not go very far, as the maximum range is only about 20 feet. This is a useful range for police work, as many encounters take place at this distance, and it provides enough reach to be out of the way of knives or other handheld weapons, but not good by military standards. The Taser, like rubber bullet guns, is also a single shot weapon. These limitations mean that the Taser has not been considered useful by the military.

As with the PEP, what the military want is a long-range, multi-shot version of the Taser. One way of doing this is with the 'Sticky Shocker', which is essentially a round like a rubber bullet with a built-in battery. When it strikes the target it sticks and delivers a series of electric shocks. The Sticky Shocker could be fired from a range of existing launchers including vehicle-mounted multiple launchers, but it does not go far beyond current capabilities.

An alternative approach is to create a conductive channel in the air and use it to send an electric shock to the target rather than relying on wires. There are several different programs underway in the US and elsewhere using 'wireless Tasers'. These spray an aerosol of fine particles or liquid droplets into the air, lowering its resistance. They are limited by the range of the spray, but lab tests by Rheinmetall show a bolt of lightning being conducted 10 metres to a target.

The USMC is funding a wireless Taser from a company called XADS. Although this will initially have a very short range – only about 10 feet – its big selling point is that it can keep spraying out electric charge 'like water from a hose'. This would make it far more effective against multiple targets, and for many situations long range is not essential. Pete Bitar, President of XADS, points out that the vast majority of police encounters are at close range, and military operations like house clearance and roadblock duty will also involve short range.

A more ambitious scheme is being carried out by HSV Inc, who are working on a 'tetanizing beam weapon', which might be loosely described as man-made lightning. Natural lightning occurs when a storm cloud builds up a high enough electrical potential to ionize the air. The ionization, in which atoms lose their outer electrons, proceeds in a narrow, zigzag channel towards the ground or another cloud. Unlike normal air, the ionized gas is a very good electrical conductor, so as soon as the ionization channel reaches the ground, the cloud is 'earthed' through the channel. An enormous electric current flows briefly through the channel, lighting it up and sending out the familiar flash and boom of lightning and thunder.

HSV's device will use a laser rather than a high electric potential to ionize the air. A laser working at the ultraviolet end of the spectrum can ionize air, and, unlike lightning, the ionization trail will be a straight line. The device will use two beams to make two channels to the target; once contact is made, an electric shock will be passed down one channel, using the other for the return current. The ionized channels, which are visible as glowing green beams, perform exactly the same function as the two wires of the Taser, except they can be much longer.

Scientists in Quebec have succeeded in created a laser-ionized conducting channel some 200 metres long. HSV believe that this could be extended to over 2 km by using a laser with the appropriate wavelength.

Rather than delivering a single powerful jolt, HSV envisage using a weak current with a high voltage. This will be modulated so that it interferes with the nervous system and paralyses the skeletal muscles, immobilizing the victim for as long as they are in contact with the beam. This effect is known as 'tetanizing', hence the name of the weapon.

The USAF are also involved in creating laser ionization channels in the air. The secret appears to be the use of a laser which has extremely high power but has a very short pulse duration: a million million watts for a million millionth of a second. This means that the overall amount of energy used is low – much too low to damage the target – but the power level is sufficient to break down the atmosphere. It is mainly a matter of making the equipment small enough. With the advent of a technology called Chirped Pulse Amplification, such lasers have become far more practical.

'A few years ago these type of lasers were the size of an office desk, but at the moment they are the size of a suitcase,' says Peter Schlesinger, President of HSV. 'Because of advancements in miniaturization processes, we expect that next year lasers will be the size of an attaché case or smaller. Hopefully in a few years time our non-lethal device will fit into a ladies' purse.'

Schlesinger has had plenty of interest in his device from around the world, including from Britain and an Australian group who were considering it as a means of catching stray kangaroos – the beam has much greater range and accuracy than a tranquillizer dart.

It remains to be seen whether the developments in this area will result in a viable weapon at an affordable price. Currently Schlesinger is talking about a price in the hundreds of thousands of dollars range for each weapon. But even at that price there will be takers; a single weapon mounted in a helicopter could be used to zap any number of rioters at high speed. The psychological effect of man-made lightning raining down might be enough on its own, but the physical effect can also be tailored to the situation.

Schlesinger is keen to emphasize that his company is only working on a non-lethal weapon, but another company in the field says that their weapon can be lethal or non-lethal at the flip of a switch.

One of the big issues with electric-shock weapons is the potential for abuse. A pocket-sized device which allows a soldier to inflict pain without damage could easily be used to punish, bully and intimidate a civilian population. In the hands of a dictatorship it would be truly horrific.

This is only beginning to emerge as a weapons technology, and the applications go far beyond the military arena. The original intention is to convey an electric shock to a target, but in principle ionization beams could be used to create a conductive channel anywhere. In effect, they will allow for the wireless transmission of high-voltage electric current. One early possible use is in the power industry, where such beams could be used to control dangerous short circuits in high-voltage equipment, providing a safe way of conducting to ground.

The costs will initially be high, but as we have seen with other technologies, they may come down quite rapidly. When this happens, there will be great opportunities for wireless power transmission. Just as the Bluetooth standard means that mobile phone headsets and other equipment no longer have to be physically plugged in to make contact, wireless transmission means that, for example, it could be possible to recharge a device simply by placing it near the recharger. A simple safety mechanism could cut off the power in a microsecond if the beam is interrupted, for example by someone putting their hand in the way.

Again, the only limits to what the technology might ultimately achieve are those of imagination. What starts as a weapon may, like the laser, go on to become a ubiquitous part of everyday life.

## Sounds Dangerous:
## Acoustic, Infrasound and Ultrasonic Weapons

Sound makes an effective weapon, as anyone whose neighbours have a bass-boost system will testify. The power of rock music to disturb and disorientate has been harnessed by the military, for example during attempts to force the Panamanian dictator General Noriega out of the Vatican Embassy where he was sheltering. The results were inconclusive, and although some of the senior offices wondered how anyone could tolerate the discordant noise for so long, some of the younger US troops reportedly thought that the music was 'pretty cool'.

But why stop at sound levels that vibrate the windows and can be felt through the floor? Make the sound loud enough and you might be able to injure or even kill.

This is another technology dating back to the anarchic days of Nazi science, and the Soundkanone was another failed prototype. There are also stories of a device built by a Dr Gavreau in the 1950s (or 1960s or 1970s, according to different versions of the tale). Gavreau researched the effects of infrasound – sound waves with a frequency too low to be heard by the human ear. According to Lyall Watson's account in his book *Supernature*, during an experiment with a giant whistle one of Gavreau's team was killed by the sound, 'his internal organs . . . mashed into an amorphous jelly by the vibrations'.

Attempts to trace Gavreau's experiments have ended in failure and the whole episode is more than a little dubious. However, claims have been made for the dangerous effects of a whole range of different sounds.

- Infrasound is claimed to cause disorientation, vomiting, nausea and bowel spasms. According to one celebrated Pentagon briefing, such a weapon could 'liquefy their bowels and reduce them to a quivering diarrhoeic mess'. At higher intensities it is said to cause resonance of the internal organs and severe injury or death.
- Audible sound at high power levels causes discomfort, pain and temporary deafness. At high levels it causes permanent deafness.
- Ultrasound (sound at frequencies too high to hear) can cause paralysis, nausea and blindness.

A recent US military study of the literature on the effects of sound at all frequencies found many of the claims implausible. However, there are a whole range of acoustic weapons under development.

At the large end, these include the jeep-mounted Acoustic Blaster made by Primex Physics International. This is intended as a crowd-dispersal system, and like all such devices it suffers from the problem of how to generate enough sound. Deafening someone 10 metres away is not too difficult, but the intensity of sound falls away with the square of distance. In other words, at 20 metres away the sound is only a quarter of the loudness at 10 metres, and at a 100 metres – a useful range for crowd dispersal – it is only a hundredth as loud. It also needs to be considered that the troops manning the acoustic weapon are likely to be closer to it than the crowd.

The prototype Acoustic Blaster features an array of four explosively-driven sound projectors, and produces 175 decibels at 15 metres – enough to cause severe chest pain.

The next stage is a more advanced version in which the combined output forms a 'mach disk'. This is a type of standing wave effect in which the sound intensity falls off much more gradually. The 'acoustic bullet' from the advanced version is expected to have an effective range of 100–200 metres.

Sometimes long range is not necessary. The events of September 11 led to an increased interest in weapons that could be used to deter hijackers without risking damage to the aircraft. One approach is an acoustic weapon called the 'Directed Stick Radiator', a tube a metre long and four centimetres in diameter, housing a series of piezoelectric disks. Each disk acts as a speaker; a pulse starts from the first disk and is amplified by the second, amplified further by the third and so on, reaching the exit nozzle at over 140 decibels.

'You could virtually knock a cow on its back with this,' said spokesman Elwood Norris, who tested a scaled-down version of the DSR on himself. His company's official claim is more modest, being only that it causes pain and disorientation. It is far from clear whether this would be an effective way of stopping a hijacker, particularly an armed one or one ready to detonate a bomb.

Like the Acoustic Blaster, the DSR is still very much a brute force approach. It may be possible to use sound as a weapon without resorting to such extreme power levels. The US Army's Tactical Command has built a prototype called the Aversive Audible Acoustic Weapon. This comprises a backpack unit connected to a tube like a bazooka which projects a highly directional beam of sound. The audible sound is a carrier for 'complex waveforms' in different frequency ranges which have a direct effect on the nervous system.

According to Tactical Command, earplugs are no protection from the sound which causes 'non-injurious behaviour modification' of an aversive nature – in other words, anyone caught in the beam will immediately try to get away from it. It is not clear how the sound does this.

Scientific Applications and Research Inc are the leading suppliers of acoustic systems to the US military, and they go even further in their claims. They use a technology called 'Pulsed Periodic Stimuli'. According to company literature:

> Under certain frequency and modulation formats, pulse acoustic waveforms potentially have the ability to interfere with nervous systems, causing disorientation, or inducing a passive state within the targeted subject. These effects may be produced at significantly lower intensity levels . . . significantly reducing the required device output power.

At only 110 decibles (no louder than some power tools) they claim PPS can produce strong disorientation, perceptual changes and 'forced calm/lethargic states', ideal for subduing rioting mobs.

Again, how PPS is supposed to work is a secret. There is no published research on this technology.

At the low-frequency end, the claims made for infrasound weapons have been downgraded. Suppliers no longer talk about uncontrolled defecation and vomiting, let alone pulping internal organs. At most they offer 'physical discomfort and disorientation' causing 'strong interference/stoppage of activities'.

Infrasound is useful because it can penetrate buildings and other structures. There has been particular interest from the US Special Forces Command in what they call an 'Infrasonic/Ultrasonic Cannon', a portable weapon capable of affecting a target inside a building through the walls and 'remotely incapacitate or disarm occupants of compartments of a ship or rooms of a building'.

Another use of such a weapon would be to force people out of buildings, bunkers or other structures. This would reduce the need for commandos to carry out hazardous operations like storming buildings, when the risk of casualties from friendly fire is particularly severe. No details are available on this 'Infrasonic/Ultrasonic Cannon' program, and what the effects of the weapon are likely to be.

As ever, the high level of secrecy will delay any possible civil application of this sort of technology. There are already suggestions that military acoustic devices using multiple beams can produce a loud noise at a localized point in the distance – this could open the way for 'virtual headphones' from a sound system that follows your movement and beams sound for your ears only.

## On the Civilian Side

Military non-lethals are of growing importance with the rise of 'operations other than war' including various forms of peacekeeping. Although none of the technologies discussed here is mature, there is no doubt that there will soon be alternatives to the rubber-bullets-and-tear-gas alternatives. Although new weapons are invariably over-hyped – selling a project involving untried technology to sceptical generals can involve a certain amount of exaggeration – they may turn out to be as effective as intended.

A directed-energy non-lethal would have the same impact on riot control as the machine-gun did on infantry warfare. In the nineteenth century, soldiers still gathered in massed ranks to exchange rifle fire at each other. The machine-gun (among other weapons) made this type of warfare impossible. A directed-energy crowd control device would make the stone-throwing mob as vulnerable as a formation of Zulu warriors. Rioters who come out in the open to attack will be 'zapped' efficiently and accurately from long range before they can present a threat.

The cunning troublemaker will still seek to take cover behind peaceful protesters or other human shields, but directed-energy non-lethals such as acoustics and VMADS will be able to drive away such shields at will. And, if the technology works, they will be able to do so in a way that does not provide any pictures of baton-swinging brutality. The suppression of riots, potential riots, and in fact any sort of illegal or undesirable gathering, will become simple and straightforward.

Few would argue that 'the right to riot' is a civil liberty, but such technology can easily be abused, especially by regimes with few scruples about human rights. The massed gatherings that preceded the downfall of regimes in the Eastern bloc, for example, might have been broken up before they became effective opposition.

Much depends on the health effects of such directed-energy non-lethals and whether they will be acceptable in the civilian world. There are still questions about the Taser, for example, and whether it is entirely safe. The makers do not recommend using it on anyone with a history of heart trouble, who might suffer from heart problems if they received a powerful electric shock. In the real world, it is not always possible to establish a suspect's medical history, especially if they are coming at you with a baseball bat. This applies even more so if they are one of a mob of dozens of stone-throwers on a dark street.

Such factors might mean that some non-lethals are restricted to military use only, at least as far as the US and UK are concerned. However, there is a definite trend for such weapons to become more acceptable over time. Both tear gas and baton rounds were used in Northern Ireland but not in mainland Britain as a matter of policy for over thirty years. But demands by the police for more effective tools led to a change in this policy. CS gas was used during the Toxteth riots in 1981, and since 1996 it has been cleared for use by several regional constabularies. Baton rounds followed, and in 2002 police in Wales used a plastic bullet to help subdue a violent suspect.

The argument from the police is that without such weapons their only recourse is to use firearms. Given a choice between lethal and non-lethal weapons, they would much prefer to have the latter. As technology advances, and if it proves to be effective and does not result in too many fatalities, it is increasingly likely that new devices will become available for police use.

But the most important area is the export market. Dictators and repressive regimes may find that soldiers will not bomb or fire on their own people, not with live ammunition anyway. But powerful non-lethal weapons give them a way of putting down protests without facing revolt by the armed forces, without the world's media being full of stories of atrocities and massacres, and without any of the outrage that follows such brutality.

Used together, the weapons described above would give future regimes an unprecedented amount of power. Calling the people on to the streets has little effect if the crowds can be swept back off the streets as easily as with a leaf-blower. Occupying the parliament building is useless if the place can be 'neutralized' and the ringleaders captured without a fight and without martyrdom.

When talking of *Nineteen Eighty-four* and the technology of repression, surveillance and information control are usually mentioned. But even without these measures, a totalitarian state with the means to quietly and harmlessly smother dissent with a feather pillow of non-lethal weapons is going to be harder to dislodge than one which relies on the bludgeon. A media-friendly means of quashing demonstrators may be bad news for the chances of democracy across the world. Any other applications in the civilian world may prove to be of lesser significance.

# CHAPTER 10

# Vortex Rings

*Mmm – doughnuts!*
Homer Simpson

Vortex rings are so exotic that we don't even know one when we see one, even though they are closely related to the familiar smoke ring. Vortex ring technology looked like one of those 'might-have-beens', like jet packs or rocket planes. Having been around since the Second World War, nothing seemed to have come of it for decades, either military or civilian. But in recent years there has been real progress and there are a wide range of products in the pipeline.

While on holiday in Sicily a few years ago, I noticed an unusual cloud. It was small and seemed to be shaped like a disc, but what had attracted my attention was that it was rising and moving against the wind. It left a faint, wispy trail behind it. Looking through binoculars I could see that it was not a perfect disc, but more like a doughnut. Clouds do not normally make neat geometric shapes; this one did.

The mystery lasted for some time, until I realized that this strange cloud was over in the direction of Mount Etna. The great volcano, then going through one of its active phases, was blowing smoke rings from a new crater called Buocca Nueva.

Strictly speaking it was blowing steam rings. I later found that these were being studied by the vulcanologists at Etna. The rings changed as they progressed, starting out thick and squat like a doughnut and gradually spreading out into a thin ring with a diameter of some 200 metres. Dr Jug Alean and Dr Marco Fulle, the two scientists who recorded them, noted that some lasted as long as ten minutes and rose to an altitude of several thousand feet. They had never seen such rings over Etna before.

It was an impressive demonstration that the physics behind smoke rings works on a gigantic scale as well as on a small one. Smoke rings and steam rings are both examples of what is known as a toroidal vortex or ring vortex. A simple vortex is like a tornado or a whirlpool, a rotating cylinder of gas or liquid. If you bend the vortex cylinder around until the ends meet, you have a vortex ring. You could think of it as a self-contained tornado, a twister twisted into a ring.

Vortex rings have some curious properties, one of which is their unusual stability. A cloud or puff of smoke quickly loses its shape, but formed into a vortex ring its longevity increases dramatically. Part of the stability is from momentum – once something starts spinning it will tend to keep going, but there are also other factors. They are more stable if the ring is less dense than the surrounding medium. Smoke and steam, being warm, are less dense that the air around them, and can form very stable rings. However, it is possible to make a vortex ring out of thin air.

This stability is so great that a vortex ring will survive an impact with a solid object, bouncing off it like a rubber ball. When it does so, the impact can involve significant momentum, especially if the vortex is moving at high speed, striking with the force of a solid projectile.

Vortex rings can travel at surprising speed through the air. Moving air is usually slowed down by drag; even an extremely strong wind, like that from an explosion, is stopped after a few tens of metres. But the airflow around a ring makes it effectively frictionless, and it glides along with very little resistance even at high speed. But they can also travel very slowly. Even though the air within a vortex ring may be circulating at very high velocity, the overall speed of the ring itself may be very low – like a tornado whirling around at 200 miles an hour that proceeds across the landscape at walking pace.

Another feature is the integrity of the ring. As a vortex ring travels it is always the same molecules moving. A smoke ring is a way of transporting a quantity of smoke from one side of the room to the other. The movement defines a surface which has an inside and an outside which cannot easily be crossed, so smoke inside the ring will not diffuse out as long as it keeps moving. The ring only breaks up when it slows down.

**VORTEX RING**

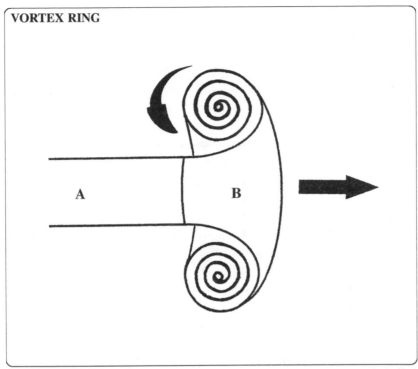

Vortex Ring formation: a stream of gas (A) passing through an aperture is 'rolled up' into a doughnut shape (B). The fast-moving centre of the stream forms the outside of the doughnut while the slower-moving edges of the stream form the core of the doughnut.

A vortex ring is formed in a gas or liquid when a fast-moving stream passes through a circular opening. As the stream comes out of the opening, the edges of the stream are slowed down by friction, but the middle of the stream continues at high speed. The effect is to roll up the stream into a doughnut shape: the fast-moving centre of the stream forms the outside of the doughnut while the slower-moving edges of the stream are rolled into the core of the doughnut.

The low speed at the heart of the ring and the high speed at its edges create a pressure gradient which will hold the ring together. An effect called 'boundary layer separation' means that there is effectively a barrier between the inside and the outside.

We can assume that the steam rings from Etna were formed because Buocca Nueva has a large chamber with a circular mouth, and as puffs of steam were explosively ejected from the mouth they would be shaped into vortex rings, with their surprising ability to keep their shape and momentum, and move against the clouds.

The renowned British physicist Lord Kelvin explored the science of vortex rings in 1867. The notion that a fluid medium could be rolled up into a stable ring that would then act like an independent body intrigued him. He even theorized that this might be the basis of matter, that the solid particles that we see as atoms might be microscopic vortex rings formed from an all-pervasive ether. Certainly vortex rings can behave like elastic solids, but the theory disappeared when the existence of the ether was finally disproved, and the work did not find any practical applications for decades.

## Vortex of Evil

To a military scientist, the vortex ring is a means of transmitting the force of an explosion over a long distance, and this concept was explored by the Germans in the Second World War. An Austrian scientist, Dr Zippermeyer, invented a device known as the 'wind cannon' intended to shoot down Allied bombers. The situation in Nazi Germany was rather chaotic as regards funding projects; given sponsorship from a suitable patron, an idea might be pursued which might not have had a hearing elsewhere. If you could persuade a senior member of the leadership of the possibilities of an invention it would be funded whatever its technical merit. This benefited seemingly eccentric ideas like Zippermeyer's.

The *Windkanone* generated an explosion in a combustion chamber and shaped the hot exhaust gases into a vortex ring using a special nozzle. The apparatus was the size of a building, with a combustion chamber 10 metres long, but it worked: Zippermeyer showed how it could break inch-thick wooden planks at a range of 200 metres. However, while breaking planks using nothing but thin air may be a good trick, it is not militarily useful: a small anti-aircraft gun has much greater range and destructive power. The Allied aircraft were way out of reach of Zippermeyer's invention even in theory, and the program was shelved.

In the anarchic world of Nazi research, other similar projects were also being undertaken at the same time. There was the 'atmospheric whirlwind gun' (*Luftwirbelkanone*), 'vortex gun' (*Wirbelkanone*), and the 'whirlwind gun' (*Wirbelringkanone*). In fact, it seems that any enthusiastic inventors with an ability to sell their ideas to the Nazi party could get funding for military development.

Few made any real progress, but the 'whirlwind gun', made by Professor Maas, also reached the stage of experimental hardware. This was more compact than the *Windkanone*, based on a cylinder 2.5 metres in diameter and of the same length with a smaller, longer tube inside it to shape the torus. Hydrogen was introduced into the outer cylinder and electrically ignited. It could send a vortex ring some 450 metres. Mass believed that a bigger version using a cylinder 30 metres in diameter would be able to bring down an aircraft at 5 km away, but this monster gun was never built.

Another Nazi design involved several nested cylinders injected with hydrogen and oxygen and fired in sequence. Known as the *Wiefealtingwindkanone* or multiple wind cannon, it was intended as an air-to-air weapon. It would have required a fairly substantial launch aircraft because of the torque (rotational force) it produced. In theory it should have been able to fire a vortex ring 1,000 metres, and such a vortex ring might have had catastrophic effects on a bomber. In practice it did not prove as effective as a conventional cannon.

A final concept developed by Dr Zippermeyer was the *Luftwirbelkanone* (air whirlwind gun). Realizing the problems with generating large vortex rings on the ground, he designed an artillery shell which would create one in the air close to the target. The device consisted of an artillery shell with a filling of coal powder and a detonator; it was supposed to burst in the air, spreading the dust into a cloud in the shape of a hollow cylinder. The detonator ignited the cloud, sending a blast wave in the form of a vortex ring along its axis.

Although the idea of using a cloud of dust as an explosive anticipated the fuel-air explosive devices of the 1960s, producing a dust mixture which would ignite reliably (let alone one that would adopt the desired shape) was an insurmountable hurdle. In fact, modern thermobaric weapons which use a similar principle of generating and detonating an explosive cloud are still working at the limits of our technology and something like the *Luftwirbelkanone* is not yet feasible.

This assortment of exotic armament provided an interesting puzzle to the US Intelligence Technical Branch which found the apparatus at the Artillery Proving ground in Hillersleben in 1945. The plethora of German devices were all investigated after the war in a secret program called Project Squid.[1] As usual, there are claims that these were developed into secret weapons, but it seems more than likely that nothing came of them; in any case the US already had its own research program in this area.

## US Vortices

The US military were investigating the same technology for different reasons. Poison gas had been used during the First World War, and there was a strong possibility that it would be used again in the Second World War. If the Germans used it, the US needed to be able to retaliate in kind. But the problem with gas warfare, proven time and again on the Western Front, was that it was unpredictable. A change in the wind direction meant that the poison gas would blow right back over friendly troops. What was needed was a way of delivering gas right on the target.

Inventor Thomas Shelton came up with a solution: use an apparatus which would form the poison gas into a vortex ring. The ring retains its integrity over a considerable distance, as we have seen, so it should be able to deliver a measured quantity of chemical agent from long range.

Shelton built a demonstration prototype with a bell-shaped chamber powered by a shotgun cartridge. When fired, it sent a 45 cm smoke ring a distance of 50 metres with 'an eerie howling sound'. The production model would have been much larger and mounted on a vehicle.

Rather than using tons of gas and indiscriminately blanketing the battlefield, the vortex ring gas projector would just fire tiny quantities at specific targets – trenches, buildings or bunkers, like a chemical 'smart bomb'. The noise of the vortex as it approached would certainly add to its psychological effect.

Shelton's device was never used – it is not clear how far the idea was progressed before the war ended – but he maintained his interest in vortex rings. He invented a toy pistol that would produce a ring capable of knocking over a paper target at a range of a few metres. This was the golden age of science fiction and ray guns, and the toy, dubbed the 'Flash Gordon Air Ray Gun' was a big hit. *Popular Mechanics* voted it 'Toy of the Year' in 1949.

Vortex ring toys reappeared several times in subsequent decades, including the popular Whammo Air Blaster from 1965, right up to a modern version, the Airzooka. They are great toys – from a child's point of view the advantage is that they never run out of ammunition, while for parents not having plastic darts, pellets or other projectiles appearing under the sofa and in the fish bowl is an asset. (Of course, it also means that you cannot simply confiscate the ammunition to stop the little darlings from shooting everything in sight.)

## Modern Vortex Weapons

Vortex ring weapons are not just toys. In the 1990s the US military started to become interested in expanding the scope of its non-lethal weapons. As we have seen, the traditional baton rounds and tear gas have their limitations. In particular, rubber bullets which were not intended to kill could be lethal, especially if fired at the head. One high-tech option was to use directed energy (see Chapter 8), another was to use vortex rings.

In conjunction with Pennsylvania State University and a number of contractors, the US Army Research Laboratory designed a non-lethal vortex ring gun. The underlying principle is the same as Zippermeyer and Shelton's, but of a size between the two. The weapon is a modified 40mm grenade launcher with a muzzle adapter over the barrel to form a cylindrical chamber. It fires 'blank' rounds which generate a flash and a bang but no projectile. A chemical incapacitating agent (such as CS or CN tear gas or pepper spray) can be injected into the chamber, or it can fire just air.

In 1998 the device was shown knocking down a 70 kg mannequin at 10 metres. The makers reckoned that this could be increased considerably, and that 'body resonant impulses' have greater effect and could be fired at greater range. It could deposit chemical agent on to the target accurately at up to 50 metres.

The weapon is safer than rubber or plastic bullets as it does not fire a solid projectile which can break bones or crack skulls. The force is spread over a comparatively large area, unlike some plastic baton rounds which are small enough to fit into the eye socket with potentially horrific results.

Weapons firing rubber bullets can fire one round only, but multi-shot vortex ring weapons are much more straightforward. Army research labs pointed to the GL-6, a portable six-round launcher, and the Mk19 tripod-mounted launcher as weapons which could easily be converted into vortex ring operation. The larger weapon is belt-fed, so the operator can keep firing until the desired effect is achieved.

There are several other projects looking at non-lethal vortex ring weapons as well. These include one by Scientific Applications & Research Associates of California, who are marketing it to police forces. According to their advertising:

> A supersonic vortex of air hits its target at about half the speed of sound with enough force to knock them off balance. The vortex feels like having a bucket of ice water thrown into your chest. For use in riot or crowd control to stop the crowd from advancing without doing them any harm.

There were also some possible lethal applications. One of these was studied under a project called MARAUDER in 1993 at the USAF's Phillips Laboratory. The wonderfully contrived acronym stands for Magnetically Accelerated Ring to Achieve Ultra-high Directed Energy and Radiation. Utilizing the USAF's mammoth Shiva Star power system, experimenters projected toroids of superheated gas at high speeds. At high temperatures, superheated gas becomes as plasma in which the electrons are stripped from the atoms, making it electrically active.

When the toroids from Marauder hit a target the impact had what was described as 'extreme mechanical and thermal shock'. The impact also resulted in an intense pulse of electromagnetic radiation, something like a miniature E-bomb (see Chapter 14) which could knock out sensitive electronics on aircraft or missiles. Having been proved in principle, the program was given a high security classification and there have been no reports since 1993.

If successful, Marauder would be a highly effective weapon. One official report mentions a velocity in excess of a hundred kilometres a second, far more than any known missile or projectile, faster than anything except a laser beam. Marauder would be an obvious choice for missions like shooting down satellites from the ground. A laser would be affected by atmospheric haze and dust which absorb and scatter the beam, but a ring of plasma would be unaffected. A range of a few hundred kilometres would be enough to strike anything in low orbit, and the effects of even a small plasma ring would be devastating on something as fragile as a satellite. One could also envisage anti-missile systems using similar technology – the high velocity makes it ideal for intercepting even the fastest missile.

However, it seems that we are unlikely to hear anything further of this particular technology until a politically convenient moment comes to reveal it.

## Rings into Ploughshares

The work carried out on these weapons has led to a much greater appreciation of the physics of vortex rings. Modelling them requires considerable computer power, as the fluid dynamics involved are complex, but the end result is that we now have the technology to create far more stable vortex rings.

Researchers at the Naval Air Warfare Center Weapons Division at Point Magu in California have already been looking into ways in which this technology can be used in the civilian world. They have been focussing on using vortex rings as a means of directing gases to a target location. Their studies have shown that a sufficiently stable ring can survive several collisions, which means that vortex rings will be able to reach the parts that other delivery systems cannot.

One proposed application is in firefighting, using the vortex ring system to send a quantity of flame-extinguishing gas (such as halon or carbon dioxide) to be directed on to a fire from a distance. Because the vortex rings are capable of travelling through corridors or air ducts, they will be able to get to locations which are inaccessible to fire hoses. Crawl spaces and maintenance shafts, which are sometimes the site of electrical fires, are also difficult for conventional apparatus.

Micro-scale vortex rings have the same essential properties as their larger cousins, making them suitable for certain types of medical use. One application is a novel inhaler. Instead of spraying the vaporized medicine into the mouth, where there is the risk that it will get no further than the trachea, a dispenser will blow a series of tiny vapour rings. These ricochet down through the airways and find their way into the lungs where they deliver the medicated vapour to where it is needed.

It has also been suggested that the same principle could be used for scuba equipment, as it might be a more efficient way of getting oxygen into the lungs than existing systems.

The Advanced Telecommunications Research Institute in Kyoto is working on a smell cannon. This couples the ability to fire a small shot of odour with a computer-aided sensor which can follow a person and direct the smell at their face. This could be used to fire fresh coffee smells at potential customers passing a coffee shop, or baking bread to shoppers as they pass the bread counter.

If smells can be directed so accurately, so could air. Why waste a huge amount of energy providing air conditioning for the whole room when it is only the occupants who need cooling? A vortex ring version of the desktop fan could ensure that the user is kept bathed with cool air using a fraction of the energy of an old air conditioner.

Vortex rings also have possible applications in liquid. Water is a thousand times denser than air, so a vortex ring in water carries a thousand times more momentum. This is being tested as a method of clearing fouling from ships. Underwater, a high-pressure hose has limited effect, but vortex rings could be much more effective. A similar technique has been proposed for cleaning pipes: again, an increase of pressure has little effect on debris on the side of the pipe, but a vortex ring with the same diameter of the pipe could do the job.

A variant of this idea has been proposed as a way of stimulating oil wells. In the usual course of oil extraction, the well gets clogged up with paraffin wax, scale deposited by carbonates and other contaminants. At present, mechanical scrapers and hydraulic devices are used to clear the fouling. Acoustic methods using high-intensity sound have also been tried, but communicating the sound into the depths of a well has proven difficult.

Using a pulsating jet stream, a series of vortex rings can be sent down the piping. These will break up fouling and clear deposits all the way through the system; the vortex ring exerts much more force than acoustic methods, and unlike the mechanical tools it can reach into the smallest perforations.

Liquid vortex rings could also be used as a means of dispensing medicine in the bloodstream. The stability of the ring is determined by its rotation, density and velocity, so a vortex ring dispenser could launch rings programmed to break up and release their load as exactly the right location.

They could also be useful at the other end of the digestive system.

Enemas have very limited ability to deliver medication to the colon, but conditions like inflammatory bowel disease need to be treated with steroids which only have a local effect. Again, this looks like the sort of area where a dispenser based on vortex rings might be able to overcome the limitations of existing technology.

Other applications could extend into many areas where we currently use sprays or aerosols. Trying to hit a fly with spray is a challenge, especially if it is the other side of the room, but a dispenser which spat out a vortex ring containing a dose of insecticide would hit the insect precisely and without warning – a kind of domestic version of the chemical warfare proposed by Thomas Shelton. On a larger scale, crop dusting using vortex rings rather than random clouds of chemical agent would be far more precise and result in less wastage and less contamination of neighbouring areas.

Hewlett Packard, one of the world's leading manufactures of computer printers, is developing a new type of inkjet printer using vortex ring technology. Inkjet printers use something called an 'orifice plate' to shape a stream of tiny ink droplets. There are a series of firing channels from which the ink is sprayed, which are aligned with the holes (or orifices) in the orifice plate. This makes manufacture relatively difficult and complex, as the alignment has to be very precise.

Attempts to make an inkjet without an orifice plate suffer from what HP call 'poor directionality' – in other words the ink ends up all over the place. HP's new design works by creating vortex rings instead of a random spray from the ink channels. The vortex rings are more even and far more directional than bubbles, so the orifice plate is not needed. This should lead to cheaper printers and better quality printing.

Even something as prosaic as house paint spraying could benefit. A spray attachment that could send a series of accurately directed rings from a distance would reduce the need for ladders or scaffolding. Again, the stability of the rings means that they do not get blown away by the wind, like the giant steam ring I witnessed at Mount Etna, so there is less risk of the paint spray being blown all over the place and consequently less need to cover everything in dust sheets.

These applications are only the most basic use of vortex rings. At present, vortex ring technology is still very much in its infancy because the engineering models of how vortex rings behave do not exist. Industry is unlikely to put money into something with no immediate payoffs, but vortex ring weapons are driving significant investment into computer modelling by the military. A conference on non-lethal weapons in Europe in 2003 attracted five papers on different aspects of vortex rings from the UK, Germany and Russia, all pointing to an increasing understanding of how they can be formed, controlled and directed.

One area where an improved understanding of vortex rings can help is in aviation safety. Under some circumstances aircraft wakes can form vortex rings; normally wakes disperse very quickly, but obviously ring formation means that they may persist for some time, and another aircraft running into one will experience a localized patch of turbulence. This can be very hazardous, as the most likely place to encounter such a vortex would be during landing or take-off behind another aircraft.

The V-22 Osprey tilt-rotor discussed earlier is a particular victim of ring vortices, as it is prone to generate them during hover. When this happens the Osprey is trapped inside its own vortex ring and suffers from a sudden loss of lift. The reasons behind this are now understood and pilots know how to avoid causing the condition, but without knowledge of vortex rings the problem would have remained mysterious.

Vortex rings also appear in pulse jets and pulse detonation wave engines. The 'doughnuts on a rope' contrail is actually a series of vortex rings, and the vortices are more effective at carrying momentum (and so pushing the plane forward) than a normal exhaust. It is possible that there is a great deal of knowledge in this area which has yet to be made public.

In the long-term we can look forward to vortex ring technology being incorporated into machinery. Pumps, filters and heat exchangers could all be enhanced by using vortex rings, but at present designers have no understanding of how to make the phenomena work. Given more information, it will be possible to improve the designs of everyday household items from central heating pumps to coffee makers and air conditioners without adding any complex new elements or elaborate electronics, simply by using vortex rings in the flow of gas or liquid.

Although this is not a technology on the same scale as nuclear power or digital computers, vortex rings clearly have applications that are only limited by the imagination. It may become so pervasive that we hardly notice it. In years to come, the toilet cleaner that disinfects 'all the way around the bend' may work using a technique that was originally pioneered by the Nazis to shoot down Allied aircraft.

# CHAPTER 11

# Mach 10 Airliners and Space Tourism

*Faster than a speeding bullet!*
Radio introduction to
*The Adventures of Superman* (1940)

One afternoon in South London, I stood on a hillside with a gathering of other people. Many had binoculars or video cameras; all had come to witness the end of an era. As we watched, three sleek white darts appeared from the west, one after the other: Concorde airliners on their final flight. The three in turn circled around the City before dropping below our line of vision towards Heathrow.

The extraordinary number of people who turned out to see these final flights underlined the fact that Concorde was without doubt the world's best-loved airliner. It did not open up the world to the masses as the 747 had done, or provide cheap travel for all, but it was a symbol. Concorde stood for aspiration, for technical excellence, for a bright future where aircraft were not merely flying buses but dream machines carrying us around the world in hours with style.

Huge crowds turned up to air displays just to see Concorde, and thousands went on short circular flights over the Bay of Biscay simply to experience the thrill of breaking the sound barrier for themselves, without actually going anywhere. The appeal had nothing to do with utility, but everything to do with Concorde's sleek looks and unique performance.

As remarked earlier, modern aircraft look very much like their predecessors from a generation ago, and have similar performance: Concorde was the single sparkling exception to the rule.

Commercial realities were against supersonic passenger travel from the start. The US bowed out of the race at an early stage, convinced that it would never be economically viable, an analysis which proved to be correct. Even after Concorde's £1 billion development costs were written off by the British and French governments, it proved impossible to operate the aircraft at a profit. For decades it made a great flagship but a poor earner, but after the loss of a Concorde in an accident outside Paris revenues were unsustainably low.

The Russians did complete in this area, but their equivalent, the Tu-144 airliner dubbed 'Concordski', was also developed very much as a prestige project. The Tu-144 had a shorter working career than Concorde and also proved highly uneconomical. There is still talk of the aircraft having an afterlife as a test plane for NASA's supersonic experiments, so it may end up as the sole survivor of the supersonic transport era.

The chances of a successor to Concorde look slim. Speed does not have enough appeal in the modern world. Boeing had plans for a 'sonic cruiser', a fast subsonic airliner perhaps 15 per cent quicker than other planes, but were forced to drop the plan as potential customers did not find increased speed attractive compared to the additional fuel costs that it entailed. Boeing moved on to a plan for a 7E7 Dream Liner, an aircraft which would be cheaper to run than existing airliners rather than faster.

This does not mean that there will never be another supersonic airliner. Supersonic flight is still a major area of interest – it is simply that the development is all taking place in the military arena. Concorde itself was closely related to the military aircraft of the day, its swept delta wings inherited from the British V-bombers like the Vulcan. Prototypes already exist for new types of jet engine that would leave Concorde standing.

Looking even further ahead, the sky itself is not the limit. In the heyday of the space race, space tourism seemed like an inevitable development. While hopes for cheap space travel faded, the first space tourist has already flown – although the $20 million that Dennis Tito paid for his one-week experience is outside most people's reach. The real dream is for space travel that is affordable to all, and that like aviation before it, travel into space will become a mass-market. All we need is the rocket equivalent of the wide-bodied airliner. As with the airliner, this is only likely to happen as a consequence of military developments.

## The New Requirement: a Hypersonic Bomber

As we saw in an earlier chapter, plans for a rocket-powered sub-orbital bomber originated in Germany in the Second World War. The Sanger project came to nothing, though the work was closely studied in the US after the war. The current geopolitical situation has created a new demand, and if DARPA are successful, a hypersonic bomber will be ready to replace the USAF's existing fleets of B-1, B-2 and B-52 subsonic bombers by 2025.[1]

DARPA's plan is for a reusable Hypersonic Cruise Vehicle, or HCV, which will take off from a runway like an aircraft, fly to any point in the world and release a 5-ton bomb load from orbital altitude before returning to base.

The HCV would resolve one of the most difficult political problems for the US at present: the need to maintain air bases in friendly countries to carry out air strikes. The HCV would remove this restriction, giving the leadership the political freedom to strike anywhere at any time without the need for support from allies.

It would also permit a much faster reaction to world events and tactical situations. Flying from bases on mainland US, B-2 bombers flying attack missions during Operation Enduring Freedom in Iraq took twelve hours to reach the battle zone; during this time many mobile targets had moved out of the original target area. Saddam Hussein reputedly avoided death on more than one occassion because US bombs and missiles arrived some time after he had left. Fast reaction means the ability to strike before the enemy can move.

As the existing bomber fleet ages and defences become more sophisticated, bombing will be increasingly hazardous. Given the small number of aircraft involved – the entire US bomber fleet only numbers 200 – continuing losses would be serious, not just politically but also operationally. The typical loss rate of a few per cent per mission experienced in the mid-twentieth century would not be sustainable for any but the shortest of conflicts. The HCV will solve that; it is far beyond interception by anything around at present, and raises the bar so high that few countries could even attempt to build a defence system.

The HCV will only be able to carry a light load of weapons – less than a quarter of the capacity of a B-52 – but high precision and phenomenal speed will make up for it. The 'bombs' it drops may not even have any explosive content, but instead would be giant darts made of tungsten or depleted uranium, each weighing 100 kg. Travelling at mach 5 or more, these darts would be able to destroy most targets, including hardened underground bunkers, simply by force of impact. The effect would be more like a meteorite strike than any current weapon.

Alternatively each bomb could contain a cluster of smaller darts, whose speed and density gives them tremendous destructive power. A study by McDonnell Douglas concluded that darts about the size of a biro and weighing half a kilogram each would be capable of knocking out tanks, while a shower of 2 kg darts would be able to destroy bridges and other structures. A large bomb would release hundreds of these darts which would all impact at the same instant, producing as much 'shock and awe' as any high-explosive weapon.

Having established a requirement for the HCV, there is the technical issue of how such a craft could be powered. As we have seen, the options for high-speed flights are basically rockets or jets.

## Ramjets

Although rockets are inherently more powerful than jet engines, they suffer from a significant disadvantage. A rocket of any size has to carry both fuel and oxidizer, whereas a jet engine is an 'air breather' and takes its oxygen from the atmosphere. When the rocket only needs to burn for a few seconds or minutes it is not at too much of a disadvantage, but for longer ranges the oxidizer becomes a major factor.

The main tank of the Space Shuttle, for example, carries over a hundred tons of liquid hydrogen – but it is dwarfed by the tanks of liquid oxygen beside it which contain the 600 tons of oxygen needed to burn the hydrogen. Because it uses atmospheric oxygen, the amount of fuel needed by a jet is a fraction of what is needed by a rocket. Carrying less fuel means greater payload, or the same payload at greater speed and economy. A fast jet is much better than a rocket, but only if it can overcome the limitations which restrict it to comparatively low speeds.

Normal jet engines, both turbojets and turbofans, rely on a system of compressors to raise the air/fuel mix to high pressure before igniting it. Such engines have a practical limit of mach 3 or so, and even this speed can only be reached with considerable difficulty. It is unlikely that any attempt will be made to outdo the Blackbird with a turbojet.

However, at high enough speed the compressors can be dispensed with, as the sheer speed of the airflow into the intake is enough to compress the air. This type of engine is called a ramjet, and it essentially has no moving parts – all it needs is a means of injecting fuel, and airflow takes care of the rest. However, because there is no compressor, the ramjet cannot work at low speeds. It needs to be boosted to a speed high enough for the ramjet to start working.

One way of doing this is with a rocket motor, and as early as 1955 the USAF were using a combination rocket-and-ramjet missiles. The Bomarc missile (named after the makers, BOeing and MARC, the Michigan Aerospace Research Center) was one such. This was an oddity, a surface-to-air missile developed by the USAF, when all others SAMs belonged to the Army. The explanation for this is that the Bomarc was conceived as a 'pilotless interceptor' for taking out Soviet bombers rather than a missile. A liquid-fuelled rocket accelerated it to supersonic speed, at which point the ramjet cut in and the missile reached its cruising speed of mach 2.8 at 100,000 feet.

The Bomarc had a maximum range of over four hundred miles and carried a small nuclear warhead, making it an effective defence against enemy bombers. It was phased out of service from the early 1960s, having become steadily more redundant as the real threat shifted from bombers to ICBMs.

Ramjets fell out of favour with the US missile community, though the French have a ramjet-powered nuclear missile. The Russians, however, are the biggest players in this field, and this has triggered some concern in the West.

The Russian SS-N-22 anti-ship cruise missile, code-named Sunburn, has been operational since the 1980s. The French Exocet and American Harpoon are the nearest Western equivalents, both travelling at some-what less than the speed of sound. The Sunburn approaches its target at over mach 3, giving far less opportunity to shoot it down or take evasive manoeuvres.

There were worries that nations like Iran and China would buy the Sunburn, threatening US carrier task forces which might be vulnerable to its high-speed attack. Officially the US Navy is not worried, but some have suggested it could turn aircraft carriers from invincible fortresses into sitting ducks.

In fact, the US even planned to purchase a derivative of the Russian ramjet-powered Kh31 cruise missiles in the late 1990s. In the end the sale fell through due to budgetary considerations, but it would have been somewhat ironic for American warships to be armed with a Russian weapon. It may be that if the source had been different then the budgetary situation could have been resolved.

Meanwhile, Russian ramjet technology is spreading. A joint Russian–Indian developed anti-ship missile, the BrahMos, is due to enter service in 2005. According to the makers, this ramjet-powered missile is three times as fast and has three times the range of earlier missiles. Opponents describe it as a cruise missile with the capacity to strike cities in Pakistan, and there is no doubt that it represents a significant improvement in capability.

Ramjets are not the ultimate jet technology though, and the stage is set for something even faster.

## Going Hypersonic: Scramjets

Designers have long wanted to build aircraft and missiles that operate in the 'hypersonic' range – more than mach 5. Ramjets work in theory at speeds up to mach 8 or so, but at that point a new difficulty arises. The forces involved in compression in the combustion chamber are so great that the gases are heated to a point where they break up without burning completely. This puts an effective speed limit of mach 8 on ramjets.

To reach even higher speeds takes a further refinement. In a ramjet, the airflow is slowed down to subsonic speed as it passes through the engine, and this causes the gases to heat up and dissociate at high speed. But if the whole combustion process takes place while the airstream is travelling faster than the speed of sound there is less compression and the problem disappears. This type of engine, a 'supersonic combustion ramjet' or scramjet, can be run at anything up to mach 25 – in theory. Designing the inlet is a major challenge, as the turbulent supersonic airflow can simply blow out the flame.

Engineers describe the problem as being like keeping a match lit in a hurricane, a hurricane faster than any wind on earth.

Scramjets have a much higher specific impulse than other types of jet, providing much more push per unit of fuel burned, enough to make them competitive with rockets. The rapid compression in the combustion chamber increases the temperature to a level where the fuel will automatically ignite and burn completely without breaking up. With few moving parts, high efficiency and the ability to operate at very high speeds, the scramjet looks perfect for very high-speed engines, but getting it to work has proven problematical.

The USAF has undertaken a number of initiatives to build high-speed missiles based around scramjets. The most public program has been NASA's Hyper-X program, which is concentrating on the X-43 unmanned vehicle. Four metres long, the X-43 looks something like a giant surfboard; like the X-15 before it, it is carried into the air by a B-52 and released at high altitude. A rocket booster takes it up to speed before the scramjet starts working.

The flight schedule called for two flights at mach 7 and one at mach 10 over the course of 2001–2 as part of the five-year program. It was loudly trumpeted as NASA's new contribution towards aviation in the twenty-first century. After an embarassing series of mishaps in space, it was a way for the embattled agency to recover some prestige, 'something that will make it easier, safer and faster to fly across continents and into space'. The first flight should have broken the existing speed record set by the X-15, but unfortunately, during the first live flight test in June 2001 there was a malfunction in the rocket booster, the X-43 went out of control and the self-destruct mechanism was triggered, turning the test vehicle into a shower of debris over the Pacific Ocean and adding some $25 million to the cost of the program.

After an accident investigation lasting two years, the second test flight was tentatively scheduled for the end of 2003, and then delayed to 2004. To everyone's relief, the flight was a success. However, by this time a new menace loomed over the project: President Bush directed NASA to turn its attention to space again, to putting astronauts on the moon and Mars. Just as the X-15 rocket plane was sidelined when NASA was forced to put its resources elsewhere, the scramjet program, to build a full-size vehicle, may be a victim of rivalry with the more glamorous field of space exploration.

There are other scramjets programs too. In Australia, a less ambitious program has quietly been making progress. Known as HyShot, it is jointly funded by several agencies including the British Defence

Evaluation & Research Agency (DERA), and it is an altogether less dramatic production than NASA's extravaganza.

Instead of boosting their scramjet up to speed with a rocket, the Australian team launched it vertically; it gained speed as it fell back to Earth, and at the point where it reached mach 8 the scramjet was activated.

The first live flight took place in July 2001. The rocket went up successfully and HyShot came down . . . but nobody knew where. A major search operation was mounted to find it. A rocket of the right description was found which turned out to be the wrong one – nobody has yet established what it was. However, HyShot was eventually recovered, and the onboard instruments confirmed that supersonic combustion had taken place. The program is now moving on to a more advanced version which will not simply combust at supersonic speed but will provide positive thrust.

Other scramjet initiatives have been less public. The US Army's Tactical Command Advanced Systems Concepts Office (TACOM ASCO) is working to launch a scramjet-propelled projectile from a 120mm cannon. This would be a replacement for a normal anti-tank round, except that instead of just being an inert mass of depleted uranium, the projectile would accelerate after launch, powered by its own scramjet engine.

Firing a scramjet from a cannon is an easy way of starting it off at high speed, but the drawback is that it has to be able to withstand the stresses of launching from a gun barrel, which can be up to 60,000g. Tank rounds are fired at over mach 6 and are slowed in flight by air resistance. By the time they reach the target up to 6 km away they may have dropped to mach 4. A projected Scramjet weapon would have the same speed at launch and would then accelerate, and at 6 km it might be travelling at mach 7, cutting down the time of flight to the target and impacting with more than twice the energy.

There have already been successful launches of prototype scramjet projectiles in projects funded by DARPA for the US Navy, showing that they can ignite an engine and that it can maintain its velocity over a significant distance. Later versions will accelerate up to mach 7 or more.

This type of launch procedure could not be scaled up, but it does open up the possibility of launching small unmanned air vehicles at phenomenal speeds. The US Navy possesses a good supply of large cannon, and these might be put to use for firing hypersonic missiles or drones for hundreds or thousands of miles – a possible rival to the HCV for the global strike role.

Meanwhile the Russians are moving from ramjets to scramjets. An engine designed by the Central Institute for Aviation Motors was tested in the 1990s by a joint Russian and French team. This was a combined ramjet-scramjet design in which combustion started off at subsonic speed and then moved to supersonic as the speed of the missile increased. The test team claim they achieved mach 5.7 and that the combustion was supersonic. American engineers looking at their data were sceptical, suggesting that the airflow was mainly subsonic although the core may have been supersonic.

The French and Russians are continuing to collaborate in this area, with an experimental hypersonic vehicle with a combined ramjet/scramjet due to fly in 2009. This is intended to provide the technological basis for a future mach 8 missile to replace the existing French nuclear deterrent.

The US Air Force Research Laboratory, or AFRL, is spending at least $49 million on scramjet engines, with a big contract going to jet engine makers Pratt & Whitney. They are satisfied with laboratory and wind-tunnel tests of small scramjets and want to progress to much larger engines with will provide much more thrust, in fact three times as powerful as the turbojets that powered Concorde.

Their interest is partly in building vehicles like the HCV, but also in engines for missiles. As the HCV studies show, a missile at high speed becomes a much more dangerous weapon, with the ability to strike more quickly without being intercepted and hitting much harder. There are several plans for bunker-busting weapons, some intended for use against stockpiles of WMD, and these will require a scramjet engine or something like it.

Enthusiasm for scramjets runs high, and the competing US, Russian, French and Australian programs show the level of international interest. Until a full-scale one is constructed that works reliably in practice it will not be possible to say whether the technology is truly viable. However, when this occurs it is a fair bet that the civilian plan makers would start carrying out calculations on costs, markets and whether a scramjet-powered Concorde successor can be made more economically viable than the original.

A scramjet aircraft will still need some means of accelerating it to the speed where the scramjet will work. Ultimately this will probably mean an engineering solution that mates a scramjet with a turbojet so that it can take off at subsonic speeds and move through supersonic and hypersonic modes. This is likely to take decades, and early scramjets will work like the old Bomarc missile, taking off with the aid of rocket boosters. This is acceptable to the military, but not for flights taking off from Heathrow: if the noise of Concorde was unacceptable, then giant rockets are unlikely to meet with approval, however quickly they get you across the Atlantic.

The noise of the sonic boom is another matter, and applies to anything travelling faster than sound. It can be reduced significantly by shaping the aircraft, and a USAF program for a Quiet Supersonic Platform is intended to produce an airframe which will not produce a significant sonic boom at ground level. Of course, engine noise will still be an issue.

'Hypersonic plane could fly to New York in forty minutes,' proclaimed one newspaper headline before the X-43A's doomed test flight. But hypersonic passenger planes are much further off than hypersonic missiles and strike aircraft. Graham Warwick, America's editor for *Flight International* and an authority in this area, estimates that scramjet-powered missiles will not be in service until 2010–15, followed by aircraft in 2015–25. Civil applications, he believes, will only follow some years after this – if at all.

## Pulse Detonation: 'Donuts on a Rope'

Whilst scramjets have had some media attention, an alternative technology for reaching hypersonic speeds has also been progressing behind the scenes. This is a Pulse Detonation Engine or PDE, also sometimes called a pulse detonation wave engine. It is a descendant of the pulse jets which we saw earlier powering the German Second World War flying bombs.

Normal jets, including turbojets and scramjets, are 'constant pressure' engines which use a steady stream of fuel. By contrast, the PDE works in pulses. It is basically a tube with a valve at one end. The valve is closed and fuel is injected and detonated, producing a powerful single pulse. The valve is opened to admit more air, and the cycle is repeated. Each cycle takes a fraction of a second.

The PDE is similar to the pulse jet but with an important difference. In a pulse jet, the fuel burns at subsonic speed – technically known as deflagration rather than detonation. In a PDE the flame runs through the cloud of fuel in the combustion chamber at supersonic speed, perhaps as much as mach 5. This is why it is known as pulse detonation to distinguish it from the earlier version which might be described as 'pulse deflagration engines'. The difference is like the difference between a burning lump of coal and a stick of dynamite going off: the amount of energy released is the same, but the dynamite releases it much more quickly and produces far more impulse.

Detonation is much more difficult to achieve than slow burning. This is a blessing when you set light to your gas burner or Christmas pudding, when you want a gentle flame rather than a shattering blast, but it is a problem for PDE engineers.

Supersonic combustion gives the same high efficiency that is seen in the scramjet. It generates very high temperatures and pressures compared to deflagration, making it suitable for an engine with a very high power-to-weight ratio; but the engine will also have to be able to withstand these conditions without shattering or melting. The two major US engine makers, General Electric and Pratt & Whitney, are both heavily involved in PDE programs.

While PDEs are inherently more efficient than constant pressure systems, the obstacles are significant. The frequency of pulses is 'uncomfortable', according to the AFRL's chief of combustion science. In real terms this may mean that it can do structural damage to the airframe over the long term, gradually shaking it to pieces. As mentioned in an earlier chapter, the original pulse jets were very noisy and the vibration made them unsuitable for manned aircraft.

Rather than the steady roar of a jet, or the buzz of a doodlebug, a PDE will make a sound like a giant machine-gun. The AFRL is working on a PDE with a frequency of 20 hertz. This means that it detonates a small supersonic explosion twenty times a second. Successful ignitions make a sharp 'crack' sound instead of the softer 'wumpff' of deflagration, suggesting that the PDE is going to be a very noisy beast.

Future PDEs will operate at much higher frequencies, probably 400 hertz or more when the technology matures. This will involve several different approaches, including using a bundle of eight or more detonation tubes together, each pulsing many times a second. To ensure that the supersonic detonation occurs, each tube may need to have a 'predetonator' which acts like the fuse that sets off a bomb. This will consist of oxygen and ethylene fuel and will be in addition to the normal fuel.

Another area which is still in doubt is how this high rate of pulses will be maintained at a consistent rate, with all the tubes being in synchronization. The likely solution will be an electronic timing system, but the problem is far from trivial and is one of many that might become show-stoppers for the PDE program.

When and if all this can be made to work together it will lead to the Holy Grail of engine design: a highly efficient engine that works equally well at subsonic, supersonic and even hypersonic speeds. It will also have a very good thrust-to-weight ratio, high enough so that if the thrust is diverted downwards it is suitable for vertical take-off. Boeing are believed to be working on a family of VTOL aircraft using PDE engines.

Unlike the ramjet which requires a rocket or other means of propulsion to bring it up to its operational speed, the PDE is equally happy going slow or fast. The combustion speed inside the engine will be the same whatever speed the aircraft is going, and fuel economy will be equally good at any speed.

This would have a tremendous impact on both the military and civil markets. In particular though, it would combine the advantages of low fuel consumption with supersonic performance. This could lead to a new generation of airliners faster than Concorde which cost less to run than conventional airliners. More importantly for the military, it would make the hypersonic bomber a viable prospect.

Some people have suggested that such an engine may already exist. Although the AFRL's prototype PDE did not fly until 2003, rumours of PDEs have been around for some time. These have focused on the aircraft known as Aurora, supposedly a replacement for the Blackbird and capable of even higher speed.[2]

Since the 1980s there have been reports in California of curious contrails shaped like 'doughnuts on a rope'. These are sometimes linked to a curious low-frequency sound, and some aviation enthusiasts have concluded that the contrails and sounds are an indication of an aircraft using a pulse-detonation engine. The pulses may be in the range of 60 to 100 hertz, but the speed of the mystery plane is impossible to estimate and guesses of mach 12 or more are sometimes bandied about. Aviation writer Bill Sweetman has made the case for these being indications of a secret aircraft powered by a PDE.

Exactly what such an aircraft would be is an open question. It may be a prototype being used to test the engines, or it might be a super-Blackbird, a high-speed, high-altitude reconnaissance aircraft. Equally there have been rumours about a secret 'silver bullet force' of bombers, an early forerunner of the HCV. This force may have been developed during the Cold War era and tasked with nuclear strike; it could yet be adapted to a conventional role.

Meanwhile the AFRL are insistent that they have never flown a PDE aircraft before. However, their prototype PDE test aircraft is named Borealis. This suggests that, at the very least, someone has a sense of humour over the Aurora question.

Unless there is a secret aircraft already flying, it appears that PDE technology still has some way to go before it is likely to challenge other types of jet. The potential pay-off is enormous, and if successful the PDE could revolutionize air transport (not to mention military aviation) just as dramatically as the turbojet did after the Second World War. But it still has serious obstacles to overcome before it gets there.

## A Ticket to Space – Tourist Class

Back in 1968, before Concorde even entered service, Pan American Airlines started taking bookings for commercial flights to the moon. This was mainly a publicity stunt, playing on the PanAm spaceliner in the film of *2001: A Space Odyssey*, but at the time there was a genuine belief that airlines would be ferrying passengers to the moon before the turn of the century, and they took more than 90,000 'reservations' (actually just expressions of interest). Space flight was compared to aviation, with commercial airlines appearing just a few decades after the Wright brothers' first flight. Mass space travel seemed just around the corner.

Faster aircraft travel at higher altitudes where the air is thinner and there is less drag. Subsonic airliners cruise at 35,000 feet; the mach 2 Concorde cruised at 60,000 feet, so high that it was above most of the atmosphere and the sky above appeared black. The X-43 should cruise at 100,000 feet, or 20 miles, and plans for a high-speed Blackbird replacement have suggested an altitude of 40 miles. At 80 miles altitude you enter the orbit of the lower satellites and air travel has effectively become space travel.

However, thirty years after the first reservations for space travel, PanAm has ceased to exist and will not be ferrying anyone into orbit in any case. More importantly, the priorities are different.

The satellite revolution described earlier gave an additional incentive to make cheap flights into orbit a reality. At present the cost of putting a payload into orbit runs to over $6,000 per kilogram; anything that can bring that cost down will be immensely profitable. Space tourism is a nice idea, but it is cheap launches that will be the driving force.

It is worth putting the flight of the private 'SpaceShip One' into context here. This was a privately-funded venture which won the 'X Prize', awarded to the first craft to successfully reach space (defined as an altitude of 100 km) twice within a specified period. In principle, the X Prize should lead eventually space tourism.

SpaceShip One comprises a small rocket-powered vehicle carried to 50,000 feet by a special aircraft. It is the product of Scaled Composites, a small company which produces aircraft for niche markets including a number of unmanned military aircraft, and, according to rumour, some stealthy Black projects. It is a great technical achievement: for a cost of only $20 million or so, Scaled Composites have built a craft capable of reaching space.

Unfortunately, there is a big difference between reaching space and doing anything useful. Although SpaceShip One can take its passengers up a hundred kilometres, they come right down again afterwards. Getting into orbit is another matter entirely. Arthur C. Clarke expressed the energy requirements for space travel in terms of walls of different heights.

SpaceShip One can climb a wall 100 km high. Low earth orbit, the minimum for a satellite, is represented by a wall 3,228 km high, and geosynchronous orbit (the most useful) would be 5,984 km. So in effect, it only gets about 3 per cent of the way to the nearest milestone.

It could be argued that this effort, like the Wright Flyer, represents a first step towards a civilian space industry, but there are important differences. The basic design of the flyer could be scaled up into something bigger and better; the Scaled Composites design is limited by the fact that it has to be lifted by an aircraft to reach launch height.

Burt Rutan of Scaled Composites believes that his company will eventually be able to produce a craft capable of ferrying a single tourist to an orbiting space hotel, but the prospect is a remote one. The true significance of SpaceShip One and the X Prize is that it might rekindle interest in space travel. Entrepreneur Richard Branson is set to invest in building a fleet of craft based on SpaceShip One for high-altitude joyrides at $200,000 a ticket, turning space tourism into an industry rather than a series of one-offs. If people are queuing up to pay thousands of dollars for a few minutes of space flight, then the government may also see some value in giving greater funding to the manned space program. True space tourism – into orbit at the top of that 3,228 km-wall – will only be accessible to those able to pay the millions charged by the Russians.

There are other private firms interested in space travel, such as Space Exploration Technologies Corporation (SpaceX) which is developing a low-cost satellite launch system. Perhaps unsurprisingly, their first customer is the US Department of Defense. The military continues to take a strong interest in space.

The USAF has redefined itself as an aerospace force. Control of the high ground does not stop 20 miles up, but continues all the way into orbit and beyond. In order to dominate this new frontier, new vehicles and new systems will be needed.

The problem from the USAF's point of view is that the Space Shuttle did not do what it was supposed to do. Although it may be a technological marvel, the first re-usable spacecraft did not achieve the goal of cheap and frequent launches into orbit. When the program was initiated in 1972, the Shuttle program was scheduled to have a fleet of four craft by 1980, making a total of fifty flights a year between them.

When the first shuttle mission took off in 1981, two years behind schedule, it was obvious that the original expectations would not be met. In fact, in the first ten years of operation only forty launches were made, including Challenger, which exploded after launch.

The cost per launch also went up. The original plans were for a cost of around $10 million per launch. The actual launch cost depends on how you calculate it, but the lowest estimate, based on the incremental cost of a single shuttle flight, is $100 million. If you start factoring the overall costs of the NASA shuttle operation and dividing by the number of shuttle launches, a much higher figure emerges. The Futron Corporation, which provides analysis of the space industry, applies a figure of $300 million for shuttle launches throughout the 1990s, making it the most expensive way of launching heavy payloads.

This left a significant gap in the planned satellite launching capability, one which left the military short of capacity. The loss of Challenger meant that for an extended period the US could only call on a single high-resolution spy satellite.

The military still had a formidable launch capability using other rockets, notably the Delta and Atlas boosters which have been in service for decades. But by the time of the loss of a second shuttle in 2002 it was more apparent than ever that a next-generation system would be needed for military purposes.

As mentioned earlier, it is quite possible that there are classified systems used for launching small satellites, which might take the form of a small rocket launched from a very fast aircraft. But this would not be sufficient for larger payloads such as the big spy satellites.

Delta can launch satellites of up to 5 tons, but Titan IVb is the serious heavy lifter at present, with the power to hurl a 20-ton payload into low earth orbit, or 5 tons into a geosynchronous orbit – but at a cost of over $250 million a time.

In a bid to reduce launch costs, the USAF is now using commercial launch services for some of its payloads, but these are not suitable for the most sensitive ones, and cannot be relied upon during periods of political crisis. The Evolved Expendable Launch Vehicle or EELV was intended to provide a competitive means of launching payloads more cheaply. The two main contractors, Boeing and Lockheed, were persuaded to increase standardization and adopt a more common approach, as well as producing updated versions of their rockets – the Lockheed Atlas V and Boeing Delta IV.

Intended to 'increase the US space launch industry's competitiveness in the international commercial launch services market', the EELV should have been a major boost for the US as a military and commercial power in space, giving the nation the lead in low-cost launches.

Boeing and Lockheed gained in two ways: their launch capability was upgraded to compete commercially with the rest of the world, and each received over $500 million for development costs. Once complete, the two companies bid for business from the US military, with Boeing receiving the lion's share of the $2 billion-worth of contracts, getting 21 out of the 28 launches.

Both rockets became operational in 2002 and incorporate new technology. The Atlas has optional boosters that can be added to give a range of different launch capabilities, and the Delta has a new engine designed for low-cost operation fuelled by hydrogen.

In addition to the new launch vehicles themselves, the EELV includes a new infrastructure for US space launches. There is a new Spaceflight Operations Center with an impressive new Launch Operations Center and Customer Support Facility. The facility is equipped with new Mobile Launcher Platforms 60 metres tall, and automated equipment for fuelling and checking the rockets.

Unfortunately this expensive project coincided with a slump in the demand for satellite launches worldwide, and Boeing and Lockheed have to compete with other companies like Arianespace for barely profitable business. According to *Aviation Week & Space Technology* magazine, in 2002 both companies requested 'an additional $100–150 million' as an annual payment just to stay in business.

Boeing spent $1.5 billion of its own money on the Delta IV, and was not planning to make a profit for some years. But the dip in the commercial market meant that they were forced to withdraw completely until the market picks up.

This sheds an interesting light on the relationship between the US military, the companies that support it and which it funds, and the progress of technology. The matter became more entangled when industrial espionage came to light and two Boeing managers were charged with stealing secrets from Lockheed relating to the EELV. Lockheed proprietary documents detailing the costs of their bid were found in possession of the two Boeing managers, one of whom had previously worked for Lockheed.

According to the prosecution, 'By covertly using a competitor's secret information, they caused harm not only to Lockheed Martin, but also to the Air Force and taxpayers who finance government operations.'

The award of the EELV contracts is being reviewed, and several of the launches which had been awarded to Boeing have been given to Lockheed, while the Boeing launches were grounded until 2005. A further threat has come from a House Appropriations Committee report which suggests that a single supplier should be selected for full funding rather than underfunding two suppliers. Whatever the outcome, the case can only harm the long-term prospects for the US launch industry.

Meanwhile, the international launch market has been strengthened with the advent of improved Russian rockets, which have become reliable enough to be commercially attractive. In June 2004 a Russian–Ukrainian Dnepr rocket launched a payload of eight commercial satellites, three US, three Saudi and two European. This rocket is a direct adaptation of the Soviet SS-18 ICBM, designed to carry a payload of ten nuclear warheads. Other Russian space hardware has been less successful, including the Buran, the Russian equivalent of the space shuttle, but it has a thriving launch industry. It may be some time before the US can get back into the competition.[3]

The situation is even worse for US rocket engine manufacturers. Because of export laws governing strategically sensitive items – rocket motors can be used for ballistic missiles – engine makers are barred from selling to foreign customers. The lack of new strategic missile contracts means that sales are at a historic low. Some have predicted that this will lead to a critical situation in the US, with engine makers going out of business. This is unlikely; the Pentagon will need another generation of missiles some day and they will keep the suppliers afloat, but it looks as though Uncle Sam may have to finance a whole flock of lame ducks. The alternative would be a new initiative for another generation of strategic missiles.

While the US may not be firing many nuclear ballistic missiles in the near future, there is a possible conventional use for them.

## Common Aero Vehicle

While the heavy lift area may have its problems, another craft may come close to giving the USAF the orbital bomber capability that it desires, although on a smaller scale than the HCV and not as reusable. The project is called Falcon (one of those contrived acronyms: Force Application and Launch from CONtinental United States Technology Demonstration), and the vehicle it launches is called the Common Aero Vehicle. Rather than being a kind of super-aircraft like the HCV, the CAV is more like a traditional rocket.

The CAV is not a fully-fledged spacecraft like the space shuttle, but instead is the upper stage that would sit on top of a larger rocket booster, such as one of the existing Atlas or Delta rockets. Alternatively, new rockets could be developed specifically for the purpose. It looks like a miniature, streamlined version of the Shuttle.

A USAF document describes CAV as 'an unpowered, manoeuvrable, hypersonic glide vehicle capable of carrying approximately 1,000 pounds in munitions or other payload'.

Very much like Sanger's original Second World War plan, the CAV would be launched from the US into orbit; it would carry out its strike mission and then glide back to friendly territory at high speed. This may sound unlikely until you realize that this is basically what the Space Shuttle does, coming back to earth without power at speeds of around mach 20.

A DARPA project, the aim of Falcon is to be able to hit a target anywhere on Earth within two hours. It would be the equivalent of a ballistic nuclear missile, but without the nuclear warhead, which raises the question of why it is worth having a reusable manned upper stage. A disposable system, like a ballistic missile with a conventional warhead, would be slightly more expensive because there are not reusable components, but instead of just a thousand pounds it would be able to carry several times as much payload.

It is not clear whether CAV would be manned. Current plans suggest that it would be, but this is not necessarily realistic considering the sort of missions it is likely to fly. Manned aircraft have advantages over unmanned ones where the pilot can carry out target identification and other tasks, but from orbital height and at mach 12 or more it is not feasible for the pilot to pick out targets by eye. It is also unlikely that he will be able to do much 'piloting' as such; as with the space shuttle, things will be happening much too fast for human intervention. Possibly it has more to do with psychological reasons than practical ones: it is more acceptable to carry out missions with manned bombers than ballistic missiles – a high-tech bomber is much more impressive than just another missile, and it looks good for the cameras.

The idea of using ICBMs for conventional strike has also been mooted, and as these missiles approach their best-before date and obsolescence, it is highly likely that they will be converted for conventional operations. A missile that plunges to earth at mach 10 or more will make an effective bunker buster armed with the sort of kinetic energy warheads planned for the CAV.

An alternative use for these outdated missiles would be to recondition them as satellite launchers for the civil market. This might really provide low-cost transport into space, but it would be supremely unpopular with those who already have a huge investment in satellite launch systems. The last thing a depressed market needs is another source of low-cost alternatives, and it is unlikely that this particular sword will be bent into a ploughshare.

The total number of satellite launches is currently around 60 a year, compared to an average of 78 a year in the late 1990s. The US is still the biggest player, with 22 launches in 2003. Russia followed with 21, and China launched 6.

The satellite launch business remains one where military and commercial interests are closely linked. In this context the National Reconnaissance Office's new policy of 'negation' is causing some concern. Under this policy, the benefit of orbiting systems will be denied to enemies. This could include systems that belong to third parties. The French commercial satellites Spot Image provide a high-resolution space mapping service that could be of military use – the company was banned from selling pictures of Afghanistan after the US invasion. There are weapon systems that use GPS satellites for their guidance, and if it is a foreign GPS then it is likely to be neutralized. The use of satellite television has also been mentioned; given the willingness to target al-Jazeera offices on the ground, it is entirely plausible that broadcast satellites belonging to hostile media might also become a target.

In the event of a conflict, the US is developing the capability to temporarily disable any systems which are considered to be a threat. This can take several forms, from electronic jamming and interfering with control systems to physically damaging the satellite or the ground stations that control it. Mapping satellites that use optical sensors can be temporarily blinded from the ground. Naturally this will have serious consequences for other commercial users of the satellite systems.

There will be political objections, but these are likely to be over-ridden. In the end, it is space capability that will count. And as Major General Judd Blaisdell, director of the Air Force Space Operations Office, put it: 'We are so dominant in space that I pity a country that would come up against us.'

However, the military interest in space has had some unusual and potentially highly beneficial offshoots. One of these was a probe called Clementine which was intended as a demonstrator and test platform for the Star Wars (SDI) program. Clementine was supposed to orbit the moon before heading off for a rendezvous with an asteroid called Geographos. This would demonstrate the system's ability to intercept a small distant object, an important part of Star Wars' missile inter-ceptor role. Unfortunately it failed due to software problems, but not before the on-board sensors had detected something new on the moon: traces of ice in the deep shadows of lunar craters.

The significance of ice on the moon for space travel is hard to overstate. If it is present in large quantities it can be used to provide not just water, but also oxygen and hydrogen for fuel by electrolysis. Carrying water up into orbit would be a very costly exercise; if there is a supply already there it will turn the moon into a stepping stone for the planets. It would also make setting up a lunar permanent base far easier and cheaper. The prospects for a scientific station or a tourist hotel on the moon are much brighter thanks to Clementine.

# CHAPTER 12

# Information Warfare

*Throughout the world today, people are beginning to see that a modern state, whether democratic or authoritarian, cannot withstand the subterranean forces of anarchy and chaos without propaganda.*

Joseph Goebbels, 1934

Napoleon observed that in warfare 'the moral is to the physical as three to one.' This is as valid now as it was then: an army which believes that it is beaten is beaten and will surrender without a fight. In Napoleon's day, the attitude of the civilian population was not as important as the front-line troops, but now civilians are also part of the battlefield. This is true in both democracies and dictatorships – it is no use winning battles if the TV pictures back home cause the people to stop supporting the war.

This whole area has become known as information warfare. Its scope includes not only putting out information in the form of propaganda and psychological warfare, but also the gathering of data from available sources for military purposes. It is not a matter of spy planes and satellite pictures, but sifting through data from public sources. Information warfare has gained new focus with the war against terror, where the attempt to distinguish between the terrorists and the good citizen by trawling databases is reaching new heights of sophistication.

Information is both the battleground and the weapon. The right information has to get to the right people, and this covers everything from making sure that CNN shows only certain pictures to getting captives to talk. As the debate over Weapons of Mass Destruction in

Iraq has shown, not only the intelligence that is extracted, but the way it is presented and the way it is perceived are important. Both friendly and enemy governments fall within the scope of information warfare.

Information warfare tends to cause concern in human rights circles. This is because, put crudely, on one side it is a matter of developing the best possible technology for mind control, and on the other it means increasingly advanced means of spying on the population. Clearly more issues will be raised as techniques developed by the military find their way into the hands of civilian operators.

## A Short History of Public Relations

Although one could argue that its antecedents go far back in history, the public relations or PR industry is a quintessentially American creation of the modern era.

In the latter years of the nineteenth century, US business came to be dominated by a small number of powerful corporate interests which maintained virtual monopolies in their fields. Historian Matthew Josephson famously called them 'Robber Barons': industrial tycoons like Andrew Carnegie of Carnegie Steel, John D. Rockefeller of Standard Oil, the Vanderbilts and their railroads, and bankers like J. Pierpont Morgan. They were powerful and ruthless, and government did little to curb their activities.

The Robber Barons did have dangerous enemies though, a new brand of investigative journalist known as 'muck rakers'. These journalists used the new media – popular newspapers, magazines and national wire services – to publicize the Robber Barons' activities, leading to public demands for greater regulation and new laws to break the monopolies.

William Vanderbilt famously reacted to public opinion with the response: 'The public be damned!' However, even he realized that the public could not be completely ignored. This did not mean giving in to public opinion: it just meant that steps would have to be taken to reshape it into a more favourable form. The public relations industry was born, with armies of publicists and press agents being taken on to deal with the media.

Carefully staged events would ensure that the Robber Barons were photographed endowing new colleges or looking after their workers' welfare. John D. Rockefeller's publicist had him handing out dimes to poor children. Publicists ensured that the news media were well supplied with encouraging and optimistic information. If they could not get on to the editorial pages, no expense was spared in taking out advertisements putting the other side of the story.

In 1906 Ivy Lee invented the press release, distributing a story about a train accident before the reporters could get there. This was a new idea, that a news story could come directly from a company, and it has provided lazy journalists with copy ever since.

By 1914 the industry had become more sophisticated, though it was gaining a reputation as being little more than the mouthpiece of the rich and powerful. When America joined the war President Woodrow Wilson took advantage of the new industry to set up a Committee on Public Information under George Creel. Creel was a newspaper publisher, editor and sometime muck-raking journalist himself: a gamekeeper called on to turn poacher.

The Committee's job was to sell the war to America. This was a major challenge, as Wilson had run for re-election on the slogan 'He kept us out of the war'.

The Committee took on the task of recruiting soldiers, selling war bonds and raising money for the Red Cross. Creel's team produced posters and leaflets, issued several thousand press releases and produced silent movies with patriotic themes. Boy Scouts were drafted to post copies of President Wilson's addresses through letterboxes throughout the country, and over 70,000 volunteers became 'four-minute men', trained to give speeches in public places urging people to buy bonds or join up. The Committee came up with ringing slogans, including one urging people to fight in 'The War to End War'.

Creel later wrote:

> The trial of strength was not only between massed bodies of armed men, but between opposed ideals, and moral verdicts took on all the value of military decisions . . . In all things, from first to last, without halt or change, it was a plain publicity proposition, a vast enterprise in salesmanship, the world's greatest adventure in advertising.

The Committee was a great success, and the 'greatest adventure' had huge benefits for the fledgling PR industry. PR itself had some good PR, and after the war the Committee was seen as playing an important role in uniting the country and ensuring victory. As one PR textbook puts it: 'The biggest and most practical human lesson learned from the war is that nothing requiring organised effort can succeed without publicity and plenty of it.'

This success led to the founding of many new PR firms in the aftermath of the war. If it was good enough for Woodrow Wilson, then Public Relations was good enough for anyone. The industry benefited from the sudden influx of a large number of bright young men who had been trained in the arts of PR to help Uncle Sam, and who were ready to try out their new-found skills in the world of business.

One such was Edward L. Bernays, a veteran of the Committee on Public Information and sometimes called the 'father of public relations'. Bernays' innovative ideas included introducing celebrity endorsements so that famous people would put their weight behind a particular campaign. He also developed media events which were deliberately staged to create a public impression. A tobacco company found that women were embarrassed to smoke in public as it was a masculine habit. Bernays linked the right to smoke in public with the right to vote and hired suffragettes to march through the streets, carrying lit cigarettes aloft as 'torches of freedom'.

Bernays was never ashamed to describe what he did as 'propaganda'. However, the term started to take on dangerous overtones a few years later when it was used by Joseph Goebbels, the 'Minister for Public Enlightenment and Propaganda'.[1] Having been helped to life by one war, PR became a powerful weapon for the Nazi party and was a vital tool for maintaining their grip on the German psyche.

'Its task is the highest creative art of putting sometimes complicated, events and facts in a way simple enough to be understood by the man on the street,' Goebbels wrote. 'The genuine propagandist must be a true artist. He must be a master of the popular soul, using it as an instrument to express the majesty of a genuine political will.'

Goebbels had learned well from the Americans and took his inspiration from Bernays, who ironically enough was an Austrian Jew by birth. In his autobiography, Bernays recalls a meeting in 1933 with Karl von Weigand, a journalist who had just returned from Europe and had met Goebbels.

Goebbels, said Weigand, was using my book *Crystallizing Public Opinion* as a basis for his destructive campaign against the Jews of Germany. This shocked me Obviously the attack on the Jews of Germany was no emotional outburst of the Nazis, but a deliberate, planned campaign.

Goebbels made good use of new technology, in particular the radio. The spoken word was a far more effective means than the printed word of transmitting the Nazi message, and in particular Hitler's own hysterical oratorical style.

'If I approach the masses with reasoned arguments, they will not understand me.' Hitler noted in *Mein Kampf*. 'In the mass meeting, their reasoning power is paralysed. What I say is like an order given under hypnosis.'

The Nazis put a priority on increasing radio ownership, and one of their first acts in 1933 was to arrange a deal with industrialists for the manufacture of cheap radios.

The *Volksempfänger* or 'people's receiver' cost half as much as other radios and could be purchased in instalments. Between 1933 and 1939 the proportion of German homes with a radio had tripled to 70 per cent. The *Volksempfänger* was designed to have a limited range so that it could not receive foreign broadcasts, only German radio which was completely controlled by Goebbels' ministry.

The regime placed strong emphasis on community radio listening, as they found that gathering in a group created a far more effective atmosphere among the audience. Hitler's speeches were advertised in advance, and factory owners, department stores, bars and blocks of flats were ordered to put up loudspeakers 'so that the whole work force and all national comrades can participate fully in the broadcast'. Thousands of local 'radio wardens' were appointed to encourage community listening, and also to record audiences' reactions so that future broadcasts could be made as influential as possible. The process of checking on how propaganda was received proved to be highly effective.

Goebbels fully appreciated the effect of saturation radio, and his first rule for broadcasting was that it must not be boring. This was the highest principle; blaring military music, dull factual lectures and dry speeches were unacceptable, however patriotic. He ordered that radio must instruct, but had to do it in a way that was entertaining and acceptable.

The Nazis also refined their version of the big media event. The Nazi party's massed torch-lit rallies invoking a mythic Aryan past were media events par excellence, whipping up a kind of public hysteria for

Hitler ('torches of freedom' becoming even more ironic). This was the kind of support which leaders in previous centuries could only dream of.

The weekly cinema newsreel was an important part of the propaganda process, and the entire film industry was effectively drafted. Leni Riefenstahl's 1934 film *Triumph of the Will* showed that the mass appeal of the torch-lit rally could be recorded in an enduring form and distributed throughout Germany and the world.

Some have suggested that it was superiority in propaganda rather than panzers that drove the early German victories in the Second World War. The overwhelming belief in German supremacy was necessary for the country to take on seemingly impossible odds and win, conquering more countries in a few months than the Kaiser's army ever managed. Napoleon's dictum about the moral and the physical was still true in the age of mechanized warfare. In his book *Mein Kampf* ('My Struggle') published in 1925, Hitler laid out his plans for conquering the world, and up until 1940 it all came true according to plan.

The success of Goebbels' Ministry of Propaganda prompted President Roosevelt to create his own team in 1939 to fight the propaganda war for the American side. The effort was masterminded by Harold Lasswell, who was given the title of Director of War Communications Research. Given government funding and resources, Lasswell was able to develop a sophisticated 'model of communication' which was so important to the war effort that it was classified Top Secret.

This model is expressed in the simple question 'Who says what, in which channel, to whom, with what effects?' Each of the five elements – who/what/which/whom/what – could be analysed and described, and all five had to be correct for the propaganda to be effective.

Lasswell's model was used to test various propaganda messages. In one case, sales of war bonds by different campaigns were analysed. The slogan 'Help win the war' proved the most effective overall, but was not quite so popular with women. Adapting it to 'Help win the war and bring our boys home' appealed to both sexes and boosted sales further.

During the Second World War, the black art of propaganda reached new heights with the development of new approaches like Lasswell's. Both sides applied a more 'scientific' approach than ever before, diverting resources away from more conventional forms of warfare. Bombing raids to drop leaflets were common from both sides, their impact on both civilian and military morale carefully gauged by intelligence analysts.

In Britain Lasswell's model might have remained a closely guarded secret for fear of it falling into the wrong hands. In the US it was recognized as a leap forward in PR, declassified in 1948 and let loose on the world. The model was *Marketing Management*, a textbook by Philip Kotler used by hundreds of thousands of students since 1967.

One side effect of the PR war was that President Roosevelt classified Hollywood as 'an essential war industry' – a role which it has seemingly retained ever since. War movies in particular often rely on the military for assistance, and this generally assures that the US armed forces are shown in a favourable light.

In 1999 the US Army provided $50 million of funding to set up the Institute of Creative Technology. This is ostensibly tasked with creating training simulators, which may involve digitized film clips using actors, special effects and the rest of the film industry's know-how. The role of the ICT goes further than simulators; after the September 11 attacks, the ICT invited some of Hollywood's top screenwriters, producers and directors for meetings to discuss possible scenarios for terrorist attacks on the US. It is quite possible that the meetings were also used to provide a flow of ideas in the other direction, ensuring that Hollywood's A-list knew what they had to do to help the war against terror.

The Cold War saw some classic cases of PR as a weapon. For example, mention the high standard of living in Scandinavia, and someone will almost invariably chip in with a comment about the high suicide rate in that part of the world. This is a popular myth which originated in the 1950s and can be traced to a speech by President Eisenhower.

Sweden was enjoying hugely successful socialism which presented a viable alternative to the American way, so Eisenhowever tried to give a negative image of the country. In the speech he claimed that Sweden suffered from the world's highest rate of suicide. This was not true at the time, and is not true now. Sweden is currently fourteenth, behind France, New Zealand, and Australia among others.

However, few people ever checked on the story, and it struck home. Perhaps because it strikes a chord – those depressing dark Nordic winters – and perhaps because of an innate unwillingness to believe anyone is happier than we are. This is part of an effect called 'cognitive dissonance', which tends to make us reject anything that threatens our deeply held beliefs, and the belief that we are happier than others is an important part of most people's self-image. This is why very few people would change places with a random stranger, as we all perceive ourselves as being better off than most. The suicidal Swedes have long been accepted without question in the West, and the superiority of capitalism remained unchallenged.

Eisenhower was apparently not aware that he had been given inaccurate information, but it seems likely that this type of disinformation was very carefully constructed and tested before it was released.

Other examples of information warfare abound. One of the most colourful came in the form of abstract expressionism. This was a new 'modern art' movement which included the splatter-paintings of Jackson Pollock and Mark Rothko's huge dark canvases with no recognizable shapes. As Frances Saunders recounts in her book *Who Paid the Piper – The CIA and the Cultural Cold War*, the CIA funded exhibits of abstract expressionist paintings all over the world in the name of promoting free individual expression against the faceless 'social realist' school of the Soviets. It was all carefully calculated. As Sanders puts it: 'the individuals and institutions subsidised by the CIA were expected to perform as part . . . of a propaganda war.'

This particular form was adopted because other artists had a disconcerting tendency to depict naked bodies and other images shocking to the sensibilities of mainstream 1950s America. Abstract expressionists, though they attracted some mockery, were harmless enough, and there was something appealing about turning paint spatterings into million-dollar artworks. Nelson Rockefeller called Abstract Expressionism 'Free Enterprise painting'. The art world is of course sensitive to the market, and the patronage of wealthy backers helped ensure that the Abstract Expressionists triumphed while social realism lost credibility in the West. If the space race was about proving to the public who was technological top dog, there was a parallel war to show America was culturally superior too. Similar battles for hearts and minds were fought in many other arenas.

Although not technological in itself, PR shows the same pattern as the technologies discussed in this book: from modest beginnings, it enjoyed a hothouse period of growth in the hands of the military before rejoining the civilian world. And it is still developing.

The latest visible incarnation was the Pentagon's Office of Strategic Influence (OSI), which was described by the *New York Times* in 2002 as being tasked with 'developing plans to provide news items, possibly

even false ones, to foreign media organizations . . . to influence public
sentiment and policy makers in both friendly and unfriendly countries'.
A week later, the OSI was officially closed due to negative reactions to
the *Times* story

The OSI was, of course, a typical PR operation, but one backed by
the power of the Pentagon. Many have doubted whether it actually
closed or simply dropped out of view. Several months after the OSI's
apparent demise Defense Secretary Donald Rumsfeld described his
approach to the attacks on it. 'I went down the next day and said, "Fine,
if you want to savage this thing, fine, I'll give you the corpse. There's
the name. You can have the name, but I'm going to keep doing every
single thing that needs to be done, and I have."'

## The State of the Art

The modern military is still in the forefront of PR, and enjoys many
advantages over its civilian counterpart. Generally known as psy-ops
(psychological operations), it encompasses a range of techniques of overt
and covert persuasion.

One of the prime examples of the technology in use is Commando Solo,
a heavily modified Hercules transport aircraft that acts as a flying
television and radio station. Operated by the Special Operation's Command 193rd Special Operations Wing, the fleet of Commando Solo
aircraft can fly over any country or area in the world and pre-empt local
broadcasting, replacing it with whatever message is required. Of course,
given its strike capability, the USAF can generally arrange for local media
to be taken off the airwaves simply by destroying transmitters.

The aircraft is equipped with transmitters for AM and FM radio as
well as television (both the European PAL and US NTSC standards as
well as the version of SECAM used in Eastern European countries) and
can broadcast on any frequency thanks to a huge array of antennae
which give it something of the look of a flying porcupine. While it can
monitor local broadcasts and override them, it is more common to use
empty bandwidth, using leaflet drops or other means to inform the local
population of their broadcasts.

'Psy-ops got a bad name with Goebbels,' comments Colonel Ernst,
the commander of the 193rd Special Operations Wing, 'but now
they're coming into their own.'

The Commando Solo broadcasts can be highly directional and cover a specific area. Sometimes it is the civilian population who are targeted, and sometimes enemy soldiers.

The Desert Storm psy-ops involved a combination of leaflets, broadcasts and bombing. The massive air strikes by B-52 bombers were advertised in advance: the Iraqis were told that their only chance of survival was to surrender or desert. A total of 29 million leaflets of more than thirty types were dropped in the Kuwait theatre alone.

The broadcasts followed the Goebbels principle that they must not be boring. There was news, weather and sport reporting, as well as interviews with Iraqi PoWs saying how well they were being treated. Iraqi soldiers would listen in, hoping to find out if their sector was going to be bombed, and in the process they would listen to hours of US propaganda.

'Your only safety is across the Saudi Arabian border,' warned the radio broadcasts. 'That is where the bombing and the starvation stop . . . [the Coalition] offer you a warm bed, medical attention and three filling meals a day.'

An outsize bomb, the BLU-82, was also used as part of this effort. Originally designed to clear helicopter landing pads in the jungles of Vietnam, the BLU-82 is so large that it cannot be dropped from a bomber but has to be pushed out of the back of a transport plane. This makes it less than suitable as a weapon, but its immense size makes it very useful for propaganda purposes: the explosion produces a mushroom cloud like a small nuclear blast and can be seen for many miles around. Leaflet drops told the Iraqis about BLU-82 strikes in advance and warned them if they continued fighting they would be the next target.

By the time of the 2003 Iraq war, this tactic had been refined further. An even larger bomb than the BLU-82, called MOAB (Massive Ordnance Air Blast – but also Mother Of All Bombs) was built weighing more than 10 tons. A single example tested at a base in Florida, and the Defense Department released a video of the test which was distributed among news media and put on the Internet. This was a weapon designed to produce the proverbial 'shock and awe' – not on the battlefield, but in the minds of those who saw it. Simply by threatening to use it (even though there was no stock of bombs to drop) the US gained a psychological advantage. Even those who had not seen the video or the pictures would know that they were facing a terrible new weapon.

There is always a certain amount of PR associated with new weapons emphasizing how effective they are. However, with MOAB we may have reached a new level where certain weapons are 'famous for being famous': deliberately created to be as intimidating as possible, regardless of their actual effectiveness or whether they are ever used. The psychological impact of some weapons may be more important than their physical effect. This can be especially important in non-lethal weapons used for crowd control: the 'lighting guns' discussed in Chapter 9 are a case in point.[2]

Other new weapons may be unveiled more as advertising. Boeing revealed a new prototype stealth aircraft called 'Bird of Prey' in 2002, apparently mainly to persuade stockholders and others that the company was still at the cutting edge of technology. Similarly, British Aerospace announced their own prototype stealth plane called Replica; the aim was to demonstrate that they were capable of keeping up with US plane makers and to ensure their share of the multinational F-35 Joint Strike Fighter program. Neither was ever intended to result in a production aircraft. They are the military equivalent of the outlandish concept cars which car companies sometimes display in mock-up form at shows to illustrate how forward-thinking they are – but at a cost of millions.

The USAF has shown a particular interest in integrating information warfare into its battle plans. The integrated plan includes what they call 'kinetic solutions' – bombs – and 'non-kinetic options' such as leaflets and broadcasts. This will enable them to knock out the enemy's means of communication, including radio and TV broadcasting and telephone systems, and ensure that the only mass communications in the field are those from the USAF itself.

'Information operations is going to play an increasing role,' Major General Paul Lebras, commander of the Air Intelligence Agency said in *Jane's Defence Weekly*. 'Perhaps we can shape adversary behaviour prior to a conflict . . . Maybe if you can shape behaviour, you can prevent conflict.'

As the technology improves, the level of individual targeting will increase. In the 2003 conflict, information operations also included the use of email in Iraqi computer networks to get the message across. There were even reports of SMS text messages being sent to Iraqi leaders and members of their families urging them to surrender. Some such reports appear to have come from Department of Defense sources whose only motive was to mislead about their intentions and capabilities and there is doubt whether the SMS texting ever took place. This may be a case of information warfare about information warfare, which leads to another important question: who do you believe?

## Controlling the Media

Vietnam was hailed as the first television war, where the civilians at home could watch their soldiers fighting from their living-room sofa. It is often claimed that this was the reason why the public turned against the war, because of the graphic images of death and destruction played out on the evening news. The war in the Pacific in the Second World War contained more atrocities, more blood and far more American casualties, but because the public was shielded from them it was much easier to maintain morale on the home front.

A reporter with a television camera is a powerful weapon in the information war. After Vietnam, armed forces exercised an increased level of control over the media in their midst. The first example of 'news management' on a massive scale took place during the 1982 Falklands conflict. If the media wanted to be with the British Task Force, they had to accept military rules. As a result, good news was reported early, and bad news was delayed – in one case it took three weeks for film to arrive back in London, long enough to ensure that it was no longer 'news'.

There was no independent means of communication for the reporters with the Task Force, so everything had to go via Navy officers, allowing them to censor everything. The only news that came back from the war zone was what the military wanted to be reported.

Improved communications mean that it is harder to control journalists quite so tightly, but access to the front line can be easily blocked for 'security reasons'. In any war there are likely to be far more journalists than can easily be shepherded around, which gave rise to the 'pool system' during the Second World War. Under this system all the accredited journalists form a pool; a few are allowed to get involved in the action, and the rest of the pool then shares their reports. The journalists got protection and access to the front line, while the military could make sure that only favoured individuals were allowed to witness action.

During the 2003 Iraq war, the pool system received another twist with the practice of 'embedding' journalists. Each reporter was assigned to a particular military unit where they would remain. Living in close proximity with the same group for weeks or months encouraged the journalists to form bonds with their companions. Instead of reporting from an objective point of view, they were much more likely to take the side of 'their' unit.

It may be even more effective to cut out the journalists altogether. During Desert Storm, the news channel CNN took to televising the Pentagon's daily briefings. This was a gift to the military: they could ignore the journalists with their negative attitude and awkward questions and speak directly to the public. The televised briefings allowed General Schwarzkopf and his colleagues to describe their successes and show video footage of bombs striking targets with pinpoint precision.

'He was talking past the reporters and past the editors right to the public,' said one former Vietnam correspondent. 'That's not journalism that CNN was doing. It's something else.'

The ability to speak directly to the public without interference has proven invaluable, and we can expect to see an increase in direct communications of all sorts. The US military already has its own web sites and other news media, and while anything so blatant as a Pentagon-branded satellite news station is unlikely, getting the message across will become increasingly important.

## Captive Audience: Interrogation and Brainwashing

In some situations information warfare gets far more intense than subtly attempting to change someone's point of view. Sometimes the military mission is one of breaking and entering into the mind of a prisoner, usually with the aim of getting information out of them.

Torture and coercion go way back in history, but the proverbial barbarity of the medieval era and the ingenious contrivances of the Spanish Inquisition gave way in the twentieth century to more subtle tools. The first signs came during the Korean War, when American soldiers who had been captured by the North Koreans publicly sided with their captors and condemned America. Some twenty-one soldiers refused repatriation, preferring to stay in North Korea.

Forced confessions and statement signed under torture are ancient devices, but in these cases the men seemed to be acting out of their own free will.[3] Patriotic soldiers had been somehow turned into agents of the enemy, and a new word was applied to the process: brainwashing.

The term was coined by Edward Hunter, a British journalist, in his book *Brainwashing in Red China* Hunter claimed that the process of 'thought reform' used by the Chinese in Korea gave the captors almost limitless power over the prisoner, allowing them to reshape his most deeply held beliefs and transform his personality. It was impossible to resist, and once brainwashed the victim would not be a zombie but would show no signs of abnormality.

This idea of brainwashing sounds like something straight out of the movies, and it received a very thorough Hollywood treatment in the form of *The Manchurian Candidate* (1962). In this classic movie a war veteran who everyone believes to be a hero is actually an enemy agent. Without realizing it, the hapless veteran is programmed to follow the orders of his Communist masters in a plot to assassinate the President.

The US government commissioned a series of studies into brainwashing to find out the truth. They concluded that the techniques used in Korea relied on having absolute control over the subject. Once kept isolated from others and in a state of anxious uncertainty, they became more easily manipulated. Mental and physical exhaustion, aggravated by sleep deprivation, helped to reduce resistance. Continuous aggressive propaganda challenged the prisoners' habitual ways of thinking, and when put in a group of others who all accepted the propaganda they would tend to bend to peer pressure.

The studies concluded that given a situation where the captors have complete control, and where the propaganda is reasonably consistent with the prisoner's own values, it is possible to produce conformity. For example, continually bombarded with Communist literature that deals with human rights and the importance of the workers, even patriotic Americans could be persuaded to agree and say that they espoused Communist principles.

In a situation where food and blankets were in short supply and those who co-operated received better treatment, it is hardly surprising that some of the PoWs in Korea were tempted into making 'un-American' statements. However, the conformity was limited in scope and likely to be temporary; away from an atmosphere where everyone was an ardent Communist, the prisoner would tend to resume their previous attitudes.

Closer study of Korean PoWs suggested that the interrogators had managed to cultivate anti-American views that were already present. The fact that only a handful of prisoners out of the several thousand exposed to 'thought reform' were converted to communism suggests that something more individual was going on.

The media preferred to ignore this, and ever since then they have acted as though 'brainwashing' is a real effect and that human beings can be reprogrammed like robots. This belief has been especially pronounced when dealing with new religious groups, usually described as 'cults'.

Although it may not be possible to change people's minds at will, such religious groups usually show a shrewd understanding of the forces needed to convert someone to their beliefs. The key is isolation: the subject needs to be taken away from their usual surroundings into an environment which is completely controlled. Any moderating factors, such as sceptical family or friends, are removed, and they are surrounded by people who all have the same set of beliefs. Combined with sleep deprivation and an atmosphere of excitement and confusion ('Yes, the wonderful New Messiah will totally change your life forever'), resistance is gradually worn away and the subject will tend to conform with the people around them.

In spite of the media hysteria, this kind of technique is of limited effectiveness. When a person leaves the influence of the group and rejoins their family and friends, their personality can reassert itself.

As Edgar Schein said about the Korean case: 'the coercive element in coercive persuasion is paramount, forcing the individual into a situation in which he must, in order to survive physically and psychologically, expose himself to persuasive attempts.'

As an interesting sidelight, military basic training bears many of the hallmarks of classic brainwashing. Everything in it – having hair shaved and uniforms issued to ensure a uniform appearance, constant shouting and enforcement of discipline down to the most trivial level – is geared towards breaking down the recruit so they can be rebuilt as a soldier with the right set of values and attitudes. Armies are generally not keen to draw attention to these similarities.

As far as interrogation goes, further refinements have since been added to 'brainwashing'. Some of these have been revealed in declassified military manuals, though British techniques are still kept very secret. US manuals from the 1960s recommend sleep deprivation, controlling prisoners' diets and cutting off sources of light – either by blocking windows or simply putting a bag over the prisoner's head. While torture is not permitted, prisoners can be made to stand or sit in 'stress positions' for extended periods.[4] Stress positions are effective because prisoners are inflicting pain on themselves, while being aware that their captors could do much worse.

CIA manuals point out that the threat or possibility of physical pain is more effective than torture itself. The aim is to create a state of disorientation in which the subject feels helpless and afraid, and ready to clutch at straws.

In these circumstances, the introduction of an interrogator provides a relief from sensory deprivation and physical pain. When invited to sit down comfortably and talk to an amiable companion who has the power to make things better, subjects become compliant.

The techniques of interrogation are not confined to the military. In fact, there is now so much expertise in the civilian world that the US military has employed civilian interrogators, most notably at the notorious Abu Ghraib prison in Iraq, where interrogators were supplied by the company CACI (slogan 'Ever Vigilant'). In the UK, CACI's subsidiary offers 'smart solutions for intelligent marketing and information systems'.

Commercially available interrogation courses include the Reid Technique (a registered trademark). John E. Reid and Associates have trained more than 150,000 police and security professionals in a technique with several specific steps for getting information out of an unwilling informant.

The environment is one with controlled lighting and concealed observation of the informant. First there is an interview, in which an interviewer talks to the informant without confronting or accusing them. Then, after a break, a single interrogator faces the informant. Rather than getting the informant to talk, the interrogator presents them with a monologue, discouraging denials or explanations. Again, the approach is sympathetic rather than accusatory.

The monologue will offer excuses for the informant's behaviour and pave the way for them to co-operate: 'I don't think you're a bad person . . . you only stole that money because the others talked you into it.'

The interrogator asks questions carefully constructed to give the informant two choices, one of which is morally superior to the other.

Questions like 'Did you spend the money on drugs, or did you pay off your debts?' or 'Have you stolen before, or was this your first time?' will encourage the interviewee to admit to the lesser evil.

Much of the technique is involved in overcoming denial on the part of the subject, and also in assessing whether they are telling the truth using body language and noting particular verbal behaviours. The interview (known as a Behaviour Analysis Interview) at the beginning allows the interrogators a chance to assess the informant's body language and verbal behaviour in a non-stressful environment, making it easier to spot later when they are lying. This allows the interrogator to focus the questioning on important areas.

The Reid Technique does not rely on threats or deception, but on using 'active moral persuasion' to lead the informant towards the desired response.

Any sophisticated method may be all too effective at achieving the desired result. One British study suggested that around 20 per cent of people going through a full psychological interrogation process would confess, whether or not they were guilty. In 2000, a Canadian court ruled that a confession in a child abuse case should be disregarded because it had been obtained using improper interrogation techniques, specifically the Reid Technique.

This sort of technique it is not simply a tool for getting information out of prisoners, but can be used in other cases where a company suspects fraud or theft by an employee. John E. Reid Associates also offer training in the Integrity Interview (another registered trade mark), a job application interview aimed at establishing the applicant's employment history, criminal behaviour, and drug and alcohol use. Clearly this type of technique could be expanded into many other situations as well.

## Total Information Awareness

The other side of information warfare is collecting and analysing accurate information about the enemy. In the past this has meant reconnaissance, scouting and espionage. When the enemy is living among your own population though, intelligence gathering becomes a different matter. The war against terror has spawned a new type of information warfare, which involves an unprecedented level of spying on the civilian population.

In 1996 DARPA started a program aimed at drawing together intelligence from all sources as a means of tackling terrorism. The September 11 attacks caused the program to be accelerated and the Information Awareness Office was created. Its extensive mission statement says that it will 'imagine, develop, apply, integrate, demonstrate and transition information technologies, components and prototype closed-loop information systems.'

Part of the IAO effort is geared towards automated language translation and other 'intelligent' software covered in Chapter 7. But there is also what was known as the Total Information Awareness program. This has the ambitious aim of drawing together data from all the available military and civilian sources into a single architecture.

If achieved, this could, in theory, lead to a system capable of identifying terrorist threats in time to counter them by assembling all the individual pieces of the jigsaw automatically.

For example, consider the following scenario in which several apparently unrelated pieces of information are recorded on different databases:

- In California, the FBI observe a known terrorist having a meeting with an unknown tall blond man aged 30–35 with tattoos.
- Bank records show a Joe Smith unexpectedly paying off $50,000-worth of debts to loan companies.
- A stone quarry reports to local police that 91 kg of explosives are found missing in routine stocktaking.
- Next day the quarry reports that employee Joe Smith has failed to show up for work.
- Two hundred miles away the local police pull over a motorist for a suspected speeding offence; the motorist pulls out a gun and starts shooting. The licence number is a fake.
- The car make, model and colour match a vehicle sold to Joe Smith two years previously.
- A CIA informant in the Middle East says a big terrorist attack is being planned in San Francisco.
- Army records show that Joe Smith is 6' 4", 33, blond with tattoos on both arms, and has experience with handling explosives.
- A car with the fake licence number is identified going through a toll booth at the Golden Gate Bridge.

Individually, none of these pieces of information is actionable. But by pulling together data from the CIA, FBI, local police, banks and other sources, the TIA would allow counter-terrorism agencies to see all this with a single view. As soon as the fake licence number turns up again the police would be able to intercept it.

Obviously the amount of data involved would be immense, far too great for any human or even any number of humans to deal with, with reports flooding in from all quarters. The solution is a computer system which can learn from past terrorism cases and identify patterns. By doing this it would be able to 'provide focused warnings within an hour after a triggering event occurs'.

TIA is not an all-seeing eye: it is a central brain which needs to be fed external information from all the available sources. It is the intelligence behind the thousand eyes that will be able to spot potential terrorism in the background noise of everyday data. Nineteen Arabs, some with suspected terrorist connections enter the US. Some of them enrol in flight training schools, and they then book seats on four different flights going to different destinations on the same day: this would be a trigger event.

Behind the TIA are other programs, including Genisys and EELD. Genisys aims to build new types of easy-to-use database which will broaden and integrate the types of information available to intelligence agencies. It has as an adjunct, the Genisys Privacy Protection Program, which will ensure personal privacy at the same time. The government wants to know as much about you as possible, but in a democracy it has to protect your privacy. The Privacy Protection Program is intended to prevent the data gathered from being misused in any way.

EELD is Evidence Extraction and Link Discovery, a magical process for finding the relationships between people, organizations and events from unstructured sources like news reports or intelligence briefings. A prototype has already shown that it can identity distinct patterns of terrorist activity within routine reports of everyday happenings. This type of join-the-dots technology is a classic AI task.

Another IAO activity is Scalable Social Network Analysis (SSNA). As the name suggests, it is a tool for analysing social networks such as groups of friends, with the aim of being able to distinguish potential terrorist cells from law-abiding citizens.

Interestingly at some point TIA stopped meaning Total Information Awareness and changed to Terrorism Information Awareness. This name change was apparently a response to public fears that DARPA was going to be able to spy on everybody all the time.

Of course, if it did not have privacy protection built in, TIA would be wide open to abuse. It is designed to identify terrorism, but the entire system could equally well be used for other applications. Virtually everything that applies to terrorism could also be used to deal with organized crime, or drug smuggling networks. Or, of course, political movements.

Something like TIA would be a boon to the business world. Businesses already spend fortunes on tracking their customers and trying to extract the maximum information they can in order to sell more effectively in a process called 'data mining'. In one much-quoted example, a supermarket found that nappies were commonly bought by young men on Friday evenings. The supermarket moved their beer next to the nappies, and increased sales of both. (Although generally taken to be a classic example of data mining, it suffers from being apocryphal.)

The TIA program will develop a whole new set of extremely powerful data mining tools that will allow far more sophisticated analysis of databases. While business is unlikely to get its hands on the databases that lie behind the TIA, they will eventually benefit from the techniques it develops. This will allow them to target customers ever more effectively.

Genisys, which allows text and voice information to be analysed as well as more conventional data, could be used to analyse sales calls and customer correspondence, so that communications can be couched in exactly the right language pitched to a customer's age, education, socioeconomic class and preferences. Similarly EELD can be used to identify exactly what type of people are (or are likely to become) customers, while the SSNA tool offers the prospect of fishing for whole groups of people. If one person buys a cool new gadget or fashionable item of clothing, SSNA could identify their social contacts so they could be influenced, for example, by special offers: 'How would you like 15 per cent off this new camera like the one your friends might just happen to have?'

Hackers and others in the information business, not to mention those involved in illegal activities, tend to develop guerrilla warfare tactics to avoid what they see as government spying, adopting ever-more powerful encrypting to make sure their emails are unreadable, and developing new methods to make themselves invisible on the Internet. Those with anarchist tendencies everywhere prefer to avoid being on any sort of government database in the first place. In the future these tactics may become more common – not to avoid the government but simply to escape from barrages of junk mail, Spam and sales calls.

## Information Warfare for Everyone

Any promises made by database marketing gurus should be treated cautiously, but more powerful tools will mean more effective marketing. Companies will believe that they require their own version of TIA on their customers simply to stay competitive – and of course the marketing edge will also call on techniques from the persuasive side of information warfare.

While their effectiveness may be open to question, techniques which are believed to be successful may become increasingly analysed and turned into formulae for achieving the desired result from a conversation, whether it is an interrogation, a job interview, a sales pitch or a seduction. Hit-or-miss methods derived from gut feeling or personal experience will give way to 'scientific' techniques, and there may be a thriving market in 'countermeasures', learning how to counter such techniques. Information warfare may become as much a part of everyday life as advertising.

As with so much technology, there are questions over whether it works and how much impact it has. Information warfare may simply become another meaningless buzzword for marketing executives; or it might represent increasingly effective techniques for businesses and others to manipulate the public.

# CHAPTER 13

# Catching New Waves

*I'm heavenly blessed and worldly wise,*
*I'm a peeping tom techie with X-ray eyes*
Timbuk3, 'The Future's So Bright I Gotta Wear Shades'

The military have always had a keen interest in technology that allowed them to see better. In 1609, Galileo Galilei demonstrated his new invention, an improved form of telescope, to the Venetian Senate. Galileo himself was interested in the telescope as a tool for examining the heavens, although this was hardly likely to attract funding. But the new telescope had other uses too: it would give the Venetian Navy a major advantage over its opponents, allowing them to be spotted and identified long before they could see the Venetians. Accordingly, Galileo was given a position at the University of Padua, and the future of the telescope was assured.

Apart from increasing magnification, improving vision was a technical challenge. We already have some of the best vision in the animal kingdom. Our eyes are highly evolved, with good depth sense, a full range of colour perception and specialized cells for day and night vision. Equally importantly, the brain has sophisticated processing powers to handle information gathered by the eyes. We can recognize a familiar object from any angle or pick out a red bead from a tray of green ones so easily we hardly notice. The brain is always on the lookout for particular signs and ready to alert us – your name leaps out at you from a printed page, even if you are just glancing at it casually.

Our sight does have some limitations. What we perceive as visible light is just a small part of a much larger electromagnetic spectrum. New technology can open up vistas that have never been seen before.

Human vision allows us to see electromagnetic waves with a wavelength between 400 nanometres (violet) and 700 nanometres (red). Below this is ultraviolet, and at shorter wavelengths still are X-rays. Above 700 nanometres there is first infrared, then microwaves and radio waves with wavelengths that range from metres to thousands of metres. The different wavelengths behave in different ways; most people know, for example, that X-rays will pass through solid objects that are opaque to visible light. You can see through glass, but you can't get a suntan through it, because glass blocks ultra-violet frequencies.

Devices that can see different wavelengths can extend our ability to see far beyond the human range. The military are quick to take advantage of anything that can give them an edge; reconnaissance is a game of hide and seek played for high stakes. A cunning commander hides his forces, using darkness or trees and bushes, or man-made cover like smoke screens or walls. Larger targets can also be concealed from view if they are underground. Everything from command bunkers to chemical weapons factories can be hidden below ground, with only the occasional camouflaged air vent to suggest their presence.

Faced with the need to find such concealed targets, military technology has harnessed the full range of the electromagnetic spectrum to develop superhuman vision. Millimetre wave sensors can see through foliage, and spot a needle in a haystack as easily as a goldfish in a bowl. A new form of radar can see through rock and earth, and its eagle eyes can pick out a tunnel a hundred feet underground.

These types of sensor will find applications in the civilian world too. The first fruits of military vision technology are already filtering through. The man who has everything can already choose from a range of night-vision gadgets developed from military night scopes which amplify available light or see slightly into the infrared. The next few years will see technology becoming available with far more significant capabilities.

## What Lies Beneath: Ground Penetrating Radar

As mentioned earlier, radar was developed before the Second World War as a counter to the threat of air attack. In 1930, an Austrian scientist had already thought of the idea of using radio waves for sensing, except that Dr Wilhelm Stern was not interested in objects in the sky; he was looking in the other direction.

Stern knew that it was possible for radio waves to pass through ice, so he devised a means of using these waves to measure the thickness of glaciers. By firing a pulse of radio waves downwards through a glacier and measuring the time taken for the wave to be reflected from the rock beneath, he could calculate the depth of the ice. This was the first use of what was later to become known as Ground Penetrating Radar (GPR).

The underlying principle of GPR is that radio waves travel through different materials at different speeds. Like all electromagnetic waves, radio waves move at the speed of light through a vacuum, but anything else slows them down. The speed is determined by a property called of the substance called permittivity. Where there is a boundary between two substances – between air and water, or between ice and rock – there will be reflections. You can see a pool of water, for example, even though it is just as transparent as air, because of the reflections on the surface. The strength of these reflections depends on the difference in permittivity.

Ice has a permittivity of 4 and granite is 5, so Dr Stern could pick up the reflections from the boundary between them. During the Cold War, the military had more than a casual interest in measuring the thickness of the ice in polar seas. One third of the US strategic nuclear triad was the submarine fleet. The other legs of the triad – missiles in concrete silos, and nuclear bombers – were vulnerable to a first strike, but the submarines would survive even if the US itself were destroyed. Their patrols took them far out from the sea lanes, to lurk under the polar ice cap where they could not be found. When the order came they would surface, breaking a hole in the sea ice, and unleash thermonuclear retaliation. The plan relied on knowing when and where the ice was thin enough for a submarine to break through, information which was gathered partly using airborne GPR.[1]

There are differences in permittivity between different geological materials – sand is 10 and limestone ranges from 7 to 9 – so it is possible to examine buried rock formations. Measuring the thickness of ice is relatively straightforward as it only calls for a simple one-dimensional measurement. Looking at things in two or more dimensions is more complex. More importantly, buried man-made artefacts can also be detected.

The frequencies with best penetration in rock and earth are those with a frequency of 10–1,000 megahertz, which translates as wavelengths between 3 cm and 30 metres. The higher the frequency the better the resolution, which determines how sharp a picture can be obtained. Resolution is directly proportional to the wavelength, so at the top range objects can be distinguished down to about 3 cm – about the size of a 50p piece or small antipersonnel mine.

A key development in military GPR came about as a result of the conflict in the Falklands in 1982. The Argentine defenders sowed thousands of anti-personnel mines in front of their positions in an attempt to slow down the British counter-attack. It was a disorganized defence and no complete maps were made the minefields. Many of the mines were of a new, all-plastic construction, developed to foil mine detectors which can only detect metal.

De-mining the Falklands looked as though it was going to be a long and difficult task. The usual solution to plastic mines is to go back to manual methods, in which a skilled engineer prods the ground very carefully with a plastic probe until he locates a buried mine. This is time-consuming and extremely dangerous, and a better solution was sought.

GPR can detect plastic as easily as metal, since the only criterion is that it has a different permittivity to the ground it is buried in. A British company, ERA Technology, developed an apparatus for the job called SPRscan (short for 'Surface Penetrating Radar scanner'). This fits into a box the size of a car battery; as the SPRscan is moved above the ground it sends out a series of pulses; an antenna picks up the returns and converts them into an audible signal. Landmines, even small plastic ones, can easily be detected.

Landmines are usually just beneath the surface, but GPR can see much further than that. Under good conditions it can see down to about 30 metres, though in most soil and rock 10 metres is about the limit. Ice is much easier, and GPR has been used to see through over 5,000 metres of polar icecap. The hardest materials to see through are clays, which can limit GPR to less than a metre.

At the beginning of 2002, the White House released a new National Security Strategy. One of the key points of this strategy is that the main threat to security is no longer Russia: the new enemies are terrorists and rogue states, and the danger comes from the proliferation of weapons of mass destruction. Pre-emptive strikes may be called for as the only way of defending America; but you can't strike a target unless you can locate it.

The Pentagon has identified over 4,000 deeply buried installations around the world, from North Korea to Iran. Some of these are command centres; some of them may be underground factories producing enriched uranium or biological warfare agents. Their approximate location can be determined from the surface, but to get a fix within a few metres, the sort of accuracy required for bombing, will require new sensors. Normal GPR cannot reach deeply enough – what is needed is something using longer radio wavelengths. Long wavelengths means using a correspondingly long antenna – several kilometres long. Advanced Power Technologies Inc (APTI) believe they have just the thing.

Near the town of Fairbanks, Alaska, is a vast assemblage of radio antennae. It looks like a Futurist version of a forest, all right angles and sharp lines, wires and metal supports. This is the installation called HAARP, the High Frequency Active Auroral Research Program, and its function is to light up the sky. The HAARP's antennae project megawatts of radio energy into the ionosphere, heating it by several degrees over an area of many square kilometres. This heating has many different effects on the ionosphere.

According to some conspiracy theorists, the real function of HAARP is to act as an anti-missile shield, raising the ionosphere to block incoming ICBMs. Others say it is part of a grand scheme for weather modification, and the more paranoid claim that it is part of a radio mind control network. None of these ideas stands up to very much scrutiny, but it is certainly true that HAARP can influence the naturally occurring electrical currents in the upper atmosphere. APTI plan to use HAARP to modulate these currents, turning the ionosphere into a giant transmitter for very low frequency radio waves. Such waves are already used for communication with submarines, as they can penetrate water to a much greater depth than other radio waves, but they could also be used to look far beneath the surface of the ground.

The radio waves beamed down from the ionosphere would penetrate deeply into the earth, reflecting back from any interface layers such as pockets of air in the rock. Sensors in low-flying aircraft would pick up the radar reflections. The major challenge is making sense of these returns, and APTI have put a major effort into software to interpret the signals and identify and man-man features such as pipes, tunnels and bunkers beneath the surface. One demonstration has already produced a picture of a tunnel at a depth of almost 30 metres.

Other features besides air spaces can give away the presence of underground facilities. The Pentagon's budget for 2003 included $1.7 million for an Electromagnetic Wave Gradiometer for HAARP. This is an instrument to aid in imaging underground structures. A cave, mine or other underground facility is not likely to be completely empty; it will almost inevitably contain metal pipes, wiring and other electrically-conducting features such as 'leakage paths' of water. When illuminated with low-frequency waves like those produced by HAARP, these features respond by 'glowing' – re-radiating energy at the same wavelength like a tuning fork responding to a certain note. The gradiometer will allow such reradiated waves to be picked up and the source identified.

In future terrorists really will have no hiding place, even under-ground.

At first sight there is little obvious application for this kind of technology in civilian life. Nobody has much need to measure the thickness of ice, find landmines or locate Osama bin Laden's hideout. But there are already a range of civilian applications for GPR, even with the hefty price tag that comes attached – currently around £25,000 for the latest model.

ERA Technology market a civilian version of the SPRscan used for mine detection in the Falklands. Fitted to a wheeled mount it looks like a space-age lawnmower, and has won fame from featuring in the television series *Time Team*. This has showcased SPRscan's ability to detect underground features, with archaeologists using it to locate buried walls, tombs and traces of other forgotten structures. In Egypt, GPR found hidden tunnels in the Great Pyramid, and also some intriguing hollow spaces under the Sphinx which may be natural caves or ancient burial chambers (without permission to excavate, these remain mysterious).

It is not only ancient graves that show up. GPR is also being used by the police for the gruesome task of finding bodies. Sometimes a person has gone missing and murder is suspected; the police may even have arrested the murderer, but without a body the case is difficult to prove. In 1994, builder Lyle Keidel was accused of killing his wife in Phoenix twenty-eight years before. Keidel had cemented over his entire back yard shortly after his wife's disappearance. Digging up the yard would have been a monumental task, but a scan using GPR located a disturbed area three feet below the surface. Police dug it up and found Mrs Keidel's remains with a ligature still tied around her neck.

A similar scenario occurred with the serial killer Fred West, a builder like Keidel. GPR helped speed the search process, finding the bodies of several of his victims under the concrete floor of the cellar in Cromwell Road.

A case in Colorado involved a rancher believed to have buried bodies somewhere on his property which covered several square kilometres. It would have been impossible to dig up the whole ranch, but investigators trailed GPR over the area to find where the soil had been disturbed and located the burials.

The police will never be an important customer for GPR because, as Ron Ardell of ERA puts it, 'There just aren't enough murders in the UK.' However, as he points out, there are plenty of more prosaic uses for the equipment.

Roadworks are one of Britain's national sports, with some stretches in urban areas being dug up repeatedly within the space of a year. The work is carried out by several different agencies – cable television, gas, water, sewerage, electricity and telecoms – who may not be aware of the others' services. Surveys using GPR can help to prevent water mains being burst by road-menders or power losses caused by cable damage during work on gas pipes. At present the equipment is expensive and can only be used fully by trained operators, but as the technology matures it will increasingly become cheap and easier to use.

SPRscan is not an imaging system; in other words it does not produce a picture. The original system was created for mine detection, and it was made to be as similar as possible to existing mine detectors, which play an audible tone through an earpiece, with a change in pitch if metal is found. SPRscan mimics this by changing in tone whenever a change in permittivity is found, and with an hour's training operators can learn how to use it.

APTI's giant tunnel finder is an imaging system, with a considerable amount of intelligence built in. It is capable of automatically recognizing and noting features which have the signature of human constructions, building up a map of the underground landscape. This type of technology will soon feature in smaller systems. The next generation of GPR will display a three-dimensional image on a computer screen. This will require large amounts of computing power, but as fast processing chips tumble in price this is unlikely to add to the cost of the system. ERA say that at present they do not have any customers for an imaging GPR and will not go ahead with developing one until funding is forthcoming.

We may not have long to wait. 'Through-the-wall radar' has become one of the military's requirements for the war against terror. A request by the military for solicitations on this topic drew a large number of responses. While the operational details have not been released, the solicitations give some indication of what is on offer.

As one supplier puts it: 'The key goal is improving the resolution of the resulting images so the image will be able to show if a moving person is carrying a weapon, and if so allow the class of weapon to be identified.' They claim that existing systems allow detection of moving people through walls at ranges of around a hundred metres. To not only see a weapon but to distinguish between different types suggests a very sophisticated sensor indeed.

These systems are likely to be acquired by US Special Forces in the near future, enabling them to see inside buildings in urban operations.

The ultimate development of GPR would be a cheap imaging system with the processing power to translate the returned signal into a picture that anyone could understand. For example, it could show earth as grey with red cables and blue pipes. The computer would do much of the interpretation, just as the brain puts together visual information to assemble a coherent image.

Such a device would have considerable commercial value for surveying. As anyone who has bought a house will know, the small print in the survey report states that anything not immediately obvious is not covered. Surveyors do not take core samples to check the foundation, nor drill into the timbers looking for woodworm.

Mass-market GPR would allow suspected tree roots encroaching on the foundation to be inspected without having to dig up the driveway. More serious problems – such as old mine workings or forgotten waste tips underneath housing – could also be identified before they caused disaster. The extent of mole tunnels, rabbit warrens and badger setts can be mapped from the surface, as well as tracing the course of water and other pipes through properties.

For the gardeners and farmers it's a way of telling what the soil is like and how deep the water table is at any given point. The army of hobbyists who scour the landscape with metal detectors would have a new toy, one which would enable them to see what lies underground without digging it up. These treasure hunters will be able to locate ancient graves and other sites of interest which do not contain metal.

The same technology will also be important to the emergency services. Locating people trapped in burning buildings is a difficult and dangerous task, even with modern equipment like thermal imagers which can see through smoke. A through-the-wall radar would allow firefighters to scan the fifth floor without having to send anyone into the blaze.

It would be similarly useful for other rescue operations, where air pockets and bodies are conveniently visible compared to the background of rubble and debris that may surround them.

Naturally, the inquisitive will also use it to look through party walls to see what the neighbours are up to. For would-be Peeping Toms it might be a disappointing experience as the resolution is likely to be limited to several centimetres, reducing people to fuzzy, unrecognisable outlines. However, this may change. Smart image processing software can sharpen an image beyond its theoretical limits – a technology pioneered to extract data from military spy satellite pictures. The same techniques could be applied to give far more recognizable pictures from GPR.

This will make an already intrusive technology even more so. Police need a search warrant to enter your house, but what if they just stand outside and scan it with a GPR? The legal status of evidence gathered in this way will be a matter for the courts to decide.

There is already one legal precedent in the US from 1998, where police used a thermal imager to detect the waste heat from a suspect's house. This showed that he was using lights to grow marijuana plants, and provided a basis for the warrant that led to his arrest. However the suspect, Danny Lee Kyllo, challenged the legal basis for the warrant on the grounds that he had 'reasonable expectation' of privacy under US Law. His appeal was upheld, so in theory the police are not allowed to use any form of high-tech imaging if it violates privacy. How this stands in the light of new legislation passed to further the war against terrorism remains to be seen.

Mass market imaging GPR will mean that it becomes increasingly difficult to hide – not just from the police, but also from paparazzi, and anyone else who might have an interest. Professional burglars can check to see if you have anything worth stealing ('Wall safe in the study by the look of it . . . control box for the burglar alarm under the stairs'), and private detectives can get conclusive evidence in spite of walls and locked doors. A man who claims to be working late at the office could be confronted by print from a GPR screen revealing what he was up to with his female co-worker. Industrial espionage will gain a new tool, and nosy gadget freaks will be able to add a new way of spying on their neighbours to add to the telescope and parabolic microphone.

Walls and floors lined with GPR-proof material might start to look like an attractive option.

## Skin Pics: Millimetre Wave Imaging

As mentioned earlier, human vision stretches to include light of wave-lengths up to 700 nanometres or so. Above that lies infrared, a range associated with heat, as all hot bodies emit IR radiation. Between infrared and radio waves lie microwaves and millimetre waves (MMW), which, as the name suggests, have a wavelength measured in millimetres. One millimetre is a million nanometres, so this is a wavelength a thousand times greater than anything we can see with our native senses.

X-rays have a much shorter wavelength, and although Superman's X-ray vision might sound like a nice idea there are problems. There are few X-ray sources around; if you saw the world through X-ray eyes it would be pitch black. Doctors take an X-ray photograph with a flash of X-rays behind the subject and look at the shadows it casts. Constantly placing a light source behind anything you want to view would be difficult, and if you did have an X-ray lantern to go with your X-ray vision it would be a serious health hazard to anyone you shone it on.

Millimetre waves, on the other hand, have some of the advantages of X-rays in being able to go through solid matter, but without the disadvantages. Falling somewhere between radio waves and visible light, they are attractive to the military. For one thing, they offer the possibility of seeing things in much greater detail than the longer radio wavelengths. Radar is very limited in this respect; it might be able to identify an object as a tank, but distinguishing one type of tank from another is difficult.

Another advantage of millimetre wave radiation is that there is a lot of it about. Radar relies on having an emitter to illuminate the target with a radio beam, like shining a torch in the dark – and, like shining a torch on them, it may be all too obvious to the target. On the other hand, any warm object that contains water emits millimetre waves, so the world is naturally bathed in a gentle glow at this wavelength. The other useful feature is that non-conducting material are translucent to MMW, whereas conducting materials like metal are opaque. If you are looking for tanks hiding in a wood or a village, MMW is a good way to do it. Even in complete darkness, there is plenty of millimetre-wavelength light to see by.

Seeing in this wavelength is not so straightforward. Video cameras and thermal imagers use a lens to focus light and an array of charge-coupled devices (CCDs) to detect the light and convert it into electrical impulses. A video camera will have an array of thousands of CCDs, each one contributing one pixel to the final image.

Focusing MMWs is easy enough using a plastic lens, but they cannot be detected with a CCD, only with a small radio antenna. Scientists have had to miniaturize the antennae and pack them into a tight array to create a millimetre wave video camera.

An early non-military use of MMW technology has been at Euro-tunnel in Calais, where a system developed by Qinetiq is being used to detect illegal immigrants hiding in lorries. Most lorries using the tunnel are canvas-sided and can be scanned easily in seconds; metal-sided trucks still have to be searched by hand.

The human body shows up clearly in this range, glowing like a beacon. The water content of the body makes it a reflector, so people appear as silvery shapes. In fact, they are silvery, naked shapes: MMW pass right through clothing. This has led to a second application, using a MMW camera for airport security to scan passengers for hidden weapons. MMW scanners can easily pick up concealed knives or guns, and even all-plastic weapons show up clearly. (The question 'Is that a gun in your pocket, or are you just pleased to see me?' can be answered without ambiguity using MMW.)

Work is progressing on software so that the system can spot things on its own. Building intelligence into the system becomes vital, otherwise human operators will have to spend an impractical amount of time peering at images and trying to make out unfamiliar objects. Qinetiq are developing recognition software which will be capable of identifying the distinguishing features of weapons, so it can tell a firing pin from the tip of a biro.

There is always a balance between the number of false alarms that the system is likely to generate and the risk of letting a weapon through, but Qinetiq reckon that their system will scan passengers or suitcases at a rate of one per second. This is much faster than a human operator, and will get rid of the bottleneck caused by security delays.

However, an early attempt to introduce a MMW scanner for airport security was quickly dropped following concerns that passengers would object to security staff seeing their naked bodies. The system was considered suitable only for use in prisons and other situations where it is a less-intrusive alternative to strip searches, and privacy was not considered to be an issue.

Pacific Northwest National Laboratory is addressing the problem caused by MMW sensors being too revealing. Once more, the key is in the processing of the image. They have produced a refined version of the software, which identifies human skin and filters it out, instead producing a picture of a generic, sexless mannequin on which the rest of the MMW image is superimposed.

On the other hand, the ability to see through clothes may be important in some applications. Qinetiq have suggested a medical imaging device using MMW radiation, which would be used in accient and emergency situations. Normally surgeons have to cut away clothing to determine the extent of the injury, but a MMW scanner would allow them to assess the damage without the risk of aggravating it. The same system could also be used to look beneath bandages and casts to check the progress of an injury and the state of the skin.

Banks and other institutions are usually fitted with CCTV systems these days. Armed raiders will usually cover their faces – but a MMW system can see through a balaclava, mask, tinted visor or even a false beard. With computer enhancement to render the silvery image into normal flesh tones it could produce a recognisable picture of the criminal.

Security alerts are likely to be an increasingly frequent occurrence as the war against terrorism gets under way. Quite often these are caused by innocent carelessness, when someone accidentally leaves a bag on a station platform. The new imagers should make it much easier to detect harmless luggage. This will reduce delays and decrease the number of incidents where parcels of overdue library books are destroyed by a controlled explosion.

Qinetiq also have plans for a MMW vision system for pilots which will give them a clear view through the thickest fog and in complete darkness – another useful trait of millimetre waves. It would be even more useful for ships' pilots and yachtsmen who have to negotiate busy sea lanes which are blanketed in fog for much of the winter.

Currently MMW systems rely on background millimetre wave radiation. In the past, background sources represented the only available millimetre wave 'light'; engineers used to speak of the 'terahertz gap' between the means of generating radio and microwaves and the visible frequencies. There simply was no suitable material that could be stimulated to produce a usable beam of millimetre wave radiation. No natural material would do, so the solution was to build a new type of material called a superlattice, a synthetic crystal with a structure that does not appear in nature.

The first terahertz laser was demonstrated in 1994; the technology is still in its infancy but it is growing up fast. Instead of just relying on the reflections of the dim background glow, it will be possible to shine a searchlight on the world. This would make it possible to detect hidden weapons not just on someone standing next to the camera, but from a significant distance.

Since non-conductive objects like walls are not opaque to MMW, just translucent, this bright light will shine through them. A suitable source, linked to a system to pick up and interpret the reflected waves, would show a strange new world. Silver-skinned humanoids inhabit buildings like fish tanks with only the metal beams and window frames blocking the line of sight. It's the next best thing to true X-ray vision.

MMW camera systems are still fairly pricey – expect to pay in the region of £60,000 for a full system that needs a couple of suitcases to carry it around. As the technology evolves, it will follow the usual path of smaller, cheaper versions. And while there might be some demand for automatic scanners to prevent people from bringing weapons, bottles or other items into football grounds, cinemas and other public places, there will also be a demand for the X-rated, fully imaging system.

The media and press are likely to pick up on it first. The green-tinged pictures taken using nightscope-type devices are now a regular feature of TV news programmes. Crime scenes and other sites of interest will soon fall under the eye of MMW imaging as well.

Many uses are not likely to be as respectable. Hollywood stars have always been considered fair game for a press pack armed with the latest technology. MMW imaging literally has the power to lay people bare;

expect the Internet to be awash with silvery images of naked famous people.

Its X-ray-like ability to see through clothes makes MMW imaging a remarkably intrusive technology. Does conservative politician X really have an intimate body piercing? Is actress Y hiding a pregnant bulge under those loose outfits? (Or worse, cellulite?) Does sportsman Z pad his shorts? The camera never lies, and it can see everything. Given the tabloids' willingness to pay big money for scandal stories, it is easy to see how you could recoup the cost of a MMW imager fairly quickly.

Nor will it stop with bodies. The same technology effectively allows you to go through the contents of someone's pockets and handbags, search their luggage and much of their home. Does the famous red dispatch box used on Budget Day really contain the Chancellor's sandwiches? A TV personality claims to have given up smoking – but isn't that a pack of cigarettes and a lighter in their jacket? Is that a bag of white powder in someone's pocket?

It used to be said that nobody knows what goes on behind closed doors; with MMW that will no longer be the case. You can slam the door on the TV cameras, but they can still switch to terahertz imaging and see what you are up to.

The same legal questions apply to MMW as to radar. Police have limited powers of stop-and-search, but can they legally scan anyone on the street with a MMW camera looking for drugs or weapons – or is this harassment and a gross violation of privacy? And, as with the radar technology, falling prices and easy availability mean that it will end up in the hands of anyone who wants it.

New technology will allow us to see more than ever before, and it will also allow us to be seen. Close the door and draw the curtains – and Big Brother can still see you. Even the grave does not provide privacy from snooping GPR systems.

Secrets will be laid bare as never before. As always, there is one camp which says that the innocent have nothing to fear, and another that worries about the erosion of civil liberties and the advance of technologies which can be used for oppressive purposes. Inevitably, MMW imaging and GPR will change the world in subtle ways and reveal things which have never been seen before. What we cannot see is how this will affect our society.

# CHAPTER 14

# Robots in the Air

Unmanned aircraft flew long before man ever took to the air. Even without counting kites and the like, toy aircraft were well ahead of the real thing. In 1875, a Mr Wright brought one home for his two sons, Wilbur, then eight and Orville, twelve. Orville describes it:

> A small toy actuated by a rubber spring which would lift itself into the air. We built a number of copies of this toy, which flew successfully . . . but when we undertook to build the toy on a much larger scale it failed to work so well. The reason was not understood by us at the time, so we finally abandoned the experiments.

Later on in life, the Wright brothers took up the challenge again, and in 1903 Orville successfully piloted their Flyer in the first ever manned, powered, heavier-than-air flight.

The piloted flying machine has developed immeasurably since that first flight; but it has always been shadowed by its unmanned cousin. Manned aircraft dominated the first century of flight, but the next century may belong more to the pilotless aircraft. These will range from giants bigger than anything flying today to machines so small as to be barely visible. The military have been flying unmanned craft for decades, and over the next few years they are likely to make themselves felt in both the military and the civil spheres.

Unmanned aircraft are big business. Planned spending for the period 2002–2009 is estimated at $15 billion for the US, $4 billion for Europe. Worldwide, the market for unmanned vehicles for law enforcement is put at $740 million over the next ten years; even power line inspection is worth $410 million.

## The First Military Drones

The Royal Flying Corps had a bad start during the First World War. On the Western Front their aircraft were outclassed by German Fokker monoplanes, and in Southern England they were having trouble intercepting the German Zeppelins that made regular incursions to bomb British cities.

Looking for new ways to fight the war, the RFC sought advice from Professor A.M. Low. Low had been commissioned into the Army as a Lieutenant, and had worked on rangefinders. He now turned his attention to a new project: flying an aircraft by remote control. Low's work was intended to produce a new type of aircraft, an unmanned kamikaze plane carrying explosives, which could destroy a Zeppelin or a ground target by crashing into it.

Low's first creation was the Sopwith AT, made from cannibalized parts of other aircraft.[1] It was smaller than a manned plane, with a 5-metre wingspan and carrying a formidable 23-kg explosive charge. Radio-controlled servomotors operated the controls.

Unfortunately the AT was rather erratic in flight. In March 1917 a demonstration was made to the top brass, including representatives from various Allied nations. Unfortunately the crowd of dignified onlookers was forced to scatter in panic when the AT dived on them. The observers were not impressed by the lack of control or the range limitations of the radio-controlled system, and the Royal Flying Corps did not adopt it.

Events followed a similar pattern in the US, where the Navy ordered a miniature pilotless biplane, known as an 'aerial torpedo'. Operation relied on accurately judging distance to the target; a time switch was set on launching, and when it went off, the bolts attaching the wings were automatically removed so that the fuselage plunged on to the target with 130 kg of explosive. The US Army had a similar device, another biplane called the Dayton-Wright Kettering Bug. This had a range of 40 miles but a similar lack of accuracy.

After the First World War, the RAF continued experimenting with aircraft which could be controlled by radio, and were used as targets for anti-aircraft gunnery in the 1920s and 1930s.

During the Second World War, the aerial torpedo concept was revived in the form of the German V-1 flying bomb described earlier. There were also many attempts to build aircraft along similar lines as the AT, modifying existing aircraft by adding radio control and packing them with explosives. The controller would fly alongside the drone in another aircraft and guide it down to the target; although easy enough in principle this proved difficult in practice. The Germans produced a whole range of these robot kamikazes, some of them big enough to attack a battleship, but had indifferent success. Such drones were radio controlled by a remote pilot, so the pilot had to stay within visual range of the drone the entire time, putting him in as much danger as the drone.

Given these limitations it is no surprise that drones were used mainly as practice targets for anti-aircraft guns and missiles until the 1960s. Then a combination of technology and politics began to make them attractive for other missions.

New technology appeared in the form of reliable guidance systems and electronics based on the integrated circuit. These meant that an unmanned aircraft could be programmed to do more than just fly in a straight line until it failed; it could be given a flight plan to go to a distant location, take photographs and return to its start. Robot spies would not be as versatile as human pilots, but they had other advantages.

The loss of Gary Powers and his U-2 spy plane over the Soviet Union in 1960 was a diplomatic disaster, whereas the loss of an unmanned aircraft would not cause any lost sleep. A robot pilot had no family at home, could not be paraded in front of the cameras and could not give away any secrets. Within three months of the Powers incident, the Ryan company had produced a robot spy plane by modifying one of their Firebee target drones.

The first of the Ryan drones was 8 metres long with a 4-metre wingspan and weighed less than 2 tons — more on the scale of an American car of the period than an aircraft. But its compact turbojet could drive the Firebee at a cruising speed of almost 500 mph, and a combination of autopilot and radio control gave it maximum flexibility.

Ryan built more than twenty variants of the Firebee alone. These carried cameras and other sensors such as radar and radio-detection gear over enemy territory. They served in Vietnam, they flew over China, Korea and the Middle East. In due course some were shot down or crashed, but as anticipated there was no reaction from the media. The Chinese press agency produced pictures of what were clearly the remains of an American-made Firebee drone, but the American military could simply shrug and point out that they had sold similar craft to other nations.

The Israelis found a new use for drones supplied by the US. Fitted with radar reflectors to make them appear larger on radar, they flew ahead of Israeli strike aircraft, drawing anti-aircraft missiles away from the real targets.

While they were far less capable than piloted aircraft, the drones did their job and became an established part of the reconnaissance team.

## Staying Up Longer

We have already looked at the nuclear-powered aircraft program of the 1950s and 1960s, a billion-dollar experiment that ran into a dead end. But while nuclear power may have proven not to be feasible, the dream of an aircraft that could stay in the air for days or weeks on end without refuelling remained.

At first this was achieved by using highly economical engines on drones optimized to fly at extreme altitude. The Compass Cope programme of 1973 called for a drone which could fly at more than 50,000 feet for over twenty-four hours, carrying photographic and other intelligence equipment, and relaying a television image back to a remote operator. Both Teledyne Ryan, an aircraft maker specializing in drones, and Boeing met the requirements, with the latter winning the contract.

In spite of the success of the aircraft, the program was cancelled in 1977 because of problems with developing the sensor equipment. Later programs met with more success, and by the 1990s the USAF had two long-endurance reconnaissance drone programs. The small tactical

Predator drone proved its value for reconnaissance during the conflict in former Yugoslavia at the end of the 1990s, and the larger Global Hawk was first used in the 2002 war in Afghanistan.

Both the Predator and Global Hawk have endurance in excess of forty hours. Predator uses a propeller and flies at low altitude and low speed, while Global Hawk is a high-altitude jet. The Predator also found a new role: armed with a Hellfire missile, one was used by the CIA to assassinate a suspected terrorist in the Yemen in 2002. Much more advanced unmanned attack vehicles are in the pipeline, but the issue of long endurance remains.

The solution lay in dropping conventionally-powered engines altogether. Like the Wright Brothers' Flyer, the first of the new aircraft came from the backyard school of engineering. A longstanding award for the first team to achieve true man-powered flight was offered by Henry Kremer, a British industrialist, in 1958. The Kremer Prize, originally set at £10,000, was steadily increased over the years as failures mounted. It would go to whoever could fly around a figure-of-eight course with two markers half a mile apart.

Dr Paul Macready of specialist aircraft makers AeroVironment had studied the Kremer challenge. He came to the conclusion that a standard hanglider would need approximately 1.5 horsepower to maintain altitude; but a human could only provide 0.5 horsepower continuously for any length of time. In simplistic terms, all that was needed was a glider with about three times the wing area of a hanglider which would provide the requisite lift for a third of the power.

The challenge was to build a glider that size without increasing the weight. To make it as light as possible, the aircraft was made from thin aluminium tubes covered with Mylar plastic and braced with steel wires. The leading edges of the wing (crucial for shaping airflow) were fabricated from corrugated cardboard and styrene foam. It was fragile but very easy to put together again after a crash. Macready's team said that it could be rebuilt within twenty-four hours if necessary.

The result was Gossamer Condor, with a wingspan of almost 30 metres but weighing less than 32 kg. To put it another way, it weighed less than a large dog and yet had the wingspan of ten eagles. On 23 August 1977, piloted by Bryan Allen, the Gossamer Condor finally took the Kremer Prize.

Other Gossamer aircraft followed. Perhaps the most significant was the Gossamer Penguin which replaced human power with solar power.

The Penguin evolved into the Solar Challenger, which was smaller, heavier and more robust than the Condor. In 1981 the Solar Challenger scored another notable first when it flew across the English Channel on nothing but sunlight, at an average speed of 40 m.p.h.

This began to look like an aircraft with military possibilities. Given a solar power source and no pilot, there would be practically no limit to how long the aircraft could stay up. If it had a means of storing energy for the night-time, it could fly on sunlight during the day and battery power at night.

Lockheed Martin, makers of the SR-71 and U-2 spy plane, undertook a series of studies for High Altitude Powered Platforms (or HAPPs) and concluded that solar power was the most promising solution. These were purely paper studies without building any hardware, but they concluded that a HAPP with a wingspan of 100 metres would be capable of carrying a useful load (45 kg or more) up to an altitude of 80,000 feet for days on end. In fact, the only limitation on endurance would be the reliability of the components.

By 1983, on the strength of the Solar Challenger, AeroVironment were working on a highly classified aircraft known as HALSOL, which stood for High ALtitude SOLar. HALSOL was all wing, made from five 6-metre segments stuck together for a 30-metre wingspan. Some creative scavenging went into the rest of it; the main landing gear used a bicycle wheel while the outboard landing gear incorporated wheels borrowed from baby buggies.

Although it flew successfully, HALSOL showed that solar cell technology was not adequate for the task. The main problem was that it was not possible to get enough power per unit area for the aircraft to climb to high altitude, let alone have enough spare to charge up the batteries for overnight use. In any case, batteries powerful enough to store sufficient energy for a night's flying would have been too heavy to carry. The project was shelved until technology matured.

Solar cells and batteries have become progressively lighter and more efficient, and at the same time a new role emerged for an unmanned air vehicle. One of the biggest failures of the 1991 Gulf War was in locating and destroying mobile Scud missile launchers. In spite of being the highest priority target on the battlefield, not a single mobile launcher was located and destroyed, even after missiles were repeatedly fired at Kuwait and Israel. (The ill-fated 'Bravo Two-Zero' mission was part of an Allied effort to locate Scud launchers using Special Forces teams.) Spy satellites could spot missile launches, but only if they happened to be overhead at the time; they could not provide 24-hour coverage.

The new Ballistic Missile Defense Organization (BMDO), successor to the Strategic Defense Initiative/Star Wars, wanted a high-altitude, long-endurance drone. This was to be the basis of a system called Raptor/Talon. According to this plan, an unmanned aircraft called Raptor[2] patrolled a potential trouble spot, watching out for the bright flare of a missile launch. It would then track and shoot down the ballistic missile with a tiny, high-velocity Talon missile.

To meet the Raptor/Talon requirement, HALSOL was dusted off and AeroVironment carried out a series of upgrades to the propellers and the power systems, giving it lighter and more efficient solar cells.

At the same time as the classified work on Raptor/Talon, the decision was made to reveal the earlier solar-powered project. It was 1993, ten years after its first flight, that HALSOL was declassified and shown to the public for the first time. It was given a new name, Pathfinder, and turned over to NASA. Officially, the solar-powered drone was going to be used for atmospheric research and as a platform for civilian sensors. Since then, Pathfinder has grown and evolved although at a leisurely pace. In 1998 it had become Pathfinder Plus, and set a new altitude record of 80,000 feet. It was succeeded by Centurion, with a wingspan of over 200 feet, leading to the ultimate form known as Helios after the Greek sun god.

In 2001 Helios achieved an absolute altitude record for a non-rocket-propelled aircraft of 96,500 feet. It is intended to be able to carry a 100-kg payload, twice the aim of the original Lockheed study twenty years before, and enough to carry the sensors and Talon missiles envisaged by the BMDO.

In 2003 NASA completed a demonstration of the solar-powered aircraft's ability to perform useful work, which involved taking digital photographs of the Kaui coffee plantation in Hawaii. During the eleven-hour flight, the plantation was never entirely cloud free, but by taking pictures in the clear areas it produced 300 images which could then be assembled into a complete mosaic of the area. The images showed the relative ripeness of the beans in different parts of the plantation.

Disaster struck in June 2003. Helios ran into turbulence above the Pacific; for all its size, Helios is as fragile as any of the descendants of the

Gossamer Condor. It broke up, and the fragments fluttered down into the ocean, as mild an air crash as can be imagined, but a minor tragedy for NASA. Already under pressure after the Columbia shuttle disaster, the failure put the Agency's management under further scrutiny. It remains to be seen whether the program will recover.

Most of the technology used by Pathfinder has been declassified, but some elements are still secret. Some of these relate to the extremely effective rare-earth magnets used in the electric motors. It is interesting that such technology is still militarily, as well as commercially, important enough to be protected.

NASA attempted to regain some ground using a model aircraft with an array of photovoltaic cells on the underside. Although little more than a model made of balsa wood, NASA could still claim a first: flown indoors, the aircraft was powered not by sunlight but by a laser beam. The laser-powered aircraft made news headlines, although it did not actually offer anything new over the earlier models. NASA suggested that the beamed-power concept could be used for supplementing solar power, especially in situations such as winter in polar latitudes where sunlight might not be adequate.

In spite of the accident, AeroVironment remain optimistic that Helios represents a commercial opportunity, and by 2006 they plan to offer a commercial service using solar aircraft. At an altitude of 90,000 feet – 17 miles – the solar craft could act as a communications relay. In effect it would do the same job as a communications satellite, but from an altitude of 20 miles rather than 20,000 and with a corresponding decrease in the amount of power needed for transmission and the need for a large antenna or dish.

SkyTower, the AeroVironment subsidiary for communications, believes that a fleet of several aircraft will be able to provide continuous coverage over a wide area at low cost. Their killer application is the broadband Internet, what used to be known as the Information Superhighway: a single aircraft should be able to provide a bandwidth of 5 gigabits per second, enough for about 10,000 high-speed Internet connections simultaneously. 3-G mobile phones (which carry much more data than earlier phones and can be used for sending pictures and video, and broadcast TV) are also seen as significant potential markets.

Solar UAVs should be highly competitive with landlines in terms of cost, especially for more remote locations. There is an obvious parallel here with the first communication satellites and their cheapness compared to the transatlantic cable. While fibre optics have made landlines cheaper, solar UAVs will be far cheaper than satellites. A single aircraft will have a 'footprint' of area covering thousands of square miles, including locations that are not accessible to ground-based transmitters. Several aircraft can relay signals between them across continental distances to provide coverage that can be scaled up as necessary.

The solar aircraft has advantages compared to satellites too. If there are problems, the aircraft can be brought down for repair and maintenance. As technology advances, the communications gear can be upgraded to the latest standard, whereas an old or malfunctioning satellite simply becomes another piece of space junk. Aircraft are also safe from micrometeorite strikes and solar storms that sometimes interfere with satellite communications.

Some estimates put the cost of a large solar UAV as low as two or three million dollars. This would make a permanent broadband Internet connection in the sky available for just a few hundred dollars virtually anywhere on earth. It is easy to see how significant this could be, especially in the developing world.

Solar aircraft may also feature more as the demand for mobile phones increases. In areas without mobile phone coverage, a single aircraft could be used as an alternative to putting up hundreds of mobile phone masts, especially in remote locations. This becomes increasingly true as suspicions about the health effects of these masts grow and local residents and campaigning groups object to new masts being erected. A solar UAV does not have to be put in anyone's backyard.

Flexibility is also an asset. Large gatherings of people can overload an area's mobile phone capacity, as happened during the eclipse-watching in the West Country in 1999. Given a fleet of high-altitude mobile phone connections, an operator can simply rearrange them, moving the spare capacity to meet changing demands.

Other applications for long-endurance drones include acting as substitutes for weather satellites and more advanced versions of the crop monitoring carried out in Maui. Environmental work, looking at everything from deforestation to irrigation patterns to watching for sea pollution, can be carried out more cheaply and easily than from a satellite. They would be welcomed by the Coast Guard and others who have to keep an eye over large expanses of territory for extended periods.

Parallel developments involve the use of solar panels on unmanned airships. These would sacrifice a certain amount of mobility for the advantages of scale: an airship could be made large enough to carry a payload of several tons. Several companies including Lockheed Martin in the US and IAI in Israel are known to be working on large, long-endurance airships, mainly for military surveillance purposes.

Solar aircraft do not need to be huge. The Belgian technology research centre VITO is building one with a 16-metre wingspan weighing just 16 kg carrying a 2-kg payload. It is intended to fly for four to six months at a time, taking pictures that will be equal in quality to satellite images but at a fraction of the price.

A French company is building an even smaller solar UAV, with a 1.5-metre wingspan and endurance of more than a week. This would be programmed to loiter at low altitude, keeping a constant watch.

In principle, solar aircraft would give every agency the power of a spy satellite for a fraction of the cost. Since they operate at much lower altitudes than the 100-mile altitude of spy satellites they need less powerful optics and other sensors; and because they are can be directed over an area of interest, less of their time is wasted. Low satellites spend most of their time over the horizon, looping round the world; solar UAVs are always there, watching over you. For good or ill, Big Brother has a new set of eyes.

## Black Boomerangs

As a curious sideline, it is possible that solar UAVs have been flying for some years. Although the Raptor/Talon was apparently abandoned, other agencies have also shown an interest in UAVs with high endurance. While HALSOL was declassified, this is not necessarily an indication that activity on classified versions ceased. It is entirely possible that the rechristened Pathfinder was simply the 'white world' part of a larger black project.

This follows an established procedure, allowing aircraft makers to share knowledge, recruit staff and gather expertise without revealing the existence of a black program. If a company suddenly starts showing an interest in, say, methane-fuelled engines, this is a giveaway that they are working on something in that line. If they have no known programs using that technology, the existence of a classified project becomes glaringly obvious.

If there is an unclassified solar aircraft program, it can cover any traces of a classified equivalent. And there have been sightings since the early 1990s of a new type of UFO in America, known as 'black triangles'. These are generally described as being enormous, boomerang-shaped aircraft that fly silently and at very slow speed – in one incident a jogger thought he was keeping pace with one.

Some watchers immediately assumed that the flying triangles were alien spacecraft, others that they were some sort of anti-gravity device developed using secret (and probably alien) technology. A more plausible alternative is that they are actually a relation of the classified solar-powered aircraft that we have already seen.

While a solar-powered spy plane might be expected to remain permanently at high altitude like HELIOS, for some missions it might be necessary to descend to very low altitude. Some sensors, such as devices for detecting sources of radioactivity or sniffing the air for traces of substances associated with chemical weapons manufacture, need to get within a few hundred metres of the suspect site. For this role a solar-powered drone would be ideal. These aircraft are perfectly stealthy, having a very low radar signature, no infra-red emission and practically silent. It is entirely possible that such aircraft would be tested in the US, though less plausible that they would be operated close to densely populated areas, unless one assumes that they were being tested to see if anyone noticed.

Such silent spies would be kept secret, as their utility depends on opponents not being aware that the US has such a capability. This secrecy also works against them. If a secret drone finds evidence of illegal nuclear activity in a rogue state, it is impossible to reveal the evidence without giving away the source of the information.

Interestingly, there were reports from U-2 pilots during the 2003 Iraq conflict of unmanned aircraft sharing their airspace. The aircraft, of a large, unknown type, were not controlled by the USAF but were apparently working for an intelligence organization. Because they were not working within the USAF's air traffic control, the pilots were concerned about the possibility of an accident. It is possible that the black boomerangs have some other explanation entirely, but they may very well be an indication of where the billions spent on classified projects are going.

## The Unmanned Airliner

The military favour UAVs for 'dirty, dull and dangerous' jobs. As one manufacturer's advert put it, 'Can you keep your eyes open for thirty hours without blinking?' No human pilot can cope with such long reconnaissance missions, and it is far better not to risk human life when attacking surface-to-air missile sites ('it's like charging at an elephant gun with an elephant').

While most drones are purpose built, in some cases they are simple conversions of existing aircraft: the QF-4 drone, for example, is a remotely-piloted version of the F-4 Phantom.

This may be attractive for the civilian market, but for reasons that have more to do with economy. A small airliner like a Boeing 737 typically has a captain, a first officer and three cabin crew. Cabin crew are cheap, with a salary of no more than £17,000, but pilots are expensive – the Captain starts at around £80,000 and the First Officer at £50,000. A study of the running costs of one no-frills airline showed that almost 10 per cent of their outgoings went on paying the pilots. The total profit of the airline was less than 10 per cent of turnover, so basic accountancy suggests that being able to dispense with the pilots would double profit at a stroke.

Pilots are paid well for good reason. It is not that their skills are particularly hard to come by – the training is no longer than for many other occupations – it is more to do with the power that they wield within the industry. A walkout by a handful of pilots will quickly bring any airline to its knees, and pay disputes are usually settled swiftly. An automated aircraft would dispel any possible pilot problems.

The big question is whether the public could be persuaded to set foot on an aircraft piloted by a computer. Flight may be fairly routine, but it still produces a high level of anxiety in most people, and if there is one thing we have learned from our day-to-day dealings with computers it is that they are prone to crash.

In London, the Docklands Light Railway, which operated driverless trains, was viewed with great suspicion, although it has operated safely. Meanwhile human train drivers running through red lights have caused several fatal train crashes, and pressure is growing to have more automatic safety systems installed.

In the air, the military are paving the way. The Global Hawk was the first unmanned aircraft to be certified by the Federal Aviation Administration so that it can use civilian air corridors with no advance notice. Over the next few years the infrastructure will be put in place so that military drones can operate in the same airspace as civil aircraft, and such unmanned flights will become routine.

The US Air Force is already looking at unmanned concepts for its logistics fleet. One of the most advanced ideas has aircraft flying in groups, with a single manned freighter leading several unmanned followers. Such flights are likely to become commonplace in the near future.

This could soon open the way for freight airlines to abandon pilots. In the US, parcel carriers like FedEx and UPS operate their own fleets of freighters. With no passengers or cabin crew to worry about, there will be much less difficulty in making the transition. They will simply be civilian versions of the military logistics transports that fly every day. One concept put forward by the US Federal Aviation Administration is for a completely automated freight airlift operation. The customer simply delivers their freight container to a terminal, and the machines take over, scheduling, loading, flying and unloading the container at any other terminal linked to the system.

As they rack up more flying hours and establish a safety record, the argument for unmanned flight will become more complex.

The two biggest causes of fatal air crashes are 'controlled flight into terrain' and 'approach-and-landing' accidents. The first of these occurs when the aircraft, under full control of the crew, runs into terrain in the form of a hill or mountain, of which they were unaware. The instruments may be showing that there is a mountain dead ahead, but if the plane is being flown by the pilot rather than the instruments the human error can be fatal. Similarly, approach-and-landing crashes happen when the crew do the wrong thing during a landing approach. Simply pulling a lever at the wrong moment will change a smooth descent into a disaster.

The recommended remedy for both problems is more and better crew training and instruction. However, aircrew are only human, often working under the burdens of stress, long hours and a confused body-clock. It is a tribute to their professionalism that accidents are so rare, but they still have human limitations. In bad weather airliners already land on instruments only, without human intervention. If unmanned aircraft can show a consistent record they may come to surpass the safety of human pilots; then it will be more a matter of whether we dare to risk having fallible humans in charge of aircraft instead of steely-nerved machines.

One way of achieving a solution that gets the best of both worlds is to have a human pilot who is not physically located inside the aircraft. Using a virtual cockpit, a pilot can sit at Heathrow and take off and land any number of airliners all over the world. Since most of the pilot's job at present consists of waiting for take-off and landing with little to do in between, this would lead to a far more efficient use of the pilot's time.

There will be battles ahead, as both the general public and pilots' organizations are likely to be deeply opposed to any developments of this nature. But as military pilots become a thing of the past, a shift away from having a human pilot in the cockpit begins to look increasingly likely.

## Micro Air Vehicles

Unmanned air vehicles do not need to be giant solar-powered craft or conversions of existing manned aircraft. Much of the military interest has focused on micro-unmanned air vehicles, or MAVs. These are craft with a wingspan of half a metre or less – in some cases very much less. MAVs the size of a wasp have already been seriously considered, and the technology is advancing rapidly. Small and cheap, they could give the foot soldier an unprecedented ability to see what is on the other side of the hill. They could also have their uses in the civilian world.

The CIA was the operator of many of the unmanned spies during the Cold War era. Most of these were conventional, but there were some very exotic UAVs built for specific missions. In one case during the 1970s the CIA needed a means of delivering a miniature listening device to its target. Existing methods were not satisfactory, so instead they set about building a robot dragonfly for the task.

According to the CIA Museum, the 'insectothopter' had a miniature oscillating engine which powered wings that flapped in a similar way to a real dragonfly. It was remotely guided by a laser beam, but suffered from loss of control in any but the lightest of winds. The insectothopter was never used in practice.

It is hard to take the insectothopter seriously because of the difficulty of building such a robot with 1970s technology. The CIA Museum is not open to the public, making it difficult to check on how plausible the insectothopter might be. Certainly the CIA has some work to do in persuading the taxpayer of their technical skill and excellence, and it is credible that the insectothopter was attempted, given the agency's penchant for over-the-top schemes.[3]

While such schemes may have been wildly ambitious in the 1970s, by the 1990s insect-sized spies began to look technically feasible. DARPA issued a requirement for a MAV with a wingspan of no more than 15 cm and weighing no more than a hundred grammes. This would be expected to carry a video camera and relay pictures back from a distance of three miles for twenty minutes.

The soldier of the future would carry a MAV in his backpack. When reconnaissance is needed, the craft would be launched into the air using either compressed air or something like a schoolboy's catapult. Once airborne, it would be controlled using a device like a hand-held video game, with a television screen and a joystick. Once the mission is complete, both the MAV and controller would be discarded.

There is already a profusion of miniature aircraft with significant capabilities, but there are still technical issues that need to be overcome before MAVs will be truly viable.

The first issue is one of aerodynamics. Moving down from the scale of aircraft to that of large insects, the rules change. The viscosity or thickness of air, hardly noticeable on a large scale, becomes a dominant factor as the lift from airflow becomes less. Eagles and albatrosses can glide for extended periods, but smaller birds have to flap their wings in proportion to their size. A butterfly's flapping is almost leisurely compared to the rate at which a fly's wings beat, and tinier insects have to flap even harder.

Many researchers believe that scaled-down aircraft or helicopters will not be able to generate enough lift. Instead, they are going down the same route as the CIA and imitating nature with designs based on a flapping wing. Such craft, known as 'entomopters' ('insect flyers') are powered by artificial muscles and look more like wasps or dragonflies than conventional aircraft.

One of the leaders in this field is Charles Ellington, a dual-national expert whose plans were turned down by the MoD and who went to work for DARPA instead. Ellington joined Robert Michelson's team at Georgia Tech. The entomopter they are working on will be dual mode, capable not only of flight but also of crawling. Michelson's goal is not just to look over the hill, but also into the next room. He believes that entomopters will prove invaluable for operations indoors as well as out.

The entomopter could fly 'down a smokestack, or through a heating or ventilation system', says Michelson, then 'go down corridors, in search of hostages, perhaps'.

But even if the aerodynamics of microflight can be mastered, there are still the problems of packing all the necessary machinery into such a small space. Designers have to follow a strategy of making everything as multi-functional as possible. Components which would be purely structural on a larger UAV will have to have extra roles. The wings, for example, may also have to function as batteries, and the body as an antenna.

Existing MAVs already meet the initial DARPA requirements. The Black Widow made by AeroVironment (the same company that makes solar-powered aircraft) is a disc-shaped craft 15 cm across and weighing barely 60 g. It can fly at up to 43 m.p.h. and send back video for more than twenty minutes. A slightly larger vehicle (170 g) has set an endurance record of one hour forty-seven minutes for a hand-launched MAV.

Details of the most advanced MAVs are hard to come by. In 2000, the Chinese claimed to have tested a MAV the size of a wasp. There may be some truth in this; a German lab has tested a MAV less than a centimetre across, although it had very limited capabilities. If the CIA failed with their insectothopter thirty years ago, it is a fair bet that they will have built something more successful in the intervening period.

One interesting aspect of this research is that it is comparatively cheap. Developing a new airliner costs billions, but micro air vehicles can be constructed by smaller nations. Taiwan has an advanced MAV program, including one vehicle with a wingspan of 15 cm carrying a 40g-payload. South Korea has already flown an even smaller MAV called Batwing with an endurance of fifteen minutes and is working towards building completely autonomous micro-aircraft. France, Canada and Israel are known to be active in this field. In Britain, a team at the University of Bath is working on a MAV 'the size of a bee'.

In Japan, Seiko Epson have already demonstrated a mini-helicopter with camera for the civil market the size of a soft drinks can. It has an endurance of just three minutes, but with Bluetooth connectivity it is already compatible with PCs and other devices, and the company intend later versions to be able to send pictures to mobile phones.

## Smart Dust

Even smaller MAVs already exist. The tiniest under consideration are so small that they are known as 'smart dust'. These are so small that they cannot control their flight; like microscopic insects and bacteria, they just float with the wind. Released in huge numbers – thousands or even millions – they would blanket a large area. Each 'mote' of smart dust would only have a rudimentary capability, but together they would be able to build up a picture of the area. Some motes would be sensitive to vibration, others to temperature, and others to magnetic fields.

Once released, the motes would communicate with each other, building up a network and sensing their relative positions. Then they could start sending back signals to base. Anything passing through the area occupied by the smart dust would be identified by its signature. As a military technology, the first interest is being able to tell the vibration, heat signature and magnetic effects of a tank from the similar but different signature of other vehicles.

In a test in 2003, a hundred motes were used to detect the location of passing vehicles. The metal in the vehicles affects the earth's magnetic field, a disturbance detected by the motes. By using the signal from several motes, the exact location and speed of the vehicle could be calculated and tracked.

Smart dust works as a network with a large number of elements. So long as the network is maintained, then it does not matter how many motes are lost or destroyed, and the network as a whole will continue to send back information. Attacking the individual motes would be a futile exercise; the only way to tackle it will be to destroy the network itself.[4] Naturally there is a major focus on ensuring that the network cannot be jammed or otherwise interfered with.

The latest smart dust motes are 2.5mm across; succeeding generations will be much smaller and correspondingly cheaper.

It is not clear what effect these miniature spies will have in the civilian world. Certainly they will make wonderful toys. They have been suggested as a means for providing unusual camera angles for televising sporting events, and of course the emergency rescue services will find uses for them in looking for lost climbers and others. More mundane tasks – investigating traffic disturbances at particular locations, or checking for leaves on railway lines, for example – might also fall to the MAV's lot, especially when they are linked to a central computer system. Checking power lines in remote locations, a task currently undertaken by manned helicopters, is typical of the situation where MAVs are likely to prove cost effective.

They will also be a boon to the police. A larger Predator UAV has already been on a fifteen-day trial with the US Customs Enforcement Agency, and has been credited with assisting in eight arrests, detaining twenty-two illegal aliens and seizing a ton of marijuana. Few police forces can afford the cost of acquiring and running a helicopter, and even when there is one it may take several minutes to get to the trouble scene. MAVs give the possibility for every squad car to have its own aerial support sitting in the back, ready to pursue a vehicle or follow a suspect across any terrain, either operated by remote control or autonomously. Trying to lose such a pursuer would be difficult, and the psychological value of an MAV pursuing a suspect and illuminating them with a searchlight or laser pointer would be considerable.

During the 1991 Gulf War, history was made when a group of Iraqis attempted to surrender to an unmanned aircraft. It is only a matter of time before police make their first arrest by remote control.

Police MAVs may not be limited to observation. There have already been military tests with tear gas dispensers and other non-lethal devices, and as technology advances (see Chapter 9) this may prove to be a very effective combination. These days bomb disposal experts will always send out a remote-controlled robot rather than approach an explosive device themselves. In a few years time, police may choose not to expose themselves to fire, instead using remote microhelicopters to locate and take out armed suspects.

Perhaps the most obvious application for the MAV is as an extension of its natural role as spy, as a new means for newsgathering (or 'press intrusion'). At present, some television news channels have helicopters so they can be first on the scene for a breaking story. Occasionally this can lead to a scoop such as the film of OJ Simpson's pursuit by police cars, but in general these helicopters are too expensive – there is only one full-time news helicopter in the UK. If MAVs become available, every news reporter will be able to have the same equipment as the infantryman mentioned above. If there is a major fire, a chemical spill, a hostage drama or a bomb alert, the press will not be kept back behind the line of safety, but will be able to beam back images from the heart of the action.

MAVs would also be a boon to film and television producers. The long, swooping aerial shot which was previously the trademark of big productions will be available to everyone. Cranes used for overhead shots will likewise become obsolete, and other possibilities will appear, such as having a bird's-eye view flying through a city which dives through an open window to discover a drama in progress. Documentary makers will not be slow to appreciate the new medium and we can expect plenty of creative use of aerial footage when it becomes available to all.

Of course, not all filming is quite so high minded. Instead of long lenses, paparazzi would have a new technology to see beyond high walls and peer through bedroom windows. Michelson's entomopter might be going down chimneys and through air ducts in search of accident victims, but if it ever makes it to the commercial world it will be put on the trail of adulterous pop stars or drug-snorting sportsmen – or simply to get pictures of celebrity weddings which the press would be otherwise barred from. Industrial espionage would also take on a new dimension.

Smart dust has the potential to be more intrusive still. While the present version may be barely able to tell a human walking past from a camel, the progress of technology means that it is only a matter of time before individuals can be distinguished from the sound of their footsteps. Spies will be literally everywhere, and every conversation becomes accessible to those who control the clouds of smart dust that can reach every corner.

In principle, existing laws cover this kind of bugging. Neither the government nor anyone else is allowed to start wafting microphones into your house, whether they are the size of a golf ball or a dust mote. But whether this will be enforceable is another matter. The motes are likely to be cheap, and unless they are tightly controlled the technology will become widely available. Lockheed may not sell the technology, but if it is used it will be virtually impossibly to stop others from gathering samples of the smart motes and reverse engineering them to build their own copy.

There is little doubt that this kind of technology can change the world. While it may have started as a matter of providing the battlefield commander with information about what is over the next hill, it will result in a wealth of spy technology in the civilian sector.

# CHAPTER 15

# Tools for Terrorists

*'I will do such things —*
*What they are I know not, but they shall be*
*The Terrors of the earth.'*
William Shakespeare, *King Lear*

So far we have looked at the constructive uses of technology in civilian life. But technology which can be used can be misused, and nowhere is this more obvious than with terrorism, in which military technology is used against the population. Technology can fight terrorism, but it can also create opportunities for terrorism which did not previously exist. The following story is significant in this context.

At 9.49 a.m. an aircraft smashed into 79th floor of the highest skyscraper in New York City, creating a hole 20 feet across. Aviation fuel spilled out and liquid fire ran through the hallways and down the stairwells.

'The plane exploded within the building,' said one eyewitness. 'Three-quarters of the office was instantaneously consumed in this sheet of flame. One man was standing inside the flame. I could see him. It was a co-worker, Joe Fountain. His whole body was on fire. I kept calling to him, "Come on, Joe; come on, Joe." He walked out of it.'

The story may sound familiar, but the ending is not. The year was 1945 and the building was the Empire State Building. It was not badly damaged and did not collapse. A total of eleven people in the building died.

This air crash was an accident, caused when a B-25 bomber was re-routed from La Guardia to Newark on a foggy day and failed to avoid the skyscrapers of Manhattan. The bomber did relatively little damage because it weighed only about 20 tons and was travelling at little over a hundred miles an hour.

Fifty-six years later, when terrorists deliberately crashed two aircraft into the World Trader Center, they used hijacked Boeing 767s, each weighing eight times as much and flying at over four times the speed. Developments in aviation technology since the Second World War meant that passenger airliners were far bigger, and so more destructive, than the propeller-driven B-25. Buildings could be built to withstand the impact of earlier aircraft, but it would take fortress walls to stop a modern airliner.

One of the most alarming aspects of the September 11 attacks was that they showed a degree of imagination. A terrorist group had identified an entirely unexpected way of inflicting damage on the US, picking on the specific vulnerability of large skyscrapers. Note the difference in the death toll at the World Trade Center – almost 3,000 – with 100 or so killed at the Pentagon.

Terrorism is the scourge of the modern age, with a US administration bent on fighting a war against it at all costs. Regardless of the politics involved, large-scale international terrorism is likely to be a feature of the world for some years to come. And there is a strong possibility that terrorist groups will use their imagination once more to come up with ways of attacking developed countries in their most vulnerable spots.

A critical factor here is how emerging military technology may alter and add to the threat of terrorism. Two particular types of weapon are worth looking at in detail: anti-aircraft missiles and e-bombs. These are significant because they have the potential to reverse many of the benefits that we have gained through technology, in one case threatening air travel and in the other the entire base of electronic technology.

## Bringing Down Airliners

The man-portable surface-to-air missile, often abbreviated to MAN-PADS, for MAN-Portable Air Defence System, has been around since the 1950s. While heavy machine-guns were adequate against older types of aircraft, infantry had nothing that could touch jet aircraft hurtling overhead at hundreds of miles per hour.

The US Army instigated several projects using light cannon and volleys of unguided rockets, but none of these proved effective. At the same time the Convair company started a private initiative, and without government backing developed a man-portable missile to their own specifications.

Aircraft were already armed with missiles using infrared guidance (otherwise known as 'heat seeking'), which had been introduced in 1949. These were almost 3 metres long and the missile alone weighed 80 kg, but Convair were confident that the whole thing could be miniaturized to make a missile system that could be carried by one man. The end result was something like a bazooka, but with a rocket with an infrared guidance unit in the nose. This IR system gave it the name of Redeye.

In 1958, the US Army officially took the program over; by 1962 it had scored its first direct hit on a target. It took until 1967 to iron out technical problems and issue Redeye to troops.

Meanwhile the Russians were working on a similar concept. Their program started in 1960 and delivered its first finished product to the Army in 1968. This was the SA-7 (also known as SAM-7, Grail or Strela).

Redeye and the SAM-7 were very similar in general appearance and function. Both of them incorporated new features required for a system fired from the ground. For example, the rocket engine consisted of two parts, an initial booster which only burned for a fraction of a second, and a main rocket motor. The booster took the missile clear of the launcher, and the main motor did not ignite until it was at a safe distance. Without this arrangement, the soldier firing the missile would get roasted alive by the rocket exhaust.

The warhead on a man-portable missile is much smaller than designers would choose. Even a small air-to-air missile like the Side-winder carries a 10-kg warhead, which is judged to be the smallest needed to get a reasonable chance of a kill. The Redeye warhead is only slightly over a kilogram, so it is far more likely to damage than destroy unless it scores a direct hit.

The guidance system is perhaps the most important part of the missile. This is based on an infrared seeker which senses heat. Early seekers like those fitted on the first SAM-7s simply homed in on the hottest thing within their field of view. This had two serious disadvantages: they were prone to be distracted by the sun, or the sun's reflection, and they could only be fired at the hot exhaust pipe of a jet plane. This meant they could only be fired at an aircraft that had already passed overhead and strafed or dropped its bombs; missile operators would need to have cool nerves.

Speed was also a problem. Because they had to be fired from behind an aircraft, they would always be aiming at a target that was rapidly receding, and the early SAM-7 could only catch up with aircraft going less than 400 m.p.h. Other limitations included a minimum range of 800 metres and a minimum altitude of 50 metres, or the ground would interfere with the infrared guidance.

The new missiles went into service immediately, and SAM-7s were used during the 1968–70 war of attrition between Egypt and Israel. They were also distributed in Vietnam to counter America's mass use of helicopters.

In the 1973 Arab-Israeli war, the Arab nations fired a large number of SAM-7 – perhaps as many as 4,000 – but only scored a few hits, bringing down six Israeli planes and damaging a few dozen more. On the face of it this might seem like a failure, but in fact it was a triumph. Pilots could no longer strafe and bomb infantry with impunity, but were forced to drop bombs from higher altitude and take evasive action. This severely degraded their effectiveness, up to about 50 per cent according to some analysts. In other words, the missile achieved the same effect as shooting down half the Israeli air force.

In Vietnam, slow-flying helicopters made much better targets. Around 600 SAM-7s were fired with over 200 scoring hits that damaged or destroyed the target. The whole concept of helicopter operations had to be changed.

In Afghanistan the man-portable missile proved even more effective. When the Russians occupied their country in 1979, Afghan Mujahideen fought a guerrilla war against them, but were always at a disadvantage because of Russian airpower. The Russians used helicopter gunships against the guerrillas with great effect; if there was ever an attack or an ambush, the helicopters would arrive within minutes, bringing decisive firepower to bear.

In 1984, the US supplied Stinger missiles, along with a handful of British Blowpipe missiles to the Mujahideen. Although the effectiveness of the Stinger has been overstated – independent experts assess its hit rate as being more like 20 per cent than the 80 per cent claimed by the CIA – it succeeded in its main objectives of neutralizing Russian aircraft. Helicopters could no longer hover for long periods or fly low over ridges, but were forced to higher altitudes and shorter exposure times. In some cases pilots refused to get close to Mujahideen positions because of the risk, leaving ground forces to rely on artillery and mortars.

As a morale-booster the Stinger was a tremendous success. Instead of having to run away or dive into cover whenever Russian aircraft appeared, the Afghans kept fighting. It might not have been as decisive a turning point as some would like, but there is no question that the presence of these missiles made a difference.

The missiles were quickly becoming more sophisticated. Stinger, like later Russian missiles, was 'all aspect' and could be fired at an aircraft from any angle and not just from behind. Seekers became progressively smarter, and instead of locking on to any heat source they could specifically identify aircraft exhaust and ignore the sun and decoy flares. The latest generation are even more advanced, and can select the most vulnerable part of an aircraft to attack.

Later missiles also have intelligence built into the guidance system. Instead of simply chasing after where the target is now, they can compute where it is going and plot a course to intercept. Propulsion and warhead have also improved; an additional fusing system detonates any unburned rocket fuel to increase the blast, and fragmentation patterns are computer-modelled for maximum effectiveness. The modern missile is much faster and deadlier than the first models.

It was not long before the threat to airliners became apparent. A device designed purely as a way of defending the infantryman could become a lethal weapon against civilian targets. Using one of these missiles, a single terrorist could, in theory, bring down a passenger jet from several miles away. In the 1960s and 1970s, hijacking was the main fear, but after 1988 when terrorists blew up a 747 over Lockerbie it was clear that terrorists were now targeting airliners for destruction. Strict new security measures were introduced to prevent bombs from being smuggled on board, but airliners were large and easy targets to anyone with a surface-to-air missile.

The terrorist with a SAM became a staple of thriller writers, but the first attack on a civilian airliner outside of a war zone occurred in 2002, when terrorists fired two SAM-7s at an Israeli airliner in Mombasa – and missed. There is some doubt over what actually happened, and it is possible that these were not live weapons at all. However, the world woke up to the threat, which was emphasized the next year when an arms dealer was apprehended trying to buy a modern Russian SA-18 Igla missile.

This prompted varying levels of panicked response. Perhaps the most extreme was in the US, where there was pressure to pass an 'Anti-Missile Protection Bill' under which all 7,000 US passenger jets would be required to be fitted with infrared countermeasures based either on flares or lasers. Under this law, the funding for the countermeasures, which could run to fifteen billion dollars or more, would come from the Department of Transportation, though it is possible that an additional tax on air travel would have been levied to pay for it. Airline passengers are already paying for new measures under the Transportation Security Administration program, but the new bill would require much steeper charges.

There are problems with this type of reaction. To be effective, new legislation would have to include not just US aircraft but also any aircraft flying from an American airport. Secondly, such countermeasures only work against certain types of missile. Infrared guided missiles would be blocked to some degree, although the more sophisticated ones have filters and other means to negate countermeasures.

Other types of missile would be completely immune to these countermeasures. British Aerospace has produced three generations of missile which are known as beamriders or SACLOS – 'Semi-Automated Command to Line Of Sight'. The British developers rejected the idea of heat-seeking missiles entirely in favour of a missile which would be directed by the operator. With these missiles, the firer has to keep the target aircraft in the cross-hairs of a sight after he has fired, and a guidance system ensures that the missile hits whatever the cross-hairs are aimed at. Because these missiles do not rely on seeking heat, they cannot be decoyed away from the target by flares or other countermeasures. They require much greater skill and training to use, and as a result the Blowpipe was unpopular with the Mujahideen. Other irregular forces may see it as a more attractive option.

So far, SACLOS missiles like BAe's Blowpipe, Javelin and Starstreak (or the Swedish Bofors RBS-70) have not found their way into the hands of terrorists. Russian SAM-7s and other missiles are much easier to get hold of on the international arms market. However, if the Anti Missile Protection Bill went through, then they would have a good motive to seek out this type of missile. As mentioned previously, blowpipes were supplied to Afghan rebels in the 1980s, and it is entirely possible that refurbished models would find their way on to the market. If a terrorist group has sufficient funding, then the market is likely to accommodate them. Such missiles would nullify any IR countermeaures.

The threat to airliners is a real one, and it will not go away, though it is hard to assess how great the risk is. MANPADS require a much greater degree of training and organization than, say, simple suicide bombs.

The early SAM-7s had only a few per cent chance of hitting an aircraft; more recent weapons have a 40 per cent chance or more. This can be reduced by infrared countermeasures. If the plan to fit all 7,000 US airliners with such countermeasures goes ahead, it is quite possible that by the time the program is complete the threat will have moved on to something even more dangerous.

Portable missiles may be replaced altogether by more advanced weapons. According to Jim O'Halloran, editor of Jane's authorative *Ground-Based Surface To Air Missiles*, the US Army is likely to abandon their Stingers entirely by the end of the decade. The replacement will be a directed-energy weapon, probably some form of laser. Such a weapon would be ideal for terrorists: with no explosive or rocket fuel, it would not even be identifiable as a weapon, and could easily be broken down into its component parts and smuggled abroad. And there would be no way of telling where it was fired from; unlike the SAM with its tell-tale flare of launching and smoking trail, the beam weapon would give no more visible sign.

If such weapons proliferate, air travel later on in the century will become a very different matter to the carefree days of flying in the twentieth century. There would be a real risk that air travel would become unaffordable for most people: the costs of airliner defence, not to mention insurance and legal costs, would be potentially astronomical. Air travel would, of course, still be possible, but cheap air travel would be a remote memory.

## E-bomb

In the run-up to the war against Iraq in 2003, the news media were full of stories about a new type weapon: the e-bomb. The story was carried by all the major news networks.

One carried by CBS was typical, describing the new bomb as 'highly classified' and saying that it would 'fry computers, blind radar, silence radios, trigger crippling power outages and disable the electronic ignitions in vehicles and aircraft', all without injuring anyone. In spite of the secrecy surrounding the e-bomb, there were numerous diagrams and descriptions of it, and accounts of the part it would play in the Pentagon's war strategy.

By all accounts, the e-bomb had been under development for years, and the latest war would provide the perfect conditions to test it in action.[1] But once the 'shock and awe' campaign of airstrikes got underway, e-bombs did not feature. Some reports suggested that e-bombs were being use to take Iraqi television off the air, but these were later refuted. It seemed that the e-bomb was not on the menu.

E-bombs were supposed to knock out the Iraqi defence radars and neutralize deep command bunkers, all without a shot being fired. The e-bomb would, in theory, be the exact opposite of the fabled neutron bomb which killed people and left property intact in that it would destroy hardware without injuring people.[2] Military analysts were puzzled, and the media were denied the chance to see a new type of bloodless warfare.

One suggestion was that there was concern over possible damage to the Iraqi infrastructure; since the US planned to occupy the country and help rebuild it, a weapon that destroyed telecommunications including the telephone system would cause problems in the long run. Though possible, this is unlikely to be the full story. Plenty of Iraqi military facilities were located at a safe distance from civilian areas and would have made good targets without causing collateral damage.

It is entirely possible that the decision was made not to use e-bombs because it would have drawn attention to a new type of weapon that is cheap, easy to build and is especially effective against countries that rely on electronics – countries like the US. The reasons for this lie in the nature of the e-bomb itself.

All electronic devices (and electrical ones for that matter) involve electric current flowing along conductive wires. Electricity and magnetism are closely linked, which is why scientists talk about electromagnetic radiation. Every electronic device is an emitter of electromagnetic waves, mainly in the radio frequency. Some devices, like radios and mobile phones, are designed to pick up radiation from the environment. Sometimes the radiation you pick up is not what you intended: this is the electrical noise we call interference.

Interference is simply stray radiation; lots of electrical devices produce it in abundance, as you will find if you hold a radio close to some televisions. It is also produced naturally, as when a distant lightning strike produces an audible crackle on your radio. Interference is most obvious in devices like radios which are sensitive to it, but it can affect any piece of electronic equipment. This is known as 'back-door coupling', in contrast with reception through an antenna or aerial, which is 'front-door coupling'.

Interference affects equipment because the wires act as an aerial, producing an electrical current. With most interference this will be too small to have any effect, but sometimes it can be damaging. This was observed early on in radar research, when it was found that at very short ranges, the beam of radar could cause surges in electrical devices and damage sensitive components. This is why, for example, the AWACS aircraft carrying powerful early-warning radar have to turn off their radar when approaching for air-to-air refuelling. The same effect could also be used offensively; when Russian spy trawlers went to eavesdrop on NATO naval exercises, there are reports that US warships would pull alongside and sweep them with high-power radar, damaging sensitive electronic listening devices.

Even low-powered equipment can cause interference in some circumstances. In recent years new rules have been introduced banning the use of devices like laptop computers on airliners because of the risks of interference with avionics. In 1993 it was reported that a Boeing 747 experienced navigational problems from a passenger's laptop. As often happens, the problem could not be reproduced afterwards, because the exact effects of interference are very dependent on the geometry of the situation, in other words the exact alignment of the emitter and the equipment receiving the interference. Anyone who has ever tried to adjust a television aerial will be aware just how much difference the slightest movement can make.

An e-bomb then is a weapon which creates interference on a gigantic scale. Not just enough to briefly disrupt television and radio reception, but to create massive power surges in all electronic equipment, causing permanent damage. If there is a power surge in the mains the fuse protecting your domestic appliances will blow; but if the power surge is within an appliance, there is no protection. The most sensitive parts of the equipment will overheat and burn out, leaving it useless.

## Early EMP: the Nuclear Option

The difficulty lies in building a weapon with a high enough power level. As noted earlier, even the most powerful radar was only dangerous at a range of a few metres. One way of producing a powerful electromagnetic pulse or EMP is with a nuclear bomb. This was discovered by accident in 1962 when a 1.4 megaton nuclear weapon was detonated in Test Shot Starfish.

The explosion was at an altitude of 400 km above the mid-Pacific and the electromagnetic pulse blocked high-frequency radio communications across the Pacific for half an hour. Streetlights went out in Hawaii – electrical power lines make good aerials – and hundreds of burglar alarms were set off.

The USAF Intelligence Targeting Guide notes:

> Under certain circumstances EMP can severely disrupt, and sometimes damage, electronic and electrical systems at distances where all other effects are absent. In fact, a detonation above 130,000 feet can produce EMP effects over thousands of square miles on the ground.

Others have predicted that higher altitudes would increase the damage further, and that a single weapon exploded at an altitude of several hundred miles above Kansas would affect the entire continental US. No high-altitude tests have been conducted since 1963, and earlier weapons were optimized for their explosive yield rather than EMP, so there is no reliable information on just how effective a bomb tailored to produce the maximum EMP would be. Clearly though, it could do a tremendous amount of damage.

Power and telephone services would stop working – modern telephone switching systems are entirely electronic. Services like the supply of water, which rely on electrical pumping, would cease to function.

Virtually everything electronic would be affected. This would include televisions, radios, computers and mobile phones, and also the microchips that control central heating, air conditioning, fridges, cookers, washing machines and almost every other electrical appliance. At home it would be a matter of going back to candles and an open fire – for those who have a fireplace. Without any television, telephone or radio (unless you happen to have an old crystal set) there would be no way of finding out what was going on.

Any car made in the last decade relies on electronic components which would cease to function. Some aircraft might be able to land if they could find an airport without any navigation equipment, but those that depended on fly-by-wire systems like the modern Airbuses and Boeing 777 would be completely disabled. Such aircraft would not actually drop from the sky, as they are capable of flying straight and level without the electronic systems, but anything more elaborate such as landing would be out of the question.

The effect on industry would be similarly devastating. No factory which relied on complex machinery would still be working after such an attack. Anything which relies on electronic controls, from air conditioning to railway signalling, would be out of action.

If your job involves a computer, a telephone, or any type of electronics, you would have to start from scratch. In the absence of cars, trains and buses, simply getting to work would be difficult for many people.

There would be human casualties as well. Hospital equipment from life-support machines to X-rays to defibrillators would be put out of action. Anyone with a heart pacemaker could become a victim in the first few seconds.

Simple, purely electrical devices like hairdryers would continue to function, although there would be nothing to plug them into. Any business or government department which relied on computers would be in trouble – EMP will not just destroy computers but it will also scramble magnetic media like back-up disks and tapes. Only optical media like CDs would still be usable, though of course finding a usable computer would be a problem. In addition to destroying the computers themselves, the EMP would also destroy the routers and other hardware that make up computer networks. Everything would have to be replaced.

The pulse could also do tremendous damage to the military, and the threat of nuclear EMP has meant that military electronics have routinely been made with built-in protection. This protection or hardening is achieved by putting the equipment inside a 'Faraday Cage' which is simply a casing made of conducting material. The nineteenth-century physicist Michael Faraday showed that the strength of an electromagnetic field inside a conducting enclosure was zero – in other words, radiation cannot pass through conducting material.

Electromagnetic waves will not travel through openings in a surface which are much less than the wavelength of the radiation. This is why something like a satellite dish (or a radar dish) can have holes in it without affecting the signal. So the cage protecting a military computer can be made of mesh rather than sheets.

Any piece of equipment connected to an aerial or a power line is also vulnerable, because EMP would cause a power surge along these lines, and additional protection has to be installed. Protection can be at the level of individual items of equipment, or it may be at a site level – in other words, entire buildings can be lined with copper or steel panelling to protect everything inside.

How effective this shielding is remains to be seen. As mentioned above, the effect of EMP depends greatly on its geometry, not to mention the strength and wavelength of the pulse, and there are various levels of shielding.

The military standard is known as TEMPEST. This is actually geared towards preventing electromagnetic emissions getting out rather than stopping them from coming in, because it is possible for eavesdroppers to pick up the emissions from a computer to see what is showing on the computer screen from some distance away. TEMPEST prevents this,

Full shielding is very costly – one source quotes $1,000 per square metre for welded steel panelling. Others estimate an additional cost of between 10 and 50 per cent for hardening communications equipment. These costs mean that only vital equipment is protected, and there is no protection for civilian equipment, as low production costs are a priority and nuclear war has always seemed like a relatively remote possibility.

## Modern E-bombs

There are other ways of producing a powerful electromagnetic pulse. One of these is the 'explosively pumped flux compression generator'. This fearsome-sounding name is applied to a device which is little more than a normal electrical generator, but one which is driven by an explosion rather than more conventional fuel.

A gallon of fuel can run a diesel generator for hours, enough for a powerful radio transmitter. But if you use an equivalent amount of explosive instead of fuel, you can generate a signal that lasts for only a fraction of a second but which is millions of times more powerful – an electromagnetic pulse.

An explosively pumped flux compression generator typically consists of a copper tube filled with explosive, inside a much larger tube made of coiled copper wire. It works like other electrical dynamos in that it has a small current running through the coiled wire, turning it into an electromagnet. As physics teaches us, a conductor (such as a copper tube) moving inside a magnetic field generates an electric current. The explosive is detonated at one end, causing the copper tube to expand rapidly, moving at extremely high speed and with tremendous force. The 'flux compression' occurs in the space between the tube and the coiled wired, where an intense electromagnetic field is produced.

The output current from a flux compression generator can be used to power the electromagnetic coil for a second, larger generator. The system could have more than two generators in it – a cascade of generators can be constructed, each multiplying the output of the previous one.

The explosively pumped flux compression generator is basically a bomb that turns some of its explosive energy into an electromagnetic pulse. It can be made to any size, but of course the question is how much destructive power you get out of it. Such a bomb is not worth developing if the EMP effects are going to be dwarfed by the explosion.

This is a field in which hard figures are difficult to come by. The difficulty comes in trying to assess how much damage is going to be done, which depends on the exact characteristics of the bomb used and the nature of the target. One television might be vulnerable to a certain frequency but not another, and the distance at which it is likely to be burned out by EMP depends on the whether it is face-on or at an angle. Other uncertainties are caused by the fact that most of the research in this field is classified. The US military have conducted a large number of tests (in fact, they maintain some large facilities which do nothing but test the effect of radio and microwaves on targets) but they do not share the results.

Rough calculations by those in the electrical engineering field have been carried out for a 907-kg e-bomb. This is the same size as many of the laser-guided weapons used against Iraq, and is also equivalent to the weight of the warhead on a cruise missile. It appears that such a weapon would have a radius of destruction of several hundred metres. This would be much larger for vulnerable items (such as those with aerials or dishes) and much less for those with some sort of shielding. This would be a far greater area than would be affected by the explosion; an e-bomb could be exploded at an altitude of a few hundred metres so that, apart from some broken windows, the effects would be purely electromagnetic.

A few dozen such bombs would completely shut down all but the most basic functions of a modern city, cutting power, lighting and most forms of transport.

The existence of e-bomb warheads has been established for some time, but military interest has moved on to something more sophisticated. The e-bomb is a very blunt instrument indeed, indiscriminately destroying everything across a wide area and also presenting a possible risk to friendly forces. More recent research has been directed towards directed-beam weapons.

The acknowledged leaders in this field are the US and UK, but very little information has been released. Interestingly, in November 2002, the Australian Defence Science and Technology Organisation released some details of their work, which is likely to be a small-scale reflection of what the other nations are doing.

The Australian work was carried out in conjunction with US defence electronics company Raytheon, and involved fitting a radio-frequency beam weapon to a Hercules transport aircraft. The weapon is powered by a series of capacitors which are discharged together to generate a pulse lasting less than one a billionth of a second. Unlike an e-bomb which affects everything in the area, the output is channelled by a 2.5-metre antenna into a narrow beam; at a range of 2 km the impact zone was 30 metres in diameter. The exact power was not disclosed, but it was considered enough to knock out military electronics.

The emphasis now is on reusable weapons; the Australian beam weapon can be fired again as soon as the capacitors are charged up. Progress is needed in compact pulsed power systems and antenna design before it can be turned into a usable weapon, but these are no more than five years away.

Being able to direct the beam precisely will make EMP weapons much more usable. A single building belonging to the defence ministry could be attacked, for example, without damaging anything in the surrounding area; radar installations and command centres could be swiftly and silently knocked out without any risk to hospitals or other civilian electronics. Individual vehicles could also be silently disabled at will.

The USAF has included similar electromagnetic beam weapons in its long-term development strategy. Plans call for these weapons to be installed in unmanned aircraft, while other directed-energy weapons such as lasers are mounted in manned planes. This reflects concerns over just how unpredictable the effects of EMP can be. Stray reflections or unexpected effects on aircraft systems could easily fry the electronics on the carrier aircraft. Unmanned drones are expendable, but pilots are likely to be cautious about pulling the trigger if they might become collateral damage themselves.

## E-weapons on the Domestic Front

The shying away from e-bombs in favour of more focused beam weapons, and the reluctance to use e-bombs in conflicts from Serbia to Iraq, suggests that there is some unease about this technology. As long as e-bombs are not used in anger, nobody knows how effective they might be. Once the genie is out of the bottle, the risk is that everyone will want one – and e-bombs may not be that difficult to replicate.

This would include not only world governments but also every terrorist group with an interest in causing damage to places which rely heavily on electronics. And while an e-bomb might not do that much damage in Nairobi or Nepal, where there is less affluence and less dependence on electronics, American cities would represent a much more 'target rich environment'. The danger is that the e-bomb is a device which is uniquely effective against rich industrialized nations.

There have been numerous scare stories about EMP weapons. The phrase 'electronic Pearl Harbour' has been bandied about for many years by defence analysts trying to drum up support for better defences against EMP, but the threat has so far failed to materialize in a tangible form.

There have been a variety of tales centring on e-bomb-type weapons from the former Soviet Union which have supposedly fallen into the hands of criminal gangs. There was a persistent story in the 1990s that one such gang was going around the City of London blackmailing banks: they threatened that, unless they were paid off, they would completely destroy the banks' operations. In an environment where even losing a day's trading can cost millions, the story goes that the banks paid off the blackmailers and hushed up the story.

This seems highly unlikely. Banks are not in the habit of handing out money to extortionists, and it is hard to imagine that if the scheme worked once it has not been repeated many times since – unless we are supposed to imagine that the gang are now all rich men and have gone into retirement.

In 1998 a Swedish newspaper reported that the Russians had started selling a type of e-bomb that fitted into a briefcase – just the sort of thing for the modern terrorist or gangster. However, it seems that the weapons were only being sold to other accredited governments, at a cost of around $150,000.

Other scare stories have suggested that any hacker could build an e-bomb at home with supplies from their local electronics shop. This is not entirely true; many years of research and millions of dollars have gone into military e-bomb programs. It might be possible for some kind of crude weapon to be built in a garden shed; but given that a bomb can be made simply by combining ammonium nitrate fertilizer and fuel oil, it is difficult to see that an e-bomb would be a greater threat.

For the time being, there is no suggestion that terrorists or organized crime have a working e-bomb or radio-frequency beam weapon. But as technology progresses and more powerful sources become available, it becomes increasingly likely that such a weapon will turn up. This could be either from a 'friendly' source, armaments redirected by corrupt officials, or developed by a rogue state or terrorist group.

The first use of an e-bomb against civilian targets is likely to spread panic as people realize how vulnerable they are. Although the e-bomb is just another weapon in the terrorist's arsenal and considerably less deadly than, say, nerve gas or an infectious biological agent, it may have a significant effect on the way people behave. Suddenly there will be a market for shielded electronic equipment.

In some places the market already exists. Dixel electronics, an Israeli company, sells a special copper foil that can be used like wallpaper, and transparent protective mesh that goes over windows, so that worried civilians can protect their homes from EMP.

## More Terror

Other new technologies may not be as neatly tailored to their requirements as e-bombs and MANPADS, but all of the emerging technologies mentioned in this section are potentially dangerous in the hands of
terrorists, even in their civilian form. This, perhaps, is the most
important lesson we can learn from the September 11 attacks.

The military are likely to keep their monopoly on armed and
armoured robots, but if commercial versions of autonomous vehicles
become available they can be turned to destructive ends. Car bombs and
truck bombs are among the most devastating weapons used by terrorists
so far, but they rely on either being able to park near to the target or the
use of a suicide driver. Reliable truck bombers are difficult to come by,
but a robot can be instructed to take a bomb to any location and will
obey without compunction.

Human terrorists may give themselves away, or have last-minute
qualms, but robots are good foot soldiers for any extremist ideologue
with their unquestioning obedience and self-sacrifice, not to mention
their reliability in not spilling the beans when captured. In fact,
without the need to recruit cannon fodder for front-line roles like
placing bombs, terrorism becomes much easier. A solitary fanatic like
the Unabomber could carry out much worse destruction with the aid of
robotic accomplices.[3]

New sensors like through-the-wall radar could also make a
difference. The ability to observe security procedures and measures,
such as counting the number and location of guards, noting the
presence of locks and alarms, is something that those planning an
attack could only dream of. With this type of sensor they would be able
to track the exact location of their target. A bomb could be set off at the
exact moment when the intended victim walks past, even though he is
out of sight behind walls and fences.

Vulnerable targets such as government buildings would have to be
turned into radar-proof citadels, but it is not clear how far such defences
could be carried. Dixel's radiation-proof wallpaper begins to look like a
possible bestseller.

A miniature unmanned air vehicle is a useful spy, not just for the
military or the snooping paparazzi, but also for terrorists. Worse, it
could become an assassin when armed with a suitable deadly payload.

There have already been accounts of Palestinian terrorist groups experimenting with radio-controlled model aircraft carrying explosives, but new technology would take this into a different level. As an assassination weapon, a small UAV with an explosive charge would allow an attack to be carried out from miles away with no need for the terrorist to expose themselves.

In Israel, a huge 'security barrier' is being erected to stop suicide bombers, while in Iraq the Coalition forces are sheltered by tall concrete blast barriers. Any sort of defence like this can be easily overcome by small UAVs carrying bombs – or other weapons.

UAVs could also be used to deliver chemical or biological agents over considerable distances. Getting hold of a modified crop-spraying aircraft is comparatively difficult, but adapting a UAV to carry an aerosol spraying anthrax is the kind of project that could be carried out in somebody's garden shed. Again, these would greatly increase the reach and capability of anyone wanting to carry out an attack. You can stop and search everyone entering a stadium or a park where a concert is being held, but when something resembling a toy helicopter appears over the rooftops, fences and barriers become useless.

Mini-UAVs may not just be used for attacks across a few kilometres either. In theory they could have inter-continental scope. In August 2003 a group of hobbyists succeeded in flying a model aircraft across the Atlantic, a distance of 1,888 miles, from Newfoundland to Ireland. Their aircraft had a wingspan of 6 feet; more advanced aircraft will follow, with smaller wingspans, increased range and better guidance. Given GPS-aided guidance and global satellite communications via mobile phones, a group based in the Middle East could launch a stream of attacks on mainland USA using modified mini-UAVs.

An enthusiast in New Zealand decided to show how easy it would be to build this kind of thing from scratch, setting up an online diary for his DIY Cruise Missile Project. His aim was to build a missile with a range of at least a hundred miles and carrying a payload of 10 kg, all for under $5,000. The plan included a pulse-jet engine similar to that on the original V-1 flying bomb, the key difference being that the modern cruise missile builder has the advantage of several decades of advances in electronics. While the team at Peenemunde had to build their mechanical gyros from scratch, the modern-day builder can buy miniature electronic ones off the shelf. Although he was prevented by the New

Zealand government from completing the project, he showed that it was indeed feasible.

This might well promote a response just as extreme as the Anti Missile Protection Bill, and a demand that the National Missile Defence System currently being envisaged to stop attacks from ballistic missiles be extended to stop every potential threat down to small model aircraft hardly seems feasible.

Vortex-ring technology could also be another weapon in the terrorist's armoury. The original Nazi weapons were designed to bring down aircraft; given sufficient understanding of how vortex rings are formed, it would be comparatively easy to build a device which would use a small explosive charge to put an air vortex in the way of an airliner. 'Comparatively easy' is of course relative; leaving a bomb on a crowded bus will always be far simpler than any of this technical stuff. One Russian newspaper has even suggested that certain recent air crashes have been the result of such a weapon being used secretly, but without any evidence this looks rather like paranoia.

As the war against terrorism progresses, we will see many new types of weapons being used by the US and its allies. And although terrorists tend to favour unsophisticated forms of attack, they are quite capable of using MANPADs, e-bombs and other new technology. While it is true that every weapon prompts a defence, and every defence leads to the next weapon, the one thing that stays the same are the people in the middle. Unfortunately, this may be one of the most visible impacts of military technology in years to come.

# CHAPTER 16

# Nanotechnology: The Future is Small

*I will show you fear in a handful of dust*
T.S. Eliot, *The Waste Land*

Nanotechnology is currently one of the newest and most hyped areas of technology, the subject of endless speculation and also scare-mongering. It is based on the restructuring of matter at a sub-microscopic scale; the name derives from the prefix 'nano' meaning one-billionth. A nanometre is one-billionth of a metre, about ten times the diameter of an atom. Nanotechnology works at the scale of molecules. Its supporters say that it will transform the world over the next few decades; its opponents see it as a threat to the ecosystem. And while some of the interest is commercial, much of the interest and investment in this area comes from the military.

'Imagine the psychological impact upon a foe when encountering squads of seemingly invincible warriors protected by armour and endowed with superhuman capabilities, such as the ability to leap over 20-foot walls.' This imaginative scenario was put forward by Professor Ned Thomas, the Director of the Institute of Soldier Nanotechnology (ISN). The ISN was set up by the Massachusetts Institute of Technology and funded by the US Army, with the promise that nanotechnology will bring new military capabilities in the near future.

'We hope to deliver some goodies early, in the next five years,' said Thomas, speaking in 2002.

Among the ideas floated by the ISN are a uniform which will give the wearer superhero powers. Although flexible and lightweight, it will be able to stop bullets. It will have built-in power cells linked to actuators, and artificial muscles that are far stronger than mere human muscles so that the wearer can carry heavy loads or run at high speed without tiring. If the wearer breaks a leg, the uniform will be transformed into a rigid cast. The same feature could be activated at will to make a 'forearm karate glove' backed with the strength of a giant. In addition, the surface of the uniform could change colour to match the background, making the wearer virtually invisible.

While this is all very spectacular, it is a little peculiar. Why would a soldier be punching an opponent in the first place? Aren't broken limbs rather uncommon compared to, say, shrapnel injuries? Does a soldier really need powerful muscles in an age of vehicles and robotics? These are indications of a technology in its very infancy, where it is still at the stage of being a possible solution looking for a problem, and the actual advantages are likely to be very different.

Long-term development tends to come with unlimited promises which evaporate as the delivery date gets closer. However, nanotechnology is already delivering new and effective products and shows every sign of delivering more.

## Nanomaterials:
## Samurai, Dragonslayers and Bunker Busters

The simplest form of nanotechnology is simply a continuation of mankind's attempts to control the properties of materials. We have been trying to manipulate structure on a molecular level for many centuries, because the large-scale properties of a material are dependent on its structure on a very small scale.

This is fundamental to working metals. A piece of iron, for example, is always tens or hundreds of times weaker than its theoretical maximum strength, because of the presence of microscopic cracks or flaws in the structure. Under stress, these cracks quickly grow and spread, and the material fails. If you can make a slight improvement to the structure you can make a much stronger metal that is closer to its theoretical limits.

The history of metallurgy is the story of how smiths have tried to improve the metal they were working – typically, they were trying to make the hardest, strongest sword possible that would hold a sharp edge. They were guided by trial and error, and had no way of knowing the underlying physics, but they hit upon some very effective techniques.

Early iron was produced by burning iron ore in a furnace with a quantity of charcoal, a technique called the direct method. In a confined space where there was little air, the charcoal burned to form carbon monoxide, and this absorbed oxygen from the iron ore, reducing it to iron. The iron accumulated at the bottom of the furnace in the form of a spongy mass called a bloom. The bloom was heated in a furnace for hours or days to drive out impurities, making the hard metal we know as steel.

When several pieces of steel had been made in this way they could be melted together and hammered into a single piece. The workpiece was made stronger if it was repeatedly hammered out flat and folded over on itself. Several rods could be welded together using this method and some smiths found that if the rods were twisted during the process the weld was much stronger; it also produced a distinctive pattern. The process became known as pattern welding and was a closely guarded secret. Swords made this way were known as Damascus Steel and were highly prized.[1]

The final part of the swordsmith's art is 'case hardening' in the forge, again heating the blade with a small amount of carbon. This hardens the outer surface of the blade so that it can take a very sharp edge. In effect the finished product is a strong, tough, inner core with a hard surface. The core metal is too soft to hold an edge for long, and the outside metal is too brittle on its own, but married together they make the ideal sword.

In medieval Japan, swordsmiths used a method similar to the direct process, but instead of heating the bloom immediately they allowed it to cool so that it could be removed from the spent furnace and examined. Select pieces of steel were broken from a bloom which ranged from very mild steel to cast iron. The skill here was in recognizing the different types of steel, so they could be welded together and folded upon themselves repetitively. The end result was a type of steel known as Tamahagane.

These Japanese swords are held to be the best in the world; samurai could demonstrate this in combat by cutting right through their enemies' blades.

What the early swordsmiths did not know was why their techniques worked. They did not know that when the iron bloom was heated with carbon-rich material, some of the carbon migrated to the surface of the iron, forming the alloy of carbon and iron which is steel. Nor did they know that the properties of steel, its hardness and brittleness, are down to the arrangement of iron and carbon atoms in its structure.

Modern science recognizes the several different types of steel produced by different methods, and our ability to produce them to order has improved. But by a more exact control of the conditions involved, we can produce steel of a quality which would astonish the ancients.

Dr Greg Olson of Northwestern University has developed a new type of steel called Ferium 69 which is of unequalled hardness. Using computer modelling he has been able to 'direct' tight chains of microscopic carbides into the surface of the steel. The new alloy is the hardest steel ever made, and Dr Olson plans to demonstrate its power by making the ultimate sword, Dragonslayer.

Dr Olson has already tested a blade made of Ferium 69 against a Japanese Tamahagane sword, and the Japanese sword came off worst. He is convinced that modern technology will triumph over traditional craftsmanship.

It is not just samurai and dragon-killers who might have a use for ultra-hard, ultra-strong materials. The USAF is looking at ways of attacking bunkers deep underground; one approach is to hit them with a missile travelling at very high speed (see Chapter 6). One of these missiles will hit at mach 6 or more, faster than any existing missile, faster even than the depleted uranium projectiles fired by tanks. Such high-speed impacts have been studied in detail, mainly as part of anti-tank warfare research.

When a projectile strikes a hard target at five times the speed of sound or more – 'hypervelocity' – at the point of impact both the missile and target behave like liquids, flowing into each other. They do not actually liquefy, but the forces involved are so great that even solid metal flows like treacle, and the projectile and the target erode each other. Under these extreme conditions the only thing that affects penetration is whether the missile or the target is eroded away first, and this depends on their density. This is why depleted uranium, twice as dense as steel, is the favoured material for anti-tank rounds.

This presents a problem for attacking buried targets. Even at the highest speed possible, a depleted uranium missile could only penetrate two or three times its own length of reinforced concrete. Practical considerations mean that warheads are limited to a length of 20 feet or so, so targets protected by more than 60 feet of rock or concrete are virtually invulnerable to high-speed missiles. The problem can be overcome if the missile could be made of, or have a coating made of, some material that could resist the ferocious erosion of travelling through rock at supersonic speed.

No normal material is hard enough to resist this sort of force, but in theory a material could be made which would not flow during impact and the old rules would not apply. Details are hard to come by, but the description of the USAF's Affordable Rapid Response Missile System refers to 'nanolayered structured penetrators'. This would be a material even harder than Dr Olson's Ferium 69.

Unclassified reports show that these materials can already be produced, and can be applied in the form of coatings. One report on nanolayer-structured composites suggests that a synthetic crystal of boron nitride can be made in complex shapes that are both very hard and elastic. This material is almost as hard as diamond and stronger than any steel. Industrial coatings using this technology are already seven times stronger than steel, making them easily strong enough for a bunker-busting warhead.

The main difficulty is in making crystals big enough to be useful. During manufacture it must be controlled to prevent the microscopic flaws that weaken normal materials. This involves exotic techniques such as 'plasma chemical vapour deposition', building up a crystal by spraying layers of charged particles on to a surface – something like spraying paint but in a molecular scale. This sort of technique is best suited to making thin films; making up a large structure one layer of molecules at a time would be a laborious business.

A warhead would not necessarily need a very thick layer of protection. So long as the outer coating is of a sufficient hardness to withstand the force of impact, it will not be eroded significantly, any more than the hull of a ship is eroded as it travels through the water. The target material, be it rock or concrete, will flow around the warhead like liquid, and deep bunkers become easy targets.

Of course, ultra-hard materials will make ideal armour as well. Something similar happened in the days of ironclad warships, when ships with hardened steel armour could shrug off shells made from weaker material. A tank armoured with nanolayer composites would be practically invulnerable to normal shells; even a depleted uranium round striking one would simply splash off it like a paintball.

## Graded Grains Make Bigger Bangs

Nanotechnology has potential in the field of explosives and propellants. Normal explosives are granular, with the grains being irregular and comparatively large, their size measured in micrometres. This is important because the rate of reaction – how fast the explosion happens – is determined by the surface area of the grains, which in turn is determined by their size. Chop a grain into eight, and you have eight pieces with the same total weight as the original but twice as much surface area.

The increase in surface area can have dramatic effects. A large lump of coal burns slowly, but if you throw a handful of coal dust on to a fire you get a sudden flash of flame. If the coal is ground into fine powder it can be explosive. It used to be thought that explosions in coalmines were caused by leaks of methane gas, but we know now that a fine haze of coal dust mixed with the air can be just as explosive.

A forceful demonstration of the explosive power of coal dust was carried out by the Department of Labor, Mines Safety and Health Administration at Pittsburgh in 1969. Some 215 kg of coal dust was laid along the floor of a test mine for several hundred metres, and ignited. A phenomenal explosion resulted, more powerful than expected, and amplified by the effect of a temperature inversion which reflected the shockwave back to the ground. Windows were broken 11 km away and the blast was heard for 48 km away.

In fact, any fine powder of flammable material can be dangerous: flour mills and other processing plants are sometimes destroyed when their product explodes. There is a major incident of this type once a month in the US alone.

Naturally this has military applications. Aluminium oxidizes in air; scratch a piece of aluminium and you will see bright metal which slowly dulls. This oxidization is effectively a slow burn. Divide the aluminium up finely enough and you can get it to burn fast enough to explode.

Grinding up metallic aluminium to a sufficiently small size is not practical. Instead, a piece of aluminium in a safe inert atmosphere can be zapped with a powerful electric pulse. The metal is vaporized; if the cooling is controlled correctly (a trade secret), the metallic gas condenses out into particles with any desired size from 10 to 100 nanometres across. The secret process ensures that the small particles do not clump together like bunches of grapes. These particles have to be treated very carefully; any exposure to the air will cause them to explode violently.

The US Air Force has already used weapons which incorporate nano-aluminium. It forms the basis of 'solid fuel air explosive' bombs which were used against Taliban positions in Afghanistan in 2001, and reported to be far more effective, weight for weight, than conventional explosives. The energy released by aluminium powder is six times as much as is released by an equivalent weight of Semtex or other plastic explosive. Boron, another material considered for this sort of use, is twice as energetic again. But the rate of energy release is also important, and as we have seen with nanomaterials, this can be increased by a factor of hundreds or thousands.

It is not simply a matter of making the powdered explosive as small as possible. By varying composition and particle size, different effects can be achieved in terms of the light, heat and shockwave it produces.

'Flash powder' was used for stage effects and early camera flashes; a grenade loaded with the modern equivalent can temporarily blind anyone who sees it. This will make for more effective 'flash-bang' grenades which are used to disorientate and dazzle opponents at night.

High-temperature, low-pressure explosives are being developed using aluminium powder as a means of neutralizing stored stockpiles of chemical or biological weapons. Another way of using the rapid release of heat is in a 'superthermites' mixture such as a combination of nano-aluminum and an oxidizing agent. These burn at higher temperature than normal incendiary weapons and are correspondingly effective.

'Hyperbaric' (high-pressure) explosives which produce a long-lasting shockwave were used in the 2003 war on Iraq; these produce a blast which can propagate through a building or underground complex, killing the occupants without destroying the structure. Again, it is simply a matter or formulating the right type of powder.

Further developments will follow, both in the fields of explosives and propellants. Nano-aluminium also shows promise as a rocket fuel. So far, the military applications dominate, but industrial and commercial applications will follow. There are industrial processes which involve using finely powdered metal for manufacturing, such as laser sintering which allows objects to be cast in one piece from powder. The increasing availability of ultra-fine metallic powders will eventually have an impact in this area – though arguably much less impact than the weapons which will use it.

## Changing Shape

The uniform that goes from being soft, flexible fabric to bulletproof armour at will is an example of another type of nanotechnology. This utilizes materials whose properties change in an electric or magnetic field.

Researchers at the ISN are working with a type of liquid called magnetorheological fluids. These consist of a base such as silicon oil made into thick syrup with the addition of extremely fine iron powder. The particles are only a few micrometres across – smaller than red blood cells. When a magnetic field is applied to such a fluid, the iron particles all line up the same way and join up. In a fraction of a second the fluid goes from a syrup to a hard solid.

To create armour on demand, fabric will incorporate the fluid into its structure along with a fine weave of electromagnets. At the touch of a button, any part of the uniform can become a rigid shell. At present, the prototype versions of such fabrics can increase their stiffness by a factor of fifty. The strength of the armour depends on the power of the electromagnetic field, which is currently limited. Researchers believe that in five to ten years they will have body armour which can compete with existing bulletproof vests.

A key part of the technology is the iron particles themselves. At present, they are spherical, so they do not stack very effectively. Researchers would rather have the particles as flat plates or doughnuts that would fit together more tightly and create far stronger armour. This creates a new challenge for the manufacturers of nanoparticles, but one that is well worth pursuing.

The same sort of technology would have a many applications in the commercial world, ranging from seat belts and safety harnesses to casts for broken limbs to portable furniture.

The traditional folding table and chairs could be replaced by sheets of fabric which unroll themselves on command to make a rugged outdoors dining set – or a handy tent, windbreak or whatever else is required. The only limitation is programming the chip which controls the fabric.

A similar principle underlies the uniform's chameleon coating which changes colour to match the background. The technology is related to liquid crystal displays (LCDs) which incorporate a material whose properties change when an electric current is passed across them. In particular, modern displays use a type of material called a 'twisted nematic', a molecule which untwists to varying degrees depending how much current is applied. The amount of twist affects how much light can pass through the material, from transparent to opaque.

A LCD panel light in a laptop has thousands of individual elements ('picture cells' or pixels) containing the nematic material, and these pixels make up the image. Simple monochrome LCDs have pixels which darken or lighten with an electric current to provide black, white or shades of grey in between. Colour ones have three 'subpixels', which are simply monochrome pixels with a red, green and blue filter.

The chameleon fabric will incorporate a flexible coating made of millions of pixels which are far smaller than those on existing displays. Sensors will detect the colour and pattern of the background and adjust the appearance of the uniform appropriately. It will not be an exact match, but previous work on camouflage has shown that a pattern with the right colours and the right general shape is highly effective.

This type of material is still some years away, but will have plenty of applications in the civilian world. The first are likely to be computer displays and television screens that can be folded away like a handkerchief. These could be followed by billboard-sized moving displays; if sufficiently cheap, this will turn every surface into a potential movie screen.

Of course, clothing made from this type of fabric does not have to be camouflage; it can be programmed to display anything. Your T-shirt could show animations, psychedelic rippling colours, a million and one different slogans – or the six o'clock news. Whether anyone other than the most ardent gadget fan would ever want to wear one is another matter.

## Little Machines

The picture gets even more interesting when we start to look beyond simple changes in structure or colour. There is a much higher level of excitement, and hype, around Micro-Electro-Mechanical Systems or MEMS. (Arguably these should be called Nano-Electro-Mechanical Systems, but MEMS existed first and are simply getting smaller.)

The technology to build on a microscopic scale has been advanced by the computer industry. To pack more and more processing elements on to a computer chip, the industry has had to find ways of working with ever smaller electronics. This has typically involved using processes which etch paths on to silicon chips, or lay down trails of conductive materials, or build up elements layer by layer.

Improvements in our ability to manipulate matter on this level is leading to smaller and smaller circuits, and 'nanocomputers' more powerful than existing machines are more or less inevitable. With each generation, the scale gets smaller so more components can be packed into a single chip making it more powerful. The smaller size means that it takes less time for a signal to travel around the chip, making it faster. Computer power becomes ever cheaper and more plentiful.

In 1965, Gordon Moore observed that computer power for any given price doubled every eighteen months. The media called this 'Moore's Law'; it has proven to be valid for the last forty years, and is set to continue for at least another ten before fundamental laws of physics start to intrude.

The same technology used for microscopic electrical circuits can also be used to manufacture microscopic machines. A microchip is an electronic 'brain' which processes signals; MEMS technology means that we can also build 'eyes' and 'arms' that work on the same scale. Microscopic sensors can be made that will detect light, or sound, heat, magnetic fields or other stimuli. This information can be passed to a 'brain' which can then set in motion actuators, the machine's equivalent of muscles.

The best-known device which used MEMS technology is the car airbag. This incorporates a tiny accelerometer on the chip that controls the triggering of the airbag. The accelerometer is a mechanical device which is linked to the electronics. When it measures a very high acceleration – the sort caused by a crash – the electronics automatically trigger the inflation of the airbag. It is a very simple and reliable technology; the amazing thing is that it can be built on such a small scale.

The cost of chips incorporating MEMS has dropped by a factor of ten in the last decade, from £30 to less than £3. This suggests something similar to Moore's Law is at work. MEMS-based devices will get increasingly common.

The military have already incorporated MEMS into robust, reliable fusing systems and sensors. One such application is the 'micro gyro', a mechanical gyroscope small enough to fit on to a chip. This will give improved guidance to a new generation of small missiles, by allowing inertial navigation systems and gyrocompasses to be built on a microscopic scale.

The impact fuse on a bomb is similar to the airbag inflation chip: it waits until it senses massive deceleration before setting the bomb off. Smart fuses can have a built-in delay. The bunker-busting bombs discussed above, for example, will not explode on impact with a hard surface, but wait until the bomb has penetrated any layers of protection before detonating.

Having proven that mechanical devices can be built on this scale, it is all a question of how much they can do. They are not limited to having a few simple measuring devices built into electronic devices; they can be true machines capable of manipulating matter. This is the great promise of nanotechnology, or to some, the great threat.

## Nanomachinery

In 2003, Prince Charles made newspaper headlines by trying to warn the world about the threat from 'grey goo'. Few took him very seriously. The Prince's other concerns – the horrors of modern architecture, the importance of organic farming – have led to him being seen as something of an eccentric. But in fact he was only echoing fears that have been expressed in the scientific community for decades.

Nanotechnology implies an ability to build machines on a molecular scale. Such machines could themselves work with material on a molecular level and build (or destroy) literally atom by atom with a precision impossible for large, 'bulk' machinery.

Some scientists are already working with individual molecules on the nano-scale. Biologists involved with genetic manipulation have a variety of tools that allow them to manipulate the DNA molecule. There are enzymes, called 'restriction enzymes', that will cut DNA when they find a specific sequence. Other enzymes can be used to stick pieces of DNA together. By using the right enzymes in sequence, scientists can cut and splice DNA at will, building up the sequence they want. They have already succeeded in putting together a virus from scratch.

In theory, the same principle could be used to build molecules other than DNA. And while DNA does all its work through chemical means, nanomachinery could be built that is mechanical in nature.

In his classic 1986 text on nanotechnology, *Engines of Creation*, Eric Drexler compares atoms to a construction set. Atoms are like beads that can be joined together using connecting rods. The connecting rods, interatomic bonds, can be long or short, flexible or rigid. They may be capable of rotating; a chemical bond makes the perfect frictionless bearing. By putting together the right atoms with the right bonds in the right sequence, we can build a tiny machine. This machine would even have its own engine, a power-source driven by, for example, a photovoltaic junction converting light into mechanical energy, or one that can absorb fuel (such as hydrogen gas) from the environment.

The basic nanomachine that Drexler proposes is called an assembler. This is composed of about a million atoms, put together to form several thousand moving parts. Looking something like a microscopic crane, the assembler has a single manipulating arm, a set of tools, and a device that reads instruction from a molecular tape.

In response to commands read from the tape, the assembler is capable of picking up different types of atom and putting them together in specified ways. Several assemblers would be joined together, along with the tape that acts as its program. Taking his cue very much from DNA and evolution, Drexler explains how all this can be packed together into a machine the same sort of size as a human cell.

It is not immediately obvious how such a tiny machine could be of any use. It would be a gigantic task to build an assembler; any error in building the whole million-piece jigsaw puzzle would ruin the whole thing, and at the end of the day the result is much too small to see.

However, the usefulness of the assembler becomes apparent when you supply it with raw material and program it with its first task: build another assembler. Working on the nano-scale, the assembler would be much quicker than a human building from a construction kit. Ants' legs work at a much greater rate than human ones, and the bacteria wave their limb-like flagella faster still. The assembler would look like a speeded-up film of a bricklayer as it constructed a copy of itself.

Drexler estimates that, based on the speeds of biological chemical processes, an assembler could copy itself in perhaps fifteen minutes. Now you have two assemblers. Set them to reproducing, and in half an hour you have four, and in an hour sixteen.

The process of doubling increases faster than you might expect. Every ten doublings is equivalent to an increase of a factor of a thousand. After ten doublings in less than three hours, we have a thousand assemblers. After six hours we have a million, nine hours a billion. After fifteen hours there are a million billion of them – enough to be visible.

The visible speck continues to grow geometrically. By the end of the first day, we have a ton of assemblers. The microscopic machines are not attached to each other, so the pile of assemblers will resemble an extremely fine powder or a very thick liquid. The small scale of construction means it will have a greyish appearance. In effect, we have a large amount of high-tech 'grey goo'.

Now, a ton of assemblers is enough to do some serious work. On the day after having created the first assembler, we can issue the appropriate chemical command to order them to stop replicating and do something useful. Given raw material – say in the form of some scrapped cars – the assemblers can rearrange the pieces on a molecular level into a shiny new sports car.

The assemblers could build anything you wanted from those materials, from a car to an executive jet, or a computer. Drexler uses the example of creating a small rocket, made of superhard materials much lighter and stronger than metal. Drexler has in mind reaching for the stars, but as we have seen there is little difference between a space rocket and a ballistic missile.

Looking at our previous examples, the assemblers could turn dozens of small unmanned air vehicles in a matter of minutes. These would be equipped with powerful nanocomputer brains and guidance systems, armed with hypervelocity missiles fuelled by aluminium nanopowder and with a powerful warhead of the same material, inside an ultra-hard casing capable of piercing normal armour like butter.

Alternatively the assemblers could put together the sort of super-uniform described by Ned Thomas, with its super-strong actuators and nano-scale material capable of stopping bullets.

Given the designs coded into a suitable form, there is in principle no manufactured product which the assemblers could not build. If the raw material contains the right elements, the assemblers can build it. In theory, they could even turn the proverbial sow's ear into a silk purse.

Assemblers cannot turn lead into gold or uranium – that would require change at a sub-atomic level rather than just shuffling the atoms around (though assemblers would be a very efficient way of extracting these elements from ore, picking through it molecule by molecule). However, given a ton of coal, the assemblers can be instructed to break it down and rebuild it in a different crystalline form. In less than an hour, the assemblers could rearrange the carbon molecules and build up a ton of freshly minted diamonds. That should go some way towards persuading backers that the costs of the development project were well spent.

Any manufactured item could be replicated. The implications for copyright and patenting are far-reaching: this method will not just turn out a good copy of a Rolex watch, but an absolutely identical one that is indistinguishable even through a microscope.

Nanotechnology has its fervent supporters like Drexler who paint a picture of a world where material goods will be available to everyone in an unending flow. They see other advantages as well, such as nano-machinery capable of working inside the human body and aiding our health. With nanotechnology, Drexler argues, we will all be rich and live forever.

While the idea of a machine capable of making perfect 3-D copies of existing machines may seem far-fetched, it is important to remember how far technology has come along in miniaturization. In 1944 ENIAC, the original digital computer, weighed 27 tons. Within twenty years, the advent of the integrated circuit meant that the same computing power weighed less than one-thousandth as much. Ten years after that came the first microprocessor, a computer on a chip more powerful than ENIAC and less than one-millionth of the weight. These days the amount of computing power that corresponds to ENIAC occupies a space on a chip too small to see. Our technology for mechanical systems is just starting to reach the steep point of the curve, and we may see some dramatic developments over the next few years.

## The Trouble with Goo

Self-replicating machines might create more problems than they solve, a point that was raised long before nanotechnology was even considered. This came during discussions of the self-replicating machines proposed by the mathematician John von Neumann in the 1940s. Von Neumann was a genius, who made important breakthroughs in computing, economic theory and mathematics; he is also thought to be the model for Dr Strangelove.[2]

Although Hungarian by birth, von Neumann became an advisor to the US government and was influential in determining nuclear policy. He invented a new branch of mathematics, game theory, in which any game, such as chess or poker, can be reduced to strict mathematic principles. Game theory said that if the opportunity arose, the US should launch a nuclear strike against Russia if it could do so with impunity. No wonder some people felt that von Neumann symbolized the cold destructive logic of the nuclear era.

His idea for a self-replicating machine was more benign, at least initially. It consisted of two elements, a computer and a constructor. Although it was a large-scale machine rather than a microscopic one, the principle is the same as Drexler's replicating assembler. Scientists who were speculating about the possibility of exploring the galaxy picked up the idea. This is a task of mind-boggling scale: there are a hundred thousand million star systems in the galaxy, separated by many light years. In order to find the ones which have an Earth-like planet, one that might be ripe for exploration or colonization, we would have to send a probe to every star system. The number of probes required would be far too great for any conceivable manufacturing capability.

Instead of manufacturing thousands of billions of unmanned probes here on Earth and sending them out, we could build just one self-replicating probe – a von Neumann machine. This would be sent to the nearest star system, where it would make copies of itself and launch them to the next nearest stars, where the process would be repeated. Each new star system would become a manufacturing base for more probes, which in turn would go on to found their own manufacturing bases. In this way the galaxy would soon fill with von Neumann probes beaming back information of what they had found.

Shifting the manufacturing process out to increasingly distant star systems mean that the new probes would not have as far to go to reach the boundaries of the known universe. It would also solve the problem of raw materials: building a billion space probes would exhaust the earth's natural resources, but the von Neumann system allows us to mine the entire galaxy. It seemed like the ideal solution. However, one problem became evident. Over the course of tens or hundreds of thousands of years, the galaxy would be full of replicating probes, which would eventually visit every star system. And they would not stop there.

Once started, the probes would keep replicating. Like the population of nanomachines, their numbers would go from a handful, to billions, and then on to billions of billions. Each star system would soon have millions of new von Neumann probes arriving every year from all directions, each one setting up its own manufacturing plant. Whole planets would be eroded away, used up as raw materials. The galaxy would swarm with billions of billions of probes like cosmic locusts, consuming everything in their path, their numbers increasing inexorably.

While the threat from replicating von Neumann probes would take millennia to materialize, nanomachines could present the same hazard in a much shorter time span. The danger arises if an undetected error in the first assembler, or one of the copies in any succeeding generation, leads to control problems. If, instead of obeying our command to stop making more of itself, the assembler carries on its initial task, the result could be catastrophic.

When human cells start to replicate out of control, the result is known as cancer. The effect with nanomachines would be equally serious.

Looking back at our original scenario, we can see that if a single cancerous assembler appears, it will have reproduced a ton of rogue assemblers in less than a day. At this stage the growing heap will have disassembled part of the laboratory as it seeks more raw materials for the construction process. Another three hours and we have a million-ton heap of assemblers. Swelling like a monster from a bad science-fiction movie, the grey goo spreads everywhere. Given energy and raw materials, before the end of the second day it has spilled over the surface of the entire earth, assimilating everything in its path.

As Drexler points out, the largest living things from whales to redwood trees all started out as single cells. But whales and trees have limits to their growth, whereas the grey goo can go on indefinitely. If its raw material are abundant elements like carbon, oxygen and silicon (sand and dirt are mainly silicon dioxide), the only limitation would be its power supply. Assemblers which relied on external fuel like hydrogen would be easy enough to control, as they would lose power outside their own controlled atmosphere, but those which can draw on solar power would go on indefinitely.

This is a familiar theme in science fiction, where a creation turns on its maker. In Mary Shelley's *Frankenstein* the monster persuades Frankenstein to make him a mate, but Frankenstein destroys her when he realizes that the two will breed more monsters.

While monsters and large-scale machines can be dealt with by conventional means, microscopic assemblers present a different problem. Fighting nanomachines with normal weapons is like trying to take on a swarm of bees using an elephant gun: you just can't damage enough of them fast enough. Even if you managed somehow to destroy 99.9 per cent of the grey goo (nuclear weapons might be destructive enough), you have just three hours before the remaining 0.1 per cent builds itself back up to the same level as before.

Drexler suggests several methods for preventing runaway assemblers. One is to design them to be dependent on some particular material which can be controlled, like the hydrogen atmosphere. Another is to build them in laboratories that are physically isolated from the rest of the world, in orbiting laboratories. A third means is to have an 'active shield', perhaps consisting of nanomachines which attack and dismantle assemblers if they moved outside a specified area.

An alternative approach would be not to have separate assemblers in the first place. If instead they were all centrally controlled or physically attached to another unit – like the assembly line of robots in a factory – they could be managed much more easily.

Assemblers are not magical and still have to obey the laws of physics. An assembler or a group of assemblers cannot tunnel through a steel wall unless designed and programmed with the tools to do so. They are no more likely to develop the ability with a random mutation than bacteria.

However, mutations will occur. Machines on the nano scale, like our own DNA, will be affected by background radiation and other factors that could make them reproduce inaccurately. Most of the time the mutation will not be viable and the mutant assembler will not function. However, if a viable mutation does develop, it is potentially dangerous. The risk of runaway reproduction has already been mentioned, but anything which affects our ability to program the machinery could be dangerous.

Controlling assemblers is relatively straightforward if you know what they are and where they are. When the first assembler is created it will be under very tightly controlled conditions. In the event of mutant assemblers appearing, other assemblers can be programmed to identify and dismantle them.

But it is a different matter if an assembler is developed specifically as a weapon.

## Replicating Weapons

Assemblers would make a devastating weapon of mass destruction. As the debate about WMD in Iraq and elsewhere has shown, finding something as large as a chemical or biological weapon facility is not straightforward. Assemblers would be worse; once built an assembler would be literally impossible to find, as it could be microscopic in size until used. How can you stop something from getting into the country when it is the size of a single bacterium and can just blow in on the wind?

Rather than simply engaging in reproduction and spreading as grey goo, they could be programmed to reproduce up to a certain point and then attack other targets. Assemblers could be programmed to dismantle certain materials such as plastics or metal alloys, destroying the industrial infrastructure of a country. Or they could simply attack the human inhabitants, like an artificial version of the lethal, flesh-eating, necrotizing fasciitis. The difficulties of stopping a swarm of such replicators have already been discussed.

Such weapons present a danger to the country that uses them as well. Even if the attack assemblers were programmed to self-destruct after a certain delay, or if they were designed with some particular vulnerability known only to their maker (as in some old B-movie, the continent-eating monster might succumb to electricity, sea water or the common cold), the risk of mutation is still present.

This may not matter to politicians. Like nuclear weapons, the attack assembler may be built purely as a deterrent. It would make a potent reply to the threat of nuclear attack from the other side, and is much easier to protect against a first strike. A nation possessing attack assemblers could guarantee that they could destroy any aggressor, even if they were utterly destroyed first.

An arms race is entirely possible, with nations finding it necessary to have an attack assembler just in case the enemy is doing the same. Terrorist groups may see an attack assembler as the ultimate blackmail weapon – or the ultimate revenge weapon if they face extermination.

Even if assemblers were not used aggressively, their manufacturing capability would transform the military balance. At present, one nation has dominance in aircraft, missiles and other military hardware, but this becomes meaningless if other nations can churn out high-tech armaments by the million in a matter of days. The possibility of robotic armaments means that it is not even a matter of supplying arms for the human armed forces; assemblers could build virtual armies of robot tanks and aircraft.

If assemblers ever get into private hands, then gun control is a non-starter. The owner will not just be able to build as many automatic weapons as they can carry; they could have their own army in days. Weapons of mass destruction? Assemblers can build you an endless supply of ballistic missiles tipped with nerve gas or other lethal warheads. The consequences would be unimaginable.

This all has distinct echoes of the nuclear experience, with the promise of endless wealth and power and the risk of global annihilation. Will replicating nanotechnology ever be safe enough for us to consider a world in which coal can be transformed into diamonds and scrap into spacecraft? If it is, will the assemblers ever be allowed to pass into the hands of other governments, who may wish to turn the technology back to military use?

Or will this be another technology, which will have to be limited and controlled, with wars fought to prevent it falling into the wrong hands?

## Back to the Present

For the meantime, replicating assemblers remain the stuff of science fiction, and nothing exists at the moment that comes anywhere close. But there are a host of nanotechnology initiatives under way that may have significant consequences.

As we have seen, stronger materials, faster computers, more powerful explosives and fuels, and micro-machines are already on the agenda. Clearly, these can be combined to make small, smart robots – not invisibly small and not capable of replicating, but they can do much else.

In the medical field, a miniature robot the size of an aspirin tablet already exists in prototype form. This robot carries a variety of sensors and can go through the digestive system looking for ulcers, broken blood vessels and other physical symptoms. As the technology advances such robot pills will become increasingly sophisticated and could, for example, release specified types of medication when particular conditions are detected. This could be anything from a change in the blood sugar level to the pill reaching a certain point in the stomach or intestines, ensuring that, in the case of an ulcer for example, the drug is only applied when and where it is needed.

As MEMs progress, these medical robots will gain an ability to carry out physical actions – cauterizing or sealing wounds, removing tumours, breaking up foreign bodies or obstructions. Smaller robots, less than a millimetre across, could be injected directly into the bloodstream to deal with thickened arteries or blood clots (an idea prefigured by SF writer Isaac Asimov in *Fantastic Voyage*, although he used a team of miniaturized surgeons in a micro-submarine).

Nanosatellites also exist in prototype form. These weigh less than a kilogram, and later versions will be much smaller and lighter. At the moment, they are simply small versions of existing types of satellite, capable of sensing or communications relay. But again, smaller and better MEMs will give them hands and tools to work with. A swarm of nanosatellites could carry out repairs and maintenance on other satellites, reconfigure and upgrade them beyond the original design – or destroy them.

We have already seen the micro air vehicles in Chapter 13, with 'smart dust' being their smallest form at present. Of course, these too will be affected by nanotechnology, and robot spies the size of gnats or smaller are likely to appear over the next few decades. Multimode vehicles are already being considered which can fly or drift with the air currents and also crawl towards a target. The networking capabilities of smart dust means that however small the individual elements are, the network as a whole can still have significant power. This may not be limited to simply observing, especially in the case of military systems.

To take a relatively harmless example, a swarm of robot insects could be used to block air intakes to neutralize an engine, or to allow themselves to be sucked into a jet. This would be very useful from a military point of view, but would potentially create something that would cause far more collateral damage than cluster bombs.

Dr Jurgen Altmann, a German scientist specializing in advanced military technology, has already called for an embargo on certain types of development. In particular, he believes that autonomous systems on a scale of less than 20 cm, should be prohibited. This would stop development on smart mines, missiles and all sorts of miniature mobile robots. He also recommends a blanket ban on self-replicating systems. This call is not premature; Dr Altmann's studies have already identified many military projects of this type, including autonomous lethal systems.

If they are already dangerous with current technology it is easy to see how dangerous nano-scale devices could be. Even without self-replication, miniature systems could easily cause enough damage to be classed as weapons of mass destruction, with the added hazard that they cannot be neutralized as easily as biological or chemical weapons.

Whatever happens in the civil world, the military are progressing with research into nanotechnology. 'Grey goo' and the threat of nano-WMD may belong to the far future (along with the promise of infinite wealth from your own personal nano-assemblers), but other forms of nano-technology are much more imminent. New materials and micro-devices, from ultra-strong versions of metals to colour-changing fabrics, may be just around the corner. The exact effect on the environment of nano-powders and other new substances is already a matter of some concern.

While we may have some years to go before nanorobots appear, there is a risk that if they come from the military side their impact on the civilian world is not uppermost in the minds of those who are designing them – any more than the bomb makers at Los Alamos were thinking of what would become of their invention in a hundred years time.

# Conclusion

*It's tough to make predictions, especially about the future.*
Baseball manager 'Yogi' Berra

As we saw in the first part of this book, much of the technology developed during the Second World War has become familiar in the civilian world, from computers and jet aircraft to satellite communications and nuclear power. And as the second part shows, the progress of military technology continues as rapidly as ever, with many potential spin-offs in prospect. In the process we have acquired computers and cheap air travel, started wearing sunglasses and discovered water on the moon. This does not necessarily mean that military technology is a good thing. Defence spending may be seen as a necessary evil at best, but it would be optimistic to suggest that a decrease in defence spending would result in the money being spent on social programs. From our present situation defence spending and the associated developments in military technology look to be as inevitable as wars themselves. And although this is a broad and diverse field, some patterns become apparent across many different types of technology that have migrated into the civilian sphere.

## Secrecy

In Greek mythology, the war goddess Athena was born when she burst out of the head of Zeus, fully adult and armed for battle. Military technology also bursts on to the scene already fully developed, unlike other fields where there is a long and visible gestation period. However many commercial companies want to keep new products under wraps, they lack the kind of security that the military can call on. Products have to be sold, and become public as soon as they are on the market, whereas a new weapon can remain secret long after it has been built. Secret weapons such as radar and computers can stay secret even after they have entered service.

The basic theory may well be public knowledge. In 1940, the theoretical foundations already existed for space rockets, atomic energy, jet aircraft and digital electronic computers. The interested layman could have found out about what was possible in theory in each of these fields. But each development came as a surprise when it was revealed.

The V-2 attacks on London were a complete mystery to the public until the truth about German rockets finally came out. The existence of the RAF's radar was kept secret behind cover stories of pilots eating carrots to improve their night vision and other tales. German jet fighters came as a surprise to Allied pilots if not Allied intelligence, and the atomic attack on Hiroshima was the most devastating use of a secret weapon ever. ENIAC worked quietly behind the scenes, and like Colossus it might never have been revealed.

We can therefore expect that while we have some idea of what is theoretically possible and what developments are currently taking place, there will be new technologies which arrive completely out of the blue. We know broadly what the state of the art is with artificial intelligence and nanotechnology, scramjets and millimetre-wave sensors, but based on past history the military are likely to have more advanced systems than they have made public. Exactly how much more advanced is a good question.

This is why close scrutiny of any hints we may get from the black world may be valuable – and why the agencies involved work so hard to conceal and confuse the truth.

If technology is not declassified, then the civilian world does not benefit. This can lead to situations like the great loss to the British computer industry of all the knowledge and experience gained through the Colossus program at Bletchley Park, compared to the situation in the US where ENIAC fed into the civilian world. However, when it is declassified, it is not necessarily the country which develops it that benefits most. Frank Whittle's jet engine proved to be of greater benefit to American aviation than British, whereas the transistor invented in the US led to a huge electronics industry in the Far East rather than America.

The secrecy aspect of secret weapons means that there is no effective democratic accountability or oversight. Humane considerations can be overlooked – not not to mention the question of whether taxpayers millions are well spent. Time and again while researching this book I found that information was restricted or otherwise not available. In this regard, the US military establishment is a model of openness compared with the British one. Whilst snooping around and fitting together the details of various programs has been a fascinating occupation, society should not have to rely on this sort of enterprise for information about what is being done in its name.

This secrecy means that there is no possibility for public discussion or international legislation. Arms control requires openness, but new weapons such as autonomous micro-robots or radio-frequency non-lethals are being developed secretly. The prospects for an international ban on such devices are slim once they have been deployed and used. And when they proliferate in the military world, they are likely to be coming our way soon.

## The Great Leap Forward

Superheroes are a recurrent image in high tech. In the 1920s there were visions of rocket ships and ray guns, which materialized as ballistic missiles and lasers. Now we have the nanotechnological soldier leaping over tall buildings with a single bound and seeing through solid walls like Superman. There has always been an element of science fantasy involved, because these developments are about reaching beyond the limits of known technology.

Military developments are aimed at overcoming major obstacles. The commercial world is more attuned to small, step-by-step developments which bring incremental improvements. These are obviously profitable and easy to finance. However, the huge leaps forward – such as the progress from propellers to jet engines, or the exploitation of nuclear power – take a level of investment which is not often available in the commercial world. They are high risk in economic terms; but a different set of rules applies in warfare, and many projects were carried out simply because of the fear that the enemy might get there first. The cost of building the atomic bomb was tremendous, but the cost of not developing it if the Nazis pressed ahead did not bear thinking about.

Virtually all of the developments covered in this book represent significant advances rather than just slight improvements over what existed before. This applies equally to ENIAC, thirty times faster than the best analogue computer, as it does to the V-2 rocket or the revolutionary aircraft-detecting radar.

In the commercial world, a product that is good enough discourages heavy investment in an alternative which may or may not work. Piston engines and vacuum tubes were perfectly good in their day. In the present era the computer industry is happy enough with steady improvements in computer power; introducing a completely new type of computer would mean restructuring the business. At present there is not even a well-defined market for something like quantum computers, whereas customers are always clamouring for machines with more memory and more processing power.

In business, being in second or third place in the market is not necessarily a disaster. Many companies have survived for decades without becoming the leader. In warfare, there is no second prize. The price of failure is much heavier for the military and the drive to leap ahead of existing technology is much stronger.

In the US this had led to a doctrine of Full Spectrum Dominance. Under this idea that the American military should not simply be better than the rest of the world, but it should be so much better as to guarantee victory at any level, on land, sea, air or space, in cities, jungles or anywhere else that conflict can occur.

Of course this sort of technological leap comes with a hefty price tag attached, but so far there is no sign of defence spending slowing, at least in the US. The ambition is there, and so are the resources and we must expect to see more giant leaps forward.

## Synchronicity

Because the basic science underlying technology is freely available, several nations may be reaching for the same goal, and may reach it at more or less the same time.

In 1940, there were at least eight nations involved in jet engine research. Five countries had their own atomic weapons programs, and five had prototype radars, while at least three pursued computing programs during the war.

Sometimes the results could be very close. The German jets made it into the air only a few months before the British, and Russia was not far behind. Several nations brought radar into service during the course of the war. As we saw, the German atomic bomb program fizzled out, and the Russians abandoned theirs in favour of projects with faster returns, but it is easy to see how the war could have ended with all three nations with functioning or nearly-complete nuclear weapons.

Looking at the present day the situation has changed somewhat. The basic science is still there, and the possible benefits of scramjets, for example, are clear to see. But there is only one nation with deep enough pockets to pursue the full range of technology.

The US is almost invariably the front-runner simply because of the sheer quantity of money poured into defence research. In 2004, the US defence budget ran to over $500 billion, more than the defence spending of the next nine nations combined. Put another way, the US accounts for some 40 per cent of global defence spending.

According to the Organization for Economic Cooperation and Development (OECD), defence research in the US amounts to .45 per cent of GDP, far ahead even in percentage terms of the next two, .26 per cent for the UK and .22 per cent for France (the next two are Korea and Spain). Put into cash terms, the difference is even starker, with the US spending $40 billion, compared to less than 10 billion for all the Euro nations put together. This is only about 14 per cent of the total US spend on R&D, but as we have seen it has a disproportionate effect because of the dramatic results of successful projects.

This is significant because the amount of research needed is the same whether a nation needs a hundred new tanks, aircraft or rockets or a thousand. The British Army has similar requirements to the American one in terms of tanks, guns and helicopters but the Americans spend ten times as much developing their kit. This leads to armed forces which are much closer to the cutting edge, with much more high-tech gear, from radios to rangefinders, night-vision equipment to camouflage.

However, even the US has its limits, whereas high-tech can demand unlimited amounts of cash and spiralling costs are common in the defence business. The F/A-22 program was originally intended to supply 648 aircraft, but this was cut to 442 and then to 339 due to rising costs. Unfortunately, cutting the number also makes the unit cost greater, and the cuts went on – down to 276 and now only 218 planes have been funded. The new super-fighter works out at over $300 million per aircraft against an original plan of $35 million. Spin-off benefits will not be proportionate.

Of course not every area of technology needs massive investment to get results, as we have seen. In many of the fields we have looked at there are still other players, like the Franco-Russian scramjet program or Britain's millimetre-wave radar.

The situation changes once a secret weapon stops being secret. Proliferation is rapid with all but the most inaccessible technologies. For atomic power, the bar is set very high indeed, and the process of enriching uranium and assembling a warhead is a major challenge for any nation. But nations do not have to start from scratch with setting up their own Manhattan Project, and many of the basics of atomic weapons construction are now well established. What was once a huge endeavour which could only be contemplated by an assembly of some of the finest scientific brains in the world is now something that can be undertaken by a developing nation with relatively modest means.

Once it has been proven that the technology works, then many of the barriers are down. Some barriers may remain, such as the need for exotic materials like uranium or the huge capital costs in building space rockets, but even a nation with modest means like North Korea can apparently build both atomic warheads and satellite-launch vehicles.

When new technology does emerge, the breakthrough can be shared across the world. It may take several decades, as with atomic power, but the process may move much faster. If the US, for example, unveils a new solar-powered aircraft, then the rest of the world is likely to learn rapidly and follow quickly with other local versions.

In the civilian sphere, giant corporate interests like Sony, IBM and Airbus are capable of investing the necessary billions in a proven technology if the market is there. If solar aircraft can be built and are profitable, then the necessary funds will appear for them to be developed for the civilian market.

## Synergy

As technology becomes more advanced, synergy – defined as a combined effect which is greater than the sum of its parts – becomes more important. This is because technologies work together.

The intercontinental ballistic missile combined the rocket science from the German V-2 program with the atomic warhead from the Manhattan Project, and was further enhanced by the addition of digital electronics. Without each of these elements the missile would never have been built and the world would have been a different place. This unique combination made possible the ICBM and with it the potential for global devastation within minutes of the order being issued.

More constructively, the modern airliner is successful because of the combination of jet engines, swept wings and advanced control and navigation systems. Like the ICBM it is a reasonably obvious combination of elements, but in other cases the synergy is less foreseeable.

Put together digital computers powered by silicon chips, some clever communications software and global telecommunications made possible by satellites and fibre optics carrying lasers, and you have the Internet. While the great Dot.Com burst spectacularly, the full capabilities of the Internet have not been realized, although its bounds expand all the time. It is not just that you can now fill in your tax return, order groceries or bid for antique teapots online. There are whole areas of life, like medicine, education and law, which are only starting to be touched by the Internet.

Just as the technologies discussed in the first six chapters combined with sometimes unexpected results, the same is likely to happen in the future. Trying to make predictions under these circumstances is absurd, because it relies not only on guessing which technologies will work out but how they will combine and whether the market will support them. Instead of predictions, one can at least speculate.

One possible combination of quantum computers coupled with GPS and mobile phones could lead to the ultimate public transport system. At the moment, you have to wait for a bus which does not go exactly where you want to go, or hail a taxi which goes right to your door but is very expensive.

In a future system you could simply hit the 'call bus' button on your mobile phone and tell it where you want to go; the computer calculates your present location and where you are going, combines it with the needs of every other passenger and the availability of vehicles, and efficiently routes the next bus to where you are. The buses themselves might be driverless, and the system could be incredibly efficient compared to existing transport systems. Given a bit of artificial intelligence which would identify patterns of movement, the entire fleet of taxibuses would go through a tidal flow during the day, shifting to the greatest anticipated need before it happened.

A fleet of solar-powered aircraft linked with advanced computing power and vortex ring generators might also have more potential than any of the elements on their own. Weather prediction is getting steadily better as computer models improve and computing power increases. The Earth Simulator, a Japanese system running on a supercomputer, is expected to deliver far more accurate predictions than ever before by keeping track of weather patterns with a finer level of detail than has been possible previously. But we need not stop at prediction.

The 'Butterfly Effect' is famously quoted as saying that a butterfly flapping its wings in the Amazon ultimately gives rise to a tornado in Texas. While the limitations of chaotic systems mean that we are unlikely to have that level of control, our fleet of solar aircraft could roam the world, providing tiny nudges to weather systems by means of vortex rings. If the prediction system is sufficiently precise, such subtle touches might be able to influence the formation of hurricanes and storm systems billions of times more powerful, not stopping them but redirecting them and modifying them so that their energies become less destructive.

Another more threatening combination would be millimetre-wave imaging with micro air vehicles. Each of these is potentially quite effective on their own, but the combination enhances both, producing a device which effectively allows the operator to go anywhere and see everything day or night. This would be great for rescue operations, with swarms of robot insects quickly combing through burning buildings looking through smoke and partition walls to find trapped victims. It would also assist police in manhunts: rather than requiring hundreds of officers to search through woodland, a few miniature craft with sensors capable of seeing through foliage could locate a suspect within minutes.

This type of spy would also have other uses. Private detectives can already use nightscopes and intercept mobile phone calls; the micro fly spy would be the perfect means of 'bugging' someone, following them everywhere, looking through closed curtains or blinds of upper windows – not to mention seeing through bedclothes to provide video evidence of exactly what went on. If you want privacy, high walls are not enough; the only protection is impervious walls and a good anti-bug system to prevent robot insects from creeping in through chimneys or other access points.

Other possibilities are even more mind-boggling, especially those involving nanotechnology, and soon begin to sound like science fiction.

Microrobots cruising your bloodstream and providing a moment-by-moment health monitor and replicators capable of creating any item you request on demand are difficult to imagine.

Looking back at predictions made in earlier decades, it is striking how often the major developments are overlooked. Whilst researching this book I consulted a fifteen-volume work on computing from the late 1980s which makes no mention of DARPANET or the Internet; during its early stages there was little recognition that it would be anything more than a playground for technology geeks to communicate and exchange software. E-mail was for programmers, and not something that ordinary people with access to the telephone would ever be interested in.

## Late Again

Almost everything discussed in the first part of this book arrived too late. British jet fighters never flew against German aircraft; the ENIAC computer did not arrive in time to calculate the ballistic tables for wartime use as intended. Dornberger's simple epitaph for the V-2 rocket was 'too late'. Even the atomic bomb was barely completed in time, and made no difference to the actual outcome of the war. Japan was already defeated; the bomb simply provided an extra level to force the surrender sooner rather than later, saving Allied lives.

Technologists have to be optimists, or else they would not embark on research whose outcome is so uncertain. Sometimes their role demands that they make promises which may not be kept. This happens so consistently with military technology that it is more than likely that any new development intended for, say, the war against terror, may not be delivered until well after the war is over.

This long lead time has led to the spectacle of Cold War technology arriving long after the Soviet tanks have been scrapped or shared out among the new democratic Russian states. The US possesses an unrivalled collection of smart anti-tank missiles for dealing with a massed armour threat which does not exist any more. The stealth bomber, specifically designed to carry nuclear weapons through the toughest air defence network in the world is now used for dropping high explosives on guerrilla forces without a single radar between them for its stealth to evade.

The F/A-22 Raptor is set to be the world's most advanced combat aircraft, but it will be born into a world where the swarms of enemy

MiGs no longer exist. Originally intended as a pure air-to-air fighter designated F-22, it had to be recast as a combined fighter/attack (hence the F/A designation) aircraft capable of carrying bombs. It has to be said that the design, which emphasizes unprecedented tactical speed, manoeuvrability and stealth, is not what would be specified for a straightforward bomb truck.

The European equivalent, the Eurofighter Typhoon, is an altogether less sophisticated aircraft, but still highly effective – though again perhaps better as a dogfighter than a bomber. Both projects are hugely expensive, the showpieces of their respective defence establishments. The big difference is that while the US already has massive air superiority over any possible opponent, the Eurofighter is essential if European powers ever need to fight a war without the assistance of the USAF – but like the Raptor it has taken more than two decades to arrive.

Not only does it arrive late, but such technology only provides a benefit for a limited period. Maintaining a technological edge is a constant process. The jet fighter gives only an advantage until the enemy has jet fighters too. New advances are needed to ensure that one's own armed forces stay ahead of the enemy. Full Spectrum Dominance is not a simple endpoint which can be reached by developing weapons with specific capabilities, because the opposition will always be catching up.

In civilian life, a few years either way is rarely a matter of life or death. People were managing before satellite communications, barcode readers and mobile phones came along. But although the military benefits of jet engines are only temporary, the impact on civilian life is permanent. Future generations will get to experience satellite television and microwave cookery into the indefinite future.

Because the timing is less critical and the effects are permanent, the impact of technology is ultimately much greater on the civil world than on the military one.

## From the Battlefield to your Kitchen

We started out with the question 'Can military technology escape from its origins?' Clearly it can, and in many cases it has become so thoroughly civilianized that nobody recognizes it roots at all.

Military technology is particularly apt for misuse, but rockets, jet engines and computers have plenty of entirely peaceful applications.

Other technologies such as nuclear power may give rise to doubts. There are basic issues that place it in more of a grey area – the link to nuclear weapons, safety and waste disposal – but it still has potential to do good in the world.

During the Cold War, the West countered the perceived threat by producing generations of advanced missiles, as well as sophisticated tanks and aircraft to fight against superior numbers and win. Such technology rarely makes it outside the battlefield. Sixty-ton tanks and batteries of Patriot anti-aircraft missiles do not find their way into the civilian world, though particular technical features may do. For example the laser rangefinder developed for tanks is now used by golfers to estimate their drive.

The war against terror is rather different. The big spending is still on aerospace, missiles and aircraft. But there is also a significant amount of technology funding directed to areas like intelligence gathering, 'operations other than war' using non-lethal weapons, and swift and stealthy attacks on rogue states. Some of the end products, from stun guns to snooping devices, may appear directly in civilian versions, but these are likely to be exceptions.

It is worth noting in passing that military firearms have made remarkably little progress in the civilian world, even in America. In spite of a certain amount of media hysteria, military-style 'assault weapons' account for less than 2 per cent of the market. There are a number of reasons for this, one being that the most popular weapons are handguns and shotguns which are not typically military. Another factor is that the Army are limited by international conventions which ban the use of expanding bullets, including dum-dums, hollowpoints and the like, in warfare. Such bullets can cause terrible injuries, and are in common use in the civilian world.

In some ways civilians have better weapons: it is actually illegal to hunt deer in the US using 5.56mm rifle rounds, the standard US Army ammunition, because this type of bullet is considered insufficiently lethal to ensure a humane kill. However, part of the military preference for small, high-velocity rounds is their better performance against armour and cover; the 5.56mm round will go clean through a brick wall, making it something of a liability for police work and self-defence. Of course, for anti-personnel use the Army can also call on machine-guns, grenade launchers, mortars and other forms of firepower which are not legal to civilians even in the US.

But most military technology is not closely tied to killing people. As we saw with vortex rings, technology can have all sorts of unexpected applications (like firefighting and inkjet printers) which have nothing to do with the military. Other technologies may not have been very warlike in the first place. As mentioned in the introduction, both Velcro and Teflon were the result of civilian research efforts. But their successors may come from military labs.

In its quest for transformational technologies that will give US forces the edge in every aspect of combat, the Pentagon is working on a superior non-stick saucepan. While Teflon may be suitable for civilian applications, in the rugged world of the field kitchen such coatings are too fragile to last up to repeated use. US forces still have plain aluminum cookware, and all the manual scrubbing which that entails. The new program will develop 'electrocodeposited quasicrystalline coatings' which are both non-stick and durable. If successful, the new coatings should rapidly find their way into the civilian world.

Velcro fastens well enough, but makes a distinct ripping sound when it is unfastened. This makes it too risky for special forces to use: when silently patrolling through the jungle at night, the last thing you want is to give away your position every time you open a Velcro pouch. 'Stealth Velcro' is the solution, a new type of hook-and-loop fastener similar to Velcro but which is completely silent.

In fact, going through the various technology programs under way at the Pentagon it is easy to get the impression of the world's largest camping-supplies store, with everything for cooking in the field (self-heating food, ultra-mobile water coolers), and the actual weapons being less conspicuous.

Developments like non-stick pans and silent Velcro may not influence the world in such a dramatic fashion as ballistic missiles – perhaps they will be around for longer and play a bigger part in people's lives.

As we saw in the Introduction, the US Navy's cotton T-type shirt was one of the more conspicuous by-products of the Second World War. The US Army's Soldier System Centre is currently working on a new type of undergarment: the cooling shirt. This is a vest which has plastic piping embedded in it, piping that circulates a stream of cooling water from a portable unit. This type of vest is currently only used by helicopter pilots, who can plug into an external power supply, but as technology improves a mobile version is on the way.

The current Compact Vapour Compression Cooling System weighs

2 kg, is the size of a brick and requires a 1-kg battery for two hours operation. But more advanced technology using absorption cycle cooling should cut this significantly.

The absorption cycle system will be powered directly by fuel — probably military grade diesel — and a cupful will provide six hours cooling or more. Just the thing for foundry workers, miners, cooks, or indeed anyone who has to commute in hot weather.

The $70 billion spent on the F/A-22 Raptor may bring some benefits in materials science and other fields — the titanium of earlier aircraft is now being replaced with more advanced composite materials which are even stronger and lighter — but everyday benefits will also come from the projects which only cost a few million. Like mobile phones, cooling vests are a practical small-scale technology that can be made to work if the military does the development first. From an expensive brick-sized unit to something tiny and used by everyone is a small step: the big step is getting the first version working. There are thousands of other projects that might find their way into everyday life in future, if and when they can be made to work.

Similarly, the cumulative effect of small offshoots of military programs — like laser eye surgery or quartz watches — will continue to make a big contribution to the civilian world.

The military will continue to develop ever sharper, stronger and more lethal swords. New Athenas will continue to burst forth, ready to amaze the world. As the preceding chapters have made clear, there is no shortage of new ideas about how to kill, destroy and lay waste to the world, from high above the atmosphere to deep underground, not forgetting the battleground of the human mind. We now possess more power for destruction than ever before.

There is no sign yet that the world is ready to heed the prophet Isaiah's call to 'study war no more'. But in spite of our destructive tendencies, humanity survives and thrives. Athena is goddess of wisdom as well as war, and new technology has peaceful uses as well as warlike ones.

The armies march on, but as each new sword is developed it will be taken up and reshaped into a ploughshare by resourceful people.

# Notes and References

## Introduction

1 The thousands of bomblets cover the target area more effectively than a smaller number of larger warheads, but have an unintended side effect. As the fuses have a significant failure rate, every time MLRS is used dozens of unexploded bomblets are left littering the target area.

## 1  Rocket Science: V-2 to E4

1 The basic gunpowder rocket still survives in the twenty-first century. Home-made rockets, called Qassam, have been used by Palestinian separatists against Israel for many years. They have a range of up to 4 km and carry a small explosive warhead. Although several thousand have been fired, they have caused little damage and only four deaths up to 2004.

2 Other nations also worked on rockets. None attempted anything as ambitious as the V-2, but in 1933 Captain Skinner of the US Army started work with Lieutenant Uhl of the US Navy on a new type of rocket launcher. It attracted little attention until it was fitted with a new type of anti-tank warhead designed by a Swiss inventor. From 1942 the rocket launcher was a tremendous success, being copied by several other countries. Better known by its nickname – Bazooka – it gave infantrymen a fighting chance against tanks for the first time.

3 GPS has also led to a new generation of cheap guided bombs referred to as JDAMs (for Joint Direct Attack Munition). Rather than relying on laser or other guidance, they are simply programmed before dropping with the exact co-ordinates of the target, and use steering vanes to glide towards it. The JDAM guidance kit, which can be fitted to any of the standard 'dumb bombs', costs as little as $10,000 compared to $100,000 or more for the laser-guided version. JDAMs were extensively used in the 2003 war in Iraq and are likely to feature in any future conflict.

## 2   Air Raids to Airliners

1 'Boiling oil', that old Hollywood favourite, is not physically possible – oil burns before it boils. But hot oil can still be an effective weapon even if it is not boiling, and its use by defenders was recorded as long ago as AD 76 by Roman chroniclers.
2 Without the compressor the air flow through the engine would be very sluggish and there would not be enough oxygen to burn a significant amount of fuel. The more compression, the greater the airflow and the more fuel can be burned – and the greater the thrust.
3 After the war, von Ohain moved to America where he worked as a researcher for the USAF, later becoming chief scientist at the Wright-Patterson Aero Propulsion Laboratory. Although not given the same recognition as Whittle, von Ohain was one of the most important figures in the jet age.

Unlike von Braun, Willi Messerschmitt was imprisoned after the war because of his involvement with the use of slave labour. However, he was released after two years and started a new Messerschmitt company which made sewing machines and prefabricated housing. In 1958, Messerschmitt started building aircraft again, and in 1969 the company merged with others to form Messerschmitt-Bolkow-Blohm or MBB. Willi Messerschmitt was named honorary chairman, holding this position until his death in 1978. MBB is a major player in the European helicopter business, as well as manufacturing other items, including the monorails used in Disneyland.
4 This risk of confusion also taught the military some lessons in ergonomics, the science of designing equipment so that it was easy to work with. One of the early US Navy jets suffered from a high rate of accidents, with crew triggering their ejector seats when there was no reason to bale out. The cause was traced to the layout of the controls – the handle which lowered the landing gear was immediately adjacent to the ejector seat control. The plane makers changed the shape of the two handles so that they could be distinguished by touch. This prevented the confusion, and the policy of using different shapes for different controls was adopted as standard.

## 3   Computing: Codebreaking to Netsurfing

1 In English usage, Boffin is a not-unfriendly slang term for a scientist. Its origins are not clear. There are some indications that it was first used for the radar researchers, and radar pioneer Sir Robert Watson-Watt suggested that it came from combining 'puffin' with 'Baffin' (the Blackburn Baffin was a biplane), but this seems very unlikely. There is a character called Noddy Boffin in Dickens' *Our Mutual Friend*, but although he is described as a 'very odd-looking old fellow', he is a rich dustman rather than a scientist.
2 A vain hope. Even during the Second World War Bletchley Park had a resident Russian spy, John Cairncross. Information Cairncross passed to the Russians is credited with helping them to win the Battle of Kursk. Although reports were

already passed to Stalin by the Allies based on intercepted German messages, Stalin was always mistrustful of these, and placed more reliance on information straight from the horse's mouth, provided by Cairncross.

3 The Mark 1 did contribute one lasting innovation to computing. When it was displayed to the public at Harvard in 1944, IBM chairman Thomas Watson brought in a designer to make the machine look more impressive. The machinery was encased in a shiny glass and steel cabinet, against the objections of engineers who wanted easy access to all the components. Hardware engineers have been cursing this decision ever since.

4 ENIAC had a staggering number of individual components, including no less than 17,468 vacuum tubes of sixteen different types.

5 Texas Instruments was originally 'Geophysical Services Inc' and made electronics to search for oil. During the Second World War it moved to submarine detection devices for the US Navy, and became a purely military electronics company. In 1951 they were licensed by Bell to make transistors; there were some doubts about whether such a small company should be licensed.

# 4   Death Rays to DVDs

1 The Soviets were working on their own maser independently, and in the same year Nikolai Basov and Alexander Prokhorov also created a working maser. In 1964 they shared the Nobel Prize in Physics with Townes for their work.

2 Hughes' previous projects included the XF-11, a high-speed, long-range reconnaissance plane made from a type of wood impregnated with plastic, and the 'Spruce Goose', an enormous eight-engined flying boat. The XF-11 prototype crashed while Hughes was piloting it, and the Spruce Goose proved impractical and was abandoned after a single test flight.

3 By analogy with the maser and laser, this X-ray laser should really be called a 'Xaser'; even more powerful devices are possible in theory, producing gamma rays, and these have been dubbed 'Grasers'.

4 This created problems in 1985. The Star Wars name had appeared on official documents for some time; the problem was that now it was also being used by pressure groups who were opposed to SDI and who had started issuing anti-Star Wars material. Sharp-eyed lawyers working for LucasFilm pointed out that they had the rights to the name, and that they had not authorized the pressure groups to use it (although they had not objected to the official use by the SDI). The action, brought for 'Trademark Infringement', was thrown out of court.

# 5   Ultimate Power

1 Although German scientists declared that Relativity was 'Jewish physics' and not based on fact, the work on uranium went ahead. However, it was always given much less attention than rockets and other advanced weapons. Heisenberg's mistaken belief that heavy water was needed as a moderator slowed

development (the Germans lacked heavy water), but Roy Irons recounts in his book *Hitler's Terror Weapons* that the Allies also benefited from a peculiar accident. In 1941 Werner Heisenberg, the great German physicist, was to give a talk to Hitler, Goering, Himmler and other Nazi leaders about an incredibly powerful weapon that could be made using a uranium warhead. This could have been the start of a German Manhattan Project, but due to a clerical error the wrong invitations were sent out, inviting the leadership to an obscure technical seminar instead. All declined to attend, and the Nazi bomb never happened.

2  This technology has progressed and centrifuges are now important for uranium production.


# 6  Flights of Fancy

1  The 1966 film *Batman* was to have featured villains using jet packs, but this was already considered too dull and everyday. Jet packs were replaced with more colourful jet-powered umbrellas.

2  The US Army has not quite given up. They already have steerable parachutes, but these are not quite the same as flying, and there is clearly a demand for something better. Austrian stuntman Felix Baumgartner used carbon-fibre wings to glide across the Channel in 2003, and was contacted by the Army shortly afterwards. The user has to jump from a plane at high altitude, but such wings offer the prospect of Icarus-like flight, and they are being tested for their suitability in combat missions.

3  The XV-1 and XV-2 were other experimental vertical take-off aircraft. The XV-1 Convertiplane by McDonnell Douglas could operate either as a helicopter, an autogiro with unpowered rotor, or normal fixed-wing. The XV-2 was the unsuccessful Sikorsky contender in the tilt-wing competition.

4  The condition is caused by a vortex ring effect; vortex rings will be discussed in more detail in a later chapter.


# 7  UFOs and Secret Technology

1  Mogul was just one of the secret balloon programs carried out by the US. Others included the giant Skyhook balloons which gave rise to many reports of flying saucers, and also the WS-141 – the first balloon to circumnavigate the Earth, fifty years before Per Linstad. These military balloon programs gave a huge boost to balloon science, and modern high-altitude balloons owe much of their technology, such as the plastic film used to make the balloon envelope, to these military projects of the 1950s.

2  The Remington company attempted to harness the connection when they described 'the world's only shaver with the sharpness of titanium-coated trimmer blades', mentioning the Blackbird and describing the shaver as 'not so much a precision style as a distant cousin of the F-14 Tomcat', a fighter with titanium wings. This attempt to play up the macho image of titanium foundered when

Norelco, another manufacturer, took them to court on the grounds that the shaver actually used softer titanium nitride and not the metal itself.

3  Ironically enough, Ufimstev himself was completely unaware of the profound impact he had on global warfare until he visited the United States to give a lecture at UCLA in 1990. The Russians had not shown any enthusiasm for his work: 'Senior Soviet designers were absolutely uninterested in my theories,' he commented.

4  There is a long-standing Nazi flying saucer mythos. During the war, Allied bomber crews reported seeing mysterious lights, nicknamed Foo Fighters, which were assumed to be some sort of German secret weapon. They never did any damage, and after the war it was found that the Germans had also seen them and assumed they were an Allied secret weapon. Many years afterwards, pro-Nazi writers capitalized on the interest in flying saucers by claiming that they were the product of advanced Nazi technology. No hard evidence for such Nazi saucers has ever emerged. One can even fuse the alien and Nazi stories together. One variant has it that an alien craft crashed in Thuringia in Germany in the 1930s, and was promptly taken apart by Nazi scientists who used it to build their own antigravity drive . . . and you can take the rest from there.

# 9  Non-lethal Death Rays

1  Names matter. The entire field of non-lethals is a process of finding acceptable alternatives to unacceptable weapons, and giving them harmless-sounding names is important. Jelly batons, sponge grenades and bean bags sound like something from a children's party, and one needs to be aware that these 'bean bags' are actually canvas bags of lead shot. Sometimes the language is deliberately used to mislead, as with the Israeli military's 'rubber-coated bullets'. These are steel balls the size of marbles with a thin rubber coating, capable of penetrating the skull. The Israeli Army also uses 'plastic bullets' which are not baton rounds but simply high-velocity rifle bullets made of PVC.

In fact, the name of the whole area has been the subject of endless dispute; many prefer terms like 'less lethal' or 'less than lethal', but 'non-lethal' has stuck in spite of its inaccuracy.

2  OC – oleoresin capsicum or 'pepper spray', derived from chilli peppers, is used by police for short-range sprays but not in grenades or other mass applications.

3  Note the changed name. From the dangerous-sounding 'pulsed impulsive kill laser' it move to the blander 'pulsed chemical laser' and now to a name which does not even include the potentially alarming word 'laser'.

# 10  Vortex Rings

1  This was back in the days before all secret projects were given cryptic names which did not give anything away. Squids project a jet of water behind them at high speed, so anyone might guess that Project Squid involved a similar principle. Project Harvey, an early stealth program, was named after the invisible rabbit in

the film *Harvey*; slightly later on was Hopeless Diamond, a diamond-shaped stealth aircraft with 'hopeless' aerodynamics. These days all code names are generated randomly by computer to prevent any useful information being gleaned from names alone. Projects called Dipole Yukon and Tacit Gold in budget documentation give away little about what's going on behind them (a missile for knocking out WMD and an active camouflage system for aircraft in this case).

## 11   Mach 10 Airliners and Space Tourism

1  The continued survival of the B-52 bomber is worth remarking on. Having first entered service in 1955, it continues to be the mainstay of the US bomber force. Several attempts have been made to replace it. The B-70 program for a mach 3 bomber was cancelled in 1969 as too ambitious. The B-1 was cancelled, resurrected, and then redesigned before entering service, changing from a supersonic high-altitude aircraft to a low-level subsonic penetrator, absorbing billions in the process and only built in limited numbers. The B-2 stealth bomber was even more expensive, with the numbers planned dropping until only twenty-one were funded. After more than fifty years, the B-52 is still the most successful aircraft in the fleet. This history of unsuccessful replacements should give pause to anyone who thinks that a shiny new high-tech bomber is the answer.

2  The name 'Aurora' may be a red herring. It appeared in a Department of Defense budget document referring to a black aircraft program, and many assumed that this meant it was a high-speed Blackbird replacement. In his book about Lockheed's Skunk Works, Ben Rich claimed that Aurora was a code name given to part of the development of the B-2 bomber which was then highly classified. Of course, Rich's claim in itself could be a further attempt at disinformation.

3  Another symptom of this is space tourism. While the US may be the home ground of capitalism and free enterprise, any budding space tourists with $20m to spare still have to go to RKA, the Russian Aviation and Space Agency. American launch economics have not yet made paying passengers feasible even at this rate.

## 12   Information Warfare

1  The term propaganda actually has older roots. It comes from the Latin name 'Congregatio de Propaganda Fide' or 'Congregation for Propagating the Faith', an organization established by Pope Gregory XV in 1622 for the purpose of spreading the Catholic faith. Not to be confused with the 'Congregation for the Doctrine of the Faith', otherwise known as the Inquisition.

2  Another example is the net-firing gun. These were tested by various police forces in the 1990s but did not catch on for a number of reasons. However, it was found that the weapon was quite effective at subduing suspects just by pointing: the net was fired from a grenade launcher, and when a suspect found himself looking down the outsize gun barrel he tended to give up without a fight.

3 This can backfire. When he left Germany in 1933, Sigmund Freud was forced to sign a statement to the effect that he had not been mistreated by the Gestapo and had no complaints against them. He duly signed, and asked if he could add a further line: 'I can heartily recommend the Gestapo to anyone.' Oblivious to irony, the Gestapo allowed him to go. Perhaps the Allies could have made more use of Freud – he might have made a significant contribution to the psychological warfare effort.

4 Donald Rumsfeld appears to have objected to a rule that limited forcing prisoners to stand for four hours at a stretch. 'I stand for 8-10 hours a day. Why is standing limited to 4 hours?' he queried in a handwritten addition to a note on acceptable interrogation methods.

## 13 Catching New Waves

1 When declassified, this provided meteorologists with a rich source of data about global warming. Measurements of changing ice thickness through the years provided strong evidence of climate change.

## 14 Robots in the Air

1 As a cunning piece of subterfuge, the Director of Military Aeronautics, Sir David Henderson, gave Low's aircraft the name AT so that people would think it meant Aerial Target. By a curious twist of fate, the main use of unmanned aircraft after the First World War really was as aerial targets.

2 Not to be confused with the later F/A-22 Raptor, the USAF's next-generation fighter aircraft.

3 The famous 'exploding cigar' with which they supposedly tried to assassinate Fidel Castro is a popular myth. However, the CIA did attempt the task using an exploding seashell, and also considered poisoning the Cuban leader's cigars so that his beard would fall out.

4 This may result in a microscopic arms race. The counter to smart dust would be swarms of miniature smart-dust-eating robots, tasked with clearing an area of any spies. This might be countered by providing the smart dust with some kind of self-defence, and so on.

## 15 Tools for Terrorists

1 No weapon is entirely trusted until it has proven its worth in battle, and the US has taken advantage of recent wars to test out new kit. The CBU-107 cluster bomb, for example, nicknamed 'steel rain', was fielded for the first time in 2003. Just two were dropped, out of all the thousands of bombs used. A new version of the Hellfire anti-tank missile was also deployed on two occasions. It

appears that this use owes more to a desire to see if the thing worked than any particular tactical requirement. After the 1991 Gulf War, US manufacturers found that being able to describe their products as 'battle proven' was a useful aid to marketing.

2 The neutron bomb was never built. Envisaged as a counter to the threat perceived from an overwhelming Russian armoured assault across Europe in the 1970s, it was a new type of tactical battlefield nuclear weapon. Unlike previous nuclear devices which produced a combination of radiation, heat and blast, the neutron bomb was intended to produce most of its energy in the form of radiations and specifically high-speed neutrons. These will go through dense material (like tank armour) with ease, killing the crew with a lethal dose of radiation. However, the idea that the neutron bomb would not destroy property was a myth – even a small neutron bomb would still produce a detonation equal to 1,000 tons of explosive. The effective range against tanks was disappointing, and given that the Soviets could choose to respond with their own tactical nuclear weapons, the idea was quickly dropped.

3 However, in this case the possibility is a remote one. Theodore Kaczynski, who became known as the Unabomber because of his bomb attacks on universities and airliners, was an anti-technology fanatic who lived in a shack in the backwoods of Montana. His bombs were all individually hand-crafted with great skill, and he shunned a high-tech mass-production approach that might have made him far more deadly.

## 16   Nanotechnology: The Future is Small

1 Damascus was a trading centre rather than a producer. The actual metal ingots came from more distant kingdoms which are now part of Sri Lanka and India, as well as Central Asia.

2 The film is a classic black comedy about the Cold War. Dr Strangelove is a sinister scientist, apparently from Nazi Germany, who advises the President. Played by Peter Sellers, he has a quirky middle-European accent and a gloved right hand which has a mind of its own; but beneath the comedy is a man who is happy to suggest the extermination of most of the human race for political ends.

## Websites

Many hundred of websites were used while researching this book. Some of the more important are:

www.FAS.ORG
www.globalsecurity.com
howstuffworks.com

# Select Bibliography

Adams, Douglas, *The Hitchhiker's Guide to the Galaxy*, Pan Books, 1985.

Alexander, John B., *Future War – Non-lethal Weapons of the 21st Century*, Thomas Dunne, 1999.

Altmann, Jurgen, *Military Uses of Microsystem Technologies*, Agenda Verlag, 2001.

Cumpsty, Nicholas, *Jet Propulsion*, Cambridge University Press, 1997.

Darlington, David, *Area 51: The Dreamland Chronicles*, Henry Holt, 1997.

Drexler, K. Eric, *Engines of Creation*, Random House, 1986.

Dyson, George, *Project Orion: The Atomic Spaceship 1957–1965*, Penguin, 2002.

European Working Group on Non-lethal Weapons, *Non-lethal Capabilities Facing Emerging Threats*, Fraunhofe, 2003.

Farren, Mick, *The Black Leather Jacket*, Abbeville Press, 1985.

Gander, Terry, *The Bazooka*, Parkgate Books, 1998.

Gervasi, Tom, *Soviet Military Power* (the annotated and corrected version of the Pentagon's guide), Random House, 1987.

Gunston, Bill, *Future Fighters*, Salamander, 1984.

Gunston, Bill, *Modern Airborne Missiles*, Salamander, 1983.

Gunston, Bill, *The Development of Jet and Turbine Engines*, PSL, 1995.

Harbinson, W.A., *Project UFO*, Boxtree, 1995.

Hecht, Jeff, *Laser Pioneers*, Academic Press Inc, 1992.

Hyland, Gary and Gill, Anton, *Last Talons of the Eagle – Secret Nazi Technology which could have Changed the Course of WWII*, Headline, 1998.

Irons, Roy, *Hitler's Terror Weapons*, HarperCollins, 2002.

Lawrence, Jeanette, *Introduction to Neural Networks*, California Scientific Software Press, 1993.

Macksey, Kenneth (ed.), *The Hitler Options*, Greenhill Books, 1995.

McCartney, Scott, *ENIAC: The Triumphs and Tragedies of the World's First Computer*, Walker & Company, 1999.

Miranda J, Mercado P, *Strange Phenomena in the Sky*, Reichdreams, 2001.

Naughton, John. *A Brief History of the Future: The Origins of the Internet*, Weidenfeld & Nicholson, 1999.

Orlebar, Christopher, *The Concorde Story*, Osprey, 1986.

Peebles, Curtis, *Dark Eagles: A History of Top Secret US Aircraft Programs*, Presidio Press, 1995.

Peebles, Curtis, *Shadow Flights*, Presidio Press, 2000.

Ponting, Clive, *Armageddon: The Second World War*, Reed, 1995.

Price, Alfred, *Sky Battles*, Cassell, 1993.

Rich, Ben and Janos, Leo, *Skunk Works*, Little Brown & Co, 1994.

Shukman, David, *The Sorcerer's Challenge*, Hodder & Stoughton, 1995.

Smith, Michael, *Station X The Codebreakers of Bletchley Park*, Channel 4 Books, 1998.

Strathern, Paul, *Dr Strangelove's Game*, Hamilton, 2001.

Suvorov, Viktor, *Inside the Soviet Army*, Grafton Books, 1984.

Sweetman, Bill, Aurora – *The Pentagon's Secret Hypersonic Spyplane*, Motor Books, 1993.

Taylor, John W.R. (ed.), *RPVs: Robot Aircraft Today*, Macdonald & Janes, 1977.

# Index